PRAISE FOR

An Empire of Wealth

"An important new book. . . . Gordon manages to capture the majesty, vision, and drive in the scramble for (economic) empire. . . . Gordon's volume possesses a forward momentum worthy of the story he is telling and the passion he brings to it." —*Boston Globe*

"Gordon exhibits a firm grasp of political economy and finance. . . . His prose comes alive. . . . Tucked inside *An Empire of Wealth* are kernels of cautionary wisdom." —*New York Times Book Review*

"Gordon manages to please, educate, and enlighten us on every page. A deft, lively handling of an ambitious project that would have daunted almost any other writer." —Ron Chernow, author of *Alexander Hamilton*

"A fantastic, sweeping, and absorbingly informative history. . . . Riveting. . . . Gordon tells this epic tale with verve and authority, captivating any reader interested in how we arrived at where we are today." —*Forbes*

"John Steele Gordon has a rare gift—his tales of business and economic history are not only clear but enormously entertaining to read." —Jean Strouse, author of *Morgan*

"A sweeping and easy-to-read narrative with plenty of fascinating nuggets." —*Washington Post*

"Fast-moving, concise, and always lucid. . . . It is refreshing to read Mr. Gordon's nonideological assessments and cool liquidation of myth. . . . *An Empire of Wealth* is must reading." —Harold Evans, *New York Sun*

"Startling . . . an epic story. . . . John Steele Gordon is a master of the economic anecdote, of telling how people built a complicated world by solving the problem in front of their noses." —*Seattle Times*

"Engrossing and comprehensive. . . . Gordon presents even the least vivid economic history in a fascinating manner. . . . A remarkably detailed and interesting book." —*Library Journal*

"If you only read one economic history of the United States, this is the one. Gordon's narrative, from colonial times to the present, is entertaining and insightful." —*American Spectator*

"John Steele Gordon is one of our nation's most interesting financial historians, and in *An Empire of Wealth*, he does not disappoint. This fascinating book . . . belongs on the bookshelf of all American history buffs." —Jay Winik, author of *April 1865: The Month That Saved America*

"[Gordon's] grasp of the larger picture is sure and his prose bright. Those seeking an introduction to the general history of American economic power will find few better places to start." —*Publishers Weekly*

"In this elegant, lucid book, John Steele Gordon argues that the story of America's rise to global power is essentially the story of money, industry, and our seemingly endless capacity to meet any challenge with a spirit of innovation and good cheer."
—Jon Meacham, author of *Franklin and Winston*

BOOKS BY JOHN STEELE GORDON

Overlanding

The Scarlet Woman of Wall Street

Hamilton's Blessing

The Great Game

The Business of America

A Thread Across the Ocean

AN EMPIRE
OF WEALTH

JOHN STEELE GORDON

AN EMPIRE
OF WEALTH

The Epic History of American Economic Power

———

HARPER ● PERENNIAL

NEW YORK ● LONDON ● TORONTO ● SYDNEY

HARPER ● PERENNIAL

A hardcover edition of this book was published in 2004 by HarperCollins Publishers.

P.S.™ is a trademark of HarperCollins Publishers.

HarperCollins books may be purchased for educational, business, or sales promotional use. For information please write: Special Markets Department, HarperCollins Publishers, 10 East 53rd Street, New York, NY 10022.

FIRST HARPER PERENNIAL EDITION PUBLISHED 2005.

Designed by Amy Hill

Library of Congress Cataloging-in-Publication Data is available upon request.

ISBN 0-06-009362-5 (hardcover)
ISBN-10: 0-06-050512-5 (pbk.)
ISBN-13: 978-0-06-050512-7 (pbk.)

05 06 07 08 09 ❖/RRD 10 9 8 7 6 5 4 3 2 1

For

Richard F. Snow

Frederick E. Allen

Timothy C. Forbes

and Everyone at *American Heritage*

America! America!
May God thy gold refine
Till all success be nobleness
And every gain divine!

— Katherine Lee Bates
"America the Beautiful,"
1893

The march of Providence is so slow and our desires so impatient; the work of progress is so immense and our means of aiding it so feeble; the life of humanity is so long, that of the individual so brief, that we often see only the ebb of the advancing ways, and are thus discouraged. It is history that teaches us to hope.

— Robert E. Lee
Letter to Charles Marshall,
ca. 1866

We in this country . . . are—by destiny rather than by choice—the watchmen on the walls of world freedom.

— President John F. Kennedy
from the speech he did not live
to deliver in Dallas,
November 22, 1963

CONTENTS

ACKNOWLEDGMENTS

M Y FIRST THANKS, needless to say, are due to my editor, Tim Duggan; his assistant, John Williams; and my agent, Katinka Matson.

Eleanor Mikucki did an excellent job of copyediting.

I must also express my sincere thanks to the editors and publishers of *American Heritage* who have given me the opportunity for the last fifteen years to write about the American economy in all its manifold aspects. This book could not have come into being otherwise.

I would also like to thank all those who have helped in many ways, especially: Marc Arkin, Ron Chernow, Walter Olson, Jean Strouse, and William L. Zeckendorf.

THE PURSUIT OF HAPPINESS

A T THE DAWN OF THE TWENTY-FIRST CENTURY, the position of the United States in world affairs has no equivalent in history. Indeed, one has to look to the apogee of the Roman Empire, almost two millennia ago, to find even a remotely comparable situation.

Rome conquered the known world by force of arms. Its power arose from its military machine, epitomized by the legions. And every Great Power since has exercised formal political hegemony over alien peoples to advance its own interests. A century ago, the British Empire covered a quarter of the globe's land area, while a third of the world's people were subjects of King Edward VII. But only a small minority of those people spoke English or regarded themselves as British.

The United States, however, has always been, at most, a reluctant imperialist. In the twentieth century it was the only Great Power that did not add to its sovereign territory as a result of war, although it was also the only one to emerge stronger than ever from each of that century's three Great Power conflicts. Today the United States possesses only 6 percent of the world's land area and 6 percent of its people—virtually all of whom regard themselves as Americans and speak English. And yet its influence in the world is far greater than Britain's was at the height of its relative power in the mid-nineteenth century.

The reason is the American economy. For while the United States has only 6 percent of the land and the people, it has close to 30 percent of the world's gross domestic product, more than three times that of any other country. In virtually every field of economic endeavor, from mining to telecommunications, and by almost every measure, from agricultural production per capita to annual number of books published to number of Nobel Prizes won (more than 42 percent of them), the United States leads the world.

Its economy, however, is not only the largest in the world, it is the most dynamic and innovative as well. Virtually every major development in technology in the twentieth century—which was far and away the most important century in the history of technology—originated in the United States or was principally industrialized and turned into consumer products here. Its culture, therefore, from blue jeans to Hollywood movies to Coca-Cola to rock and roll to SUVs to computer chat rooms, pervades the rest of the planet. As the new technologies spread around the world, they inescapably brought with them American ways and an American perspective.

English has ever increasingly become the world's unifying language, as Latin was Europe's for centuries. Sixty percent of the students who are studying foreign languages in the world today are studying English, which is increasingly a required subject in school systems everywhere. Partly this is because, thanks to the British Empire, so many countries use English as a first or second language, but equally it is because the United States dominates the world in communications and entertainment. The Internet, the most powerful means of communication ever devised, is largely an American invention, and English is the language of more than 80 percent of the four billion Web sites now in existence.

The ultimate power of the United States, then, lies not in its military—potent as that military is, to be sure—but in its wealth, the wide distribution of that wealth among its population, its capacity to create still more wealth, and its seemingly bottomless imagination in developing new ways to use that wealth productively.

If the world is becoming rapidly Americanized as once it became Romanized, the reason lies not in our weapons, but in the fact that others want what we have and are willing, often eager, to adopt our ways in order to have them too. The relentless spread of democracy and capitalism in recent decades, to a large extent in the light of the American example, is a peaceful and largely welcomed conquest—at least by the people, if often not by the elites who have seen their own power slipping away. It is a conquest more subtle, more positive, more pervasive, and, in all likelihood, more permanent than any known before.

Thus America's is an empire of wealth, an empire of economic success and of the ideas and practices that fostered that success. Like many successes, that of the American economy is often seen, in retrospect, as inevitable, even foreordained. After all, the country has always had a vast, varied, and fecund national territory, abundant natural resources, and a large and well-educated population. But Argentina had all these assets as well, and hasn't fought a protracted war since 1870. Yet it struggles just to maintain its status as a developed nation. Its GDP is less than a third that of the United States per capita.

Much of the reason for the difference lies in politics, for Argentine politics, inherited from Spain's control-from-the-top imperial system, has all too often destroyed wealth or, even more often, prevented rather than fostered its creation. But American politics had the great good fortune to be grounded in English traditions, especially the idea that the law, not the state, is supreme. The uniquely English concept of liberty—the idea that individuals have inherent rights, including property rights, that may not be arbitrarily abrogated—was also crucial.

That England was able to develop these concepts, incorporate them into its politics, and pass them on to its children was due in large part to the most consequential geographic fact in Europe, perhaps the world:

the twenty-three miles of often treacherous water that separates the island of Great Britain from the European mainland. The English Channel is narrow enough that England was in close and continuous contact with the mainland and yet broad enough to make invasion, at best, a chancy proposition. It has succeeded only once in the last thousand years.

With no need to maintain a large and expensive army, England was a low-tax state for much of its history, able to put its economic resources into developing more resources. Further, it could afford the luxury of decentralized government; the running of local affairs was left in the hands of local people with little interference from the king. And England had the most fluid social structure of any major nation in Europe. While it had an aristocracy—indeed one legendary for its wealth and influence—it had no closed noble class. Marriage alliances between successful bourgeois families and landed families have always been far more common in Britain than on the Continent, and talent could thus rise much more easily to the top. Napoleon's description of England as a nation of shopkeepers was meant as an insult. It was taken as a compliment.

Englishmen became used to this state of affairs and deeply antagonistic to any attempts to change it. The English who began to settle in America in the early seventeenth century brought these ideas with them and applied them to the new situation in which they found themselves.

Geopolitically, that situation resembled England's, only on a much grander scale. Until the second half of the twentieth century, North America was largely immune from foreign attack, and the hand of government (and thus the tax man) lay very lightly indeed upon it for most of that time. And just as Great Britain was perfectly situated on the map to dominate the trade routes of northern Europe as that area began to dominate European and world affairs, the United States was perfectly positioned to take advantage of the emergence of a fully globalized economy. The United States is the only Great Power to front on both the Atlantic and the Pacific and the only one whose national territory sprawls across arctic, temperate, and tropical climate zones. It is at once both effectively an island, with all of an island's military security, and a continent, with all of a continent's resources.

Further, most of the people who came to what is now the United States—by no means all of them English, of course—came here precisely to be able to run their own affairs as they saw fit, both to worship their God and to better their economic situation in what has been seen for centuries as a land of opportunity. It is perhaps not a coincidence that the United States is both the most religious nation on earth and the most secular, the most devout and the most commercial.

And there can be no doubt that if the United States is famous for its get-up-and-go, that is because Americans are descended from those who got up and came. Those who chose to leave all they had ever known and come to a strange and distant land came to pursue their own ideas of happiness. Here, the great majority found conditions that allowed them to do so with less interference than anywhere else and thus gave them a better chance to find it. Even those who came as captives, rather than of their own free will, had each somehow survived an ordeal beyond modern imagination and passed that strength to their descendants. And because a national economy is nothing more than the collective economic accomplishments of the citizenry, the American economy has become, over the nearly four centuries of its existence, one the greatest wonders of the modern world and, indeed, a prime creator of that modern world.

That is not to say that the history of the American economy has been one of triumph following triumph. That would make a very dull book. At numerous points in the history of the United States, the economy was in deep, deep trouble, and that trouble could easily have spiraled out of control if the political leadership had failed, as Argentina's leadership has failed that country so often.

After the Revolution, the American economy was in profound recession. The country's products were largely excluded from the British

Empire, where most of its traditional markets had lain. Its currency, to the extent that it even amounted to a currency, was nearly worthless; its government's debts were unpaid and unpayable. In 1932 the disaster of the Great Depression was so all-pervasive that the future not only of the economy but of the Republic itself was in doubt to many.

In both cases the country managed to weather the storm and emerge stronger than ever thanks to extraordinary leadership, in the first case by George Washington and Alexander Hamilton, in the latter by Franklin D. Roosevelt. But the country's politicians have also made grave errors, including those that produced the Great Depression itself. The destruction of the Second Bank of the United States at the hands of Andrew Jackson meant that the country had no central bank for nearly eighty years. As a result, the recurrent financial panics in the nineteenth and early twentieth centuries were far worse, and the ensuing depressions far deeper, than they otherwise might have been.

Like most stories of empire, the story of the empire of wealth is an epic one, full of triumph and disaster, daring and timidity, new ideas and old prejudices, great men and utter fools. But most of all it is an epic powered by uncountable millions pursuing their self-interests within the rule of law, which is the essence of liberty. And as with all epics, it is at its heart a window into what makes us human, for I can think of no more overwhelming a refutation of William Wordsworth's notion that in "getting and spending we lay waste our powers" than the story of the American economy.

PART I

A VAST AND ROARING WILDERNESS

———

They must come into, and go through a vast and a roaring Wilderness, where they must be bruised with many pressures, humbled under many overbearing difficulties . . . before they could possess that good Land which abounded with all prosperity, flowed with milk and honey.

— The Reverend Thomas Hooker
The Application of Redemption, 1659

Chapter One

THE LAND, THE PEOPLE,
AND THE LAW

ABRAHAM LINCOLN thought that a nation—however much the whole might exceed the sum of its parts—consists of nothing more than its people, its land, and its laws.

For the nations of the Old World, the three parts are inextricably bound up together in the long individual histories of those nations. But the United States, like all nations founded by European settlers in the great expansion of Western culture that began in the late fifteenth century, has no ancient history. At the beginning of American history, there was only the land.

The land that would become the United States presented a world that was at once hauntingly familiar and quite unlike the one in which the first European explorers and settlers had grown up. Western Europe was a world of dense population, concentrated in cities, towns, and villages; intense cultivation of arable areas; limited wildlife; and limited and carefully husbanded forests.

America was located in the same temperate zone and featured often familiar trees, plants, and animals, along with some exotic new ones, such as raccoons, skunks, maize, and rattlesnakes. But beyond the rocky shore of what is now the state of Maine and the vast sandy beach that stretches nearly unbroken from New Hampshire to Mexico and far beyond, lay a wilderness, upon which the hand of its human inhabitants had lain very lightly indeed.

This wilderness was a forest larger than all of western Europe, broken only by the occasional beaver meadow, bog, swamp, rock outcropping, mountain bald, and the slash-and-burn fields of Indians. It stretched from the water's edge to well past the Mississippi. From there it extended fingerlike along river and creek bottoms into the great plains that covered the center of the continent.

This huge forest was, of course, not uniform. In the North, great stands of white pine—the preferred wood for the spars and masts of sailing ships—alternated with hardwood forests, where maples, sycamores, and ash predominated in the lowlands, oaks and hickories on the drier and higher slopes. Farther south were stretches of different species of pine along the Atlantic seaboard, and these reached inward to where they met hardwood forests at higher elevations.

The eastern shore of North America is a welcoming one. A broad coastal plain made for easy settlement. Peninsulas such as Cape Cod and Delmarva; islands such as Long Island; and the barrier beaches farther south provided shelter for the early sailing ships. A series of rivers—the Merrimac, the Charles, the Thames, the Connecticut, the Housatonic, the Hudson, the Raritan, the Delaware, the Susquehanna, the Potomac, the Rappahannock, the York, the James, the Peedee, the Ashley, the Cooper, the Savannah—provided access to the deep interior for the small and relatively shallow-drafted vessels of the day. As early as 1609, Henry Hudson, an Englishman in the employ of the Dutch, sailed his full-rigged ship *Half Moon* 150 miles up the river later named for him, reaching as far as present-day Albany. An overland trip so far inland would have required a month or more. Hudson, although moving cautiously in unfamiliar and narrow waters, covered the distance in a week.

And because these rivers had been formed when the sea level was lower than it is today, the subsequent rise drowned the rivers' mouths and provided harbors that rank among the finest on the North Atlantic. Many of the country's first cities—Boston, Newport, New London, New York, Baltimore, Norfolk, and Charleston—sprang up where these harbors are located.

The climate of North America that the first settlers encountered was, like the land, both familiar and exotic. It is temperate rather than arctic or tropical, and with abundant rain. But, being on the eastern edge of a great continent, the climate is continental in nature, whereas western Europe's is maritime, and greatly tempered by the warmth of the Gulf Stream. American winters are much colder than western Europe's while the summers are hotter. The high and low temperature records for London, located north of the fifty-first parallel of latitude, are 99 degrees and 2 degrees and only rarely approach either extreme. The records for New York, just north of the forty-first, are 106 degrees and −15 degrees, and the extremes are approached with disconcerting frequency. By European standards, New England winters and southern summers were long and brutal.

This vast area was not uninhabited. Dubbed "Indians" (*les indiens* in French, *los indios* in Spanish) through the ignorance of the first European explorers, who thought themselves on the fringes of Asia, the aboriginal inhabitants of North America lived throughout the continent. But by European standards, their population was very low relative to the size of the land. Exact figures are impossible to come by and even estimates vary widely, but the Indian population of eastern North America was probably somewhere between one and two million people at the time of Colombus. The number declined, sometimes rapidly, as increased European contact prior to settlement introduced diseases to which the Indians had no immunity.

And the Indians were anything but culturally homogeneous, even by diverse European standards. There were approximately 250 languages being spoken in North America at the beginning of the European exploration (and about 2,000 in the Western Hemisphere as a whole). Even within languages, the people of North America were divided into many small, often mutually hostile tribes. Low-level warfare was chronic among these groups.

Only the Indians of the Mississippi Valley, socially organized as chiefdoms, depended primarily on agriculture for sustenance. The Indians who lived on the eastern seaboard, mostly organized as tribes, were primarily hunter-gatherers. Less than 1 percent of the arable land of eastern North America was used for growing food crops. Using slash-and-burn methods, the Indians would grow corn, squash, and beans on a patch of land for a few years and then move to new fields as the fertility of the old ones declined.

Technologically the eastern Indians were Neolithic, using sophisticated tools but lacking metal. Their culture was a highly advanced one, however, using hundreds of different materials and techniques in what James Fenimore Cooper, two centuries later, would call "the gentle art of the forest." Developed over thousands of years of extracting a living from the land, these arts, taught to the settlers, would more than once save them from disaster and even extinction as they struggled to establish themselves in the unfamiliar New World.

The more technologically advanced culture that these settlers brought with them and traded with the Indians, however, would, in turn, destroy the latter. Once the Indians became used to the superior metal tools, cloth, and firearms of the Europeans, the skills needed to use the raw materials at hand began to disappear. Before long, the Indians had no choice but to trade for what they needed on increasingly unequal terms and, inevitably, lost their economic sovereignty. Once that was gone, their political sovereignty and the rest of their culture soon followed.

———

TWO OF THE MOST SIGNIFICANT technological developments in human history had brought the European medieval world to an end by the beginning of the sixteenth century and made the settlement of the New World possible. The printing press had greatly reduced the cost of books, and thus of knowledge. In the mid-fifteenth century there had been only about fifty thousand books in all of Europe, most of them controlled by the Church, which ran the universities. By the end of the century there were more than ten million books in Europe, on an endless variety of subjects, many of them technical and agricultural. They were largely in the hands of the burgeoning merchant class and the landed aristocracy. The Church's monopoly of knowledge was broken and, soon, so was its monopoly of religion, as the Protestant Reformation swept over much of northern Europe in the early sixteenth century, setting off more than a century of warfare as a result.

The other great invention of the late Middle Ages was the full-rigged ship, capable of making long ocean passages. As late as 1400, European ships were mostly small and single-masted, not very different from those that had been used by William the Conqueror almost four hundred years earlier to cross the channel to England. But by 1450, far larger ships with three and sometimes four masts had appeared, and they were pushing out the boundaries of the world known to Europeans.

They had need to. In 1453 the Turks had taken Constantinople, the ancient capital of the eastern Roman Empire. A Muslim power now sat athwart the trade routes to the East, extracting taxes on all goods that passed. More, the Turks were expanding into Europe itself, and by the middle of the sixteenth century would be at the very gates of Vienna. Christendom felt itself under attack as it had not since the Dark Ages a thousand years earlier.

But thanks to the full-rigged ship, western Europeans could do an end run around the Muslim control of the ancient trade routes. By the end of the fifteenth century, the Portuguese had rounded the Cape of Good Hope at the southern tip of Africa and reached India. By 1510 they had reached the Spice Islands, source of the spices, such as pepper, that yielded fabulous profits once brought to Europe.

Columbus, working on a sound theory but with a flawed idea of the size of the globe, stumbled onto the New World when he tried to reach Asia by sailing west in 1492.

Once it was clear that Columbus and other early explorers had, indeed, found a New World, explorations were funded by all the major west European maritime powers. The Spanish were the first to hit the jackpot, with the conquest of Mexico and, ten years later, Peru. Huge sums in gold, silver, and precious stones began to flow into Spain, which became the dominant power in Europe as a result. Portugal began producing sugar in Brazil by the middle of the sixteenth century, an immensely profitable crop when grown with slave labor. By the end of the century, the French had begun to use the St. Lawrence River to sail deep into the continent of North America and establish a great fur trade with the Indians who lived around the Great Lakes.

But, except for a Spanish settlement in the Chesapeake Bay that was attacked by Indians and abandoned in 1572, the eastern shore of what is now the United States was largely ignored. It was too far north for tropical crops such as sugar to be grown there, and too far south for the best furs to be found there. Nor was there any sign of precious metals.

England had funded the explorations of Giovanni Caboti (an Italian known in the English-speaking world as John Cabot) in search of a northwest passage. But it came late to the race to exploit the New World by colonization. Here was an opening. In 1585 and again in 1587, Sir Walter Raleigh had tried to establish a colony on Roanoke Island in Albemarle Sound, in what is now North Carolina. The colony vanished, leaving only a cryptic message carved on a tree trunk. But twenty years later England tried again, and this time succeeded.

THE COLONY AT JAMESTOWN was not founded by the English state; it was founded by a profit-seeking corporation.

Tangible inventions, such as the printing press and the full-rigged ship, usually get far more attention from historians, but intellectual inventions are often just as important. And two intellectual inventions of the Renaissance, double-entry bookkeeping and the corporation, proved vital to the development of European civilization in the New World and particularly in what is now the United States.

Accounting had been known since civilization arose in Mesopotamia. Indeed, writing, the defining attribute of civilization, was in all likelihood invented to help keep the books. But accounting did not advance much for several thousand years thereafter, until double-entry bookkeeping was developed in Italy in the fifteenth century. Double entry makes it much easier to detect errors and allows a far more dynamic financial picture of an enterprise to emerge from the raw numbers. Because of double-entry bookkeeping, it became possible for people to invest in distant enterprises and still keep track of how their investment was doing. Ferdinand and Isabella saw to it that an accountant sailed with Columbus on his first voyage, to ensure that they got their full share of the hoped-for profits.

The joint-stock company proved equally important. Exploring far-distant lands in full-rigged ships was both hazardous—many ships simply never returned—and extremely capital-intensive by the standards of the sixteenth century. At first, it was largely done by expeditions funded by the crown in each country. But England was a small country with a small population and lacked the financial resources available to France and Spain. The Dutch Republic as well, once it revolted from Spain in 1572, needed another method to finance these costly but potentially immensely profitable enterprises that were far beyond the financial reach of even the wealthiest individual private citizens.

Partnerships, of course, had been around since ancient days. But in a partnership, each partner is liable for the debts of the entire enterprise. Thus an investor, by investing even a small sum, might find himself bankrupt if the enterprise failed. Few were willing to take on such risks, especially in an enterprise over which they would, necessarily, have very limited control. The joint-stock company solved the problem by limiting each investor's liability to what he had invested. This, of course, shifted some of the risk to the corporation's creditors but made it possible for large sums of capital to be amassed from many small investments. Next to the nation-state itself, the joint-stock company was the most important organizational development of the Renaissance and, like the nation-state, made the modern world possible.

Several English joint-stock companies for purposes of facilitating trade in various areas were established in the latter half of the sixteenth century, the Moscow Company (1555), the Levant Company (1583), and the East India Company (1600) among them. The Dutch also established an East India company. It quickly wrested away most of Portugal's far eastern empire and made the Netherlands, a country with few natural resources, the foremost trading nation in the world and the richest country in Europe in the early seventeenth century.

The Virginia Company, established by a group of London merchants, was chartered by King James I in 1606. The charter stated that the purposes of the company were to help build up England's merchant fleet, increase the number of the country's able mariners by broadening its trade, find precious metals, found a Protestant colony in a land that was under Spanish threat, and, while they were at it, convert the heathen.

The last objective would, in fact, receive precious little attention. Indeed, the English sent no missionaries at all. Instead, they apparently intended to Christianize the Indians through a sort of economic osmosis, intending "to settle and plant our men and diverse other inhabitants there, to the honour of Almighty God, the enlarging of Christian religion, and to the augmentation and revenue of the general plantation in that country, and the particular good and profit of ourselves."

Thus, from the very beginning, the English approach to colonization was profoundly different from that of the Spanish and French. The governments of Spain and France sought to control all aspects of their subjects' activities in the New World and made concerted efforts to bring the Indians to the Catholic religion, whether they wanted to convert or not.

Further, only approved colonists were permitted to immigrate to New Spain and New France, lest heretics and ne'er-do-wells taint them. But the government of England had little at stake in the enterprise and was only too happy to get rid of its troublemakers—religious and criminal—and the unemployed, of which it had far too many.

The English economy had been going through wrenching change for most of the sixteenth century. The population had grown rapidly, from about three million in 1500 to four million a century later and five million by 1650. But employment did not keep pace. The cloth industry, the mainstay of English manufacturing since Flemish weavers had settled there in the mid-fourteenth century, had been losing ground to its continental competition.

Meanwhile, the old feudal system of landholding had been decaying rapidly. The gentry and the aristocracy—the 5 percent of the population who owned most of the agricultural land in England—had been enclosing their estates, dispossessing tenants to run far more profitable flocks of sheep with hired hands. In the century between 1530 and 1630, about half the English peasantry lost their tenancies, and many of them had great difficulty in finding other employment.

In addition, the great influx of gold and silver into the European economy from the New World, thanks to the Spanish conquests, had set off a rapid inflation, and prices rose about 400 percent in the sixteenth century.

The dispossessed peasants and unemployed cloth workers, known as "sturdy beggars" to distinguish them from the traditional beggars disabled by disease or injury, took to the road, often pushed from parish to parish by local officials who did not want to have to care for them. They tended to gravitate toward the market towns and seaports. London, by far the largest city in England already, saw its population rise from 120,000 in 1550 to 200,000 only fifty years later. By 1650, London's vast rabbit warren of narrow, crooked streets and tenements was home to 350,000 people, many of them desperately poor.

It was people from these ranks, fleeing starvation or the sheriff, whom the Virginia Company recruited, together with gentlemen adventurers, often the younger sons of gentry families. In December 1606 three ships, the *Susan Constant,* the *Godspeed,* and the *Discovery,* left England and arrived in the Chesapeake Bay on April 26, 1607, with 105 men on board (39 had died at sea). Sailing some sixty miles up the James River to make their presence less obvious to the Spanish, the three ships anchored on May 13 at the site of what became Jamestown, named, like the river, for England's king.

But other than its relative security from Spanish assault, the chosen site, on the north bank of the James and beside a swamp, had very little to recommend it. The swamp, while perhaps providing some protection from Indians, bred mosquitoes by the millions in the spring and summer, and these spread malaria through the colonists. More, the water in the shallow wells the colonists dug was often brackish, especially when the river was running low. This caused salt poisoning among the colonists as they sweated in the fierce Virginia heat and drank copiously. And, when the river ran low, the garbage and sewage thrown into it did not pass out to sea, but festered and promoted such diseases as typhoid and dysentery.

The result was a slaughter. Of the 105 original colonists, only 38 remained alive nine months later.

The basic problem was that the Virginia Company was venturing into a brand-new business—American plantations—that had been made possible by a radically new technology—the full-rigged ship. As has so often been the case since—railroads in the early nineteenth century, the Internet in the late twentieth come to mind—there was a very steep and expensive learning curve to be mastered before steady profits could be achieved under these circumstances. The commercially savvy and often very wealthy London merchants who dominated the Virginia Company simply had no idea what it took to establish a successful colony on the edge of the American wilderness, three thousand miles and three months from home.

As a result, they made mistake after mistake. By holding out the promise of gold, they made the colonists reluctant to undertake the back-breaking labor of farming in virgin soil. There was, of course, no gold to find, but the colonists found quantities of the local mica and convinced themselves that it was a rich gold ore. "There was no talke," Captain John Smith, reported in his best-selling *Description of Virginia,* published in 1612, "no hope, no worke, but dig gold." Shipped back to England, it was found to be worthless.

And the company at first retained title to the land, expecting the colonists to work it as, essentially, peasants. But neither the gentlemen adventurers nor the recruits swept up from the docks of London and Bristol were inclined to work hard for the company. Nor, indeed, did they possess the necessary skills, the gentlemen for lack of need, the "sturdy beggars" for lack of opportunity.

The result was often starvation in the winter, as the Indians had scant surplus to trade and often even scanter inclination to do so. Although the company shipped more and more colonists every year, the total number rose only slowly. In December 1609 Jamestown counted 220 English souls. By the time spring arrived, there were only 60 still alive, thanks largely to lack of food. One desperate colonist had killed and eaten his wife (and was burned at the stake for his crime).

The remaining colonists abandoned Jamestown in June 1610 and sailed for home, only to meet up with three ships at the mouth of the James River carrying three hundred new recruits. They returned once more to the tiny settlement.

Numerous attempts were made to find an export that would pay the bills and generate a profit to the stockholders. Because glass was increasingly in demand in England, but the wood needed for fuel in its production was increasingly scarce, the company tried exploiting Virginia's endless forest and abundant sand, but could not ship it profitably across the Atlantic. Iron, pitch and tar, clapboards, and sassafras likewise did not yield anything like enough revenue.

By 1616 the Virginia Company had transported more than seventeen hundred people to Virginia and invested the staggering sum of 50,000 pounds in its enterprise on the Chesapeake. To give some idea of what that sum meant in Jacobean England, a gentleman at that time might have an annual income from land of only 50 pounds. The annual crown revenues from customs duties, a major source of the king's income, averaged about 75,000 pounds. But for all this money, all the company had to show for it was a rickety fort on the edge of the James River that was home to 350 people, many of them sick and hungry.

After nine years, the English toehold on the North American continent was still threatened with extinction by many means, Indian attack, Spanish assault, disease, and starvation among them. But by no means the least of the threats to its existence was the fact that the Virginia Company had not yet learned how to make its colony generate more wealth than it consumed.

————

THE ANSWER TO THE PROBLEM turned out to be a common plant native to the Americas, *Nicotiana tabacum*. Originally found in what is now Peru and Ecuador, tobacco had been cultivated for thousands of years before the coming of Europeans. It is not known when smoking the dried leaves for pleasure first began, but the addictive nature of the practice must have been revealed not long afterward. By the time Columbus arrived in the New World, the habit had spread to all temperate parts of the Western Hemisphere where tobacco could be grown and beyond.

Columbus brought tobacco back to Europe from his first voyage. In the next century the habit spread rapidly in the Old World, and the cultivation of tobacco had begun to spread around the Mediterranean

Basin. The Spanish began growing it in the West Indies as well, for export. The habit soon swept through Britain. King James loathed tobacco, which he regarded as an instrumentality of the devil, and he wrote and published a pamphlet entitled *A Counterblaste to Tobacco*. Not surprisingly, his subjects paid no attention whatever to the royal opinion, and smoking continued to increase in popularity.

But Britain's cool, rainy climate did not favor growing tobacco commercially, and the country had to import most of its needs from Spain, a country with which it was frequently at war.

The local Indians of eastern Virginia were also addicted to tobacco, but the variety they grew was not popular with the English colonists who had appeared in their midst. The colonists preferred the tobacco produced by the Spanish in the West Indies. Then, in 1612, a man named John Rolfe brought some seeds he had obtained there, probably from Trinidad, and planted them. They grew abundantly in Virginia's hot, humid climate, and with the help of the local Indians (in 1614 Rolfe married the Indian princess Pocahontas), he learned the exacting art of growing and curing tobacco.

In 1616 he took the first commercial crop to England, along with his wife. They both caused sensations, although the English climate soon killed Pocahontas. When Rolfe returned to Virginia in 1617, Virginia celebrated the first American Thanksgiving because that year's tobacco crop was safely in and promised commercial salvation for the colony.

The first American economic boom was under way. Captain John Smith, back in England, reported that when the new governor arrived that year, he found Jamestown more or less of a wreck, but "the marketplace, and streets, and all other spare places planted with tobacco."

In 1618 twenty thousand pounds of tobacco were grown in Virginia and shipped to England. Four years later—despite an Indian attack that year that killed one-third of the colonists, including, most likely, John Rolfe—the crop had tripled. By 1627 it amounted to five hundred thousand pounds, in 1629 it was one and a half million pounds. By 1638

Virginia was exporting three million pounds of tobacco to Britain every year and had become the major source for western Europe, outstripping the West Indies.

By no means the least of the reasons that tobacco production increased so quickly was that the Virginia Company in 1616 changed its land policy. Instead of colonists being expected to work the land for the benefit of the company, colonists could now own their own land. Also, in hopes of getting more men to immigrate, it gave land away free to new settlers under a system known as head rights. Every man who could pay his own way was awarded fifty acres, with another fifty acres for every relative he brought and for every servant whose way he paid. These "indentured servants," who agreed to work for a term of years to pay off the cost of their passage, were also entitled to fifty acres of land once their indentures were paid off. Of course, they had to survive to collect, and about 25 percent of immigrants died in their first year in the Chesapeake in these early decades of English settlement.

Still, despite the dangers, the prospect of owning a hundred, two hundred, or more acres of land free and clear was a powerful inducement. In an age when agriculture was the foundation of all national economies, wealth was measured less in pounds sterling than in acres. In land-starved Europe, two hundred acres of prime agricultural land made one rich. The Virginia Company, by giving it away, was exploiting, very effectively, what has always been one of America's most profound comparative advantages: its nearly inexhaustible supply of land.

And the tobacco-growing colonists were only too glad to pay for the passage of these indentured servants. Tobacco is a labor-intensive crop, and if the amount available for export was to grow quickly, the population of Virginia had to grow quickly as well. Already, what would prove another enduring characteristic of the American economy was manifest: a shortage of labor.

The tobacco boom saved Virginia, but it could not save the Virginia Company. With the company already deeply in debt, the great Indian attack of 1622 and its costs proved more than it could bear. In 1624 King James revoked the bankrupt company's charter and took Virginia into his own hands as a crown colony. And while he hated tobacco and the habit of smoking it, he had no qualms whatever about taxing it. As Virginia tobacco production began to soar and its trade with England and Europe to increase rapidly as a result, James established a monopoly to closely regulate this quickly growing trade. Within a generation, tobacco would provide fully a quarter of the customs revenue of the crown.

———

THE YEAR 1619, with tobacco already changing Virginia's landscape and economy, would prove a fateful one in the history both of Virginia and of the country of which it would one day be a part.

Since the colony's founding twelve years earlier, Jamestown's population had been overwhelmingly male as the Virginia Company struggled to establish a viable colony in the New World. But in 1619 the company brought the first shipload of women to the colony, ninety in all, and the bachelor settlers, many long starved of female companionship, snapped them up as wives at the price of 125 pounds of tobacco each. Immediately, a subtle shift in Virginia society began. The masculine atmosphere reminiscent of a mining camp or military bivouac faded slowly away, to be replaced, equally slowly, by something far more like the normal human society the new Virginians had left behind in England.

And that year as well, the first representative assembly in the Western Hemisphere was constituted. The great parliamentarian Sir Edwin Sandys was elected treasurer of the Virginia Company that year, effectively making him the chief executive officer. He soon sent out a new governor, Sir George Yeardley, with instructions for him to form a representative assembly, called the House of Burgesses, with the governor and his council sitting as an upper house of the new legislature. This miniature Westminster met for the first time on July 30, 1619, in the church at Jamestown.

While seen as momentous in retrospect, it was not thought so at the time. The English had long enjoyed the least intrusive government in Europe and the greatest personal rights, the concept of which—bound up in the word *liberty*—had been evolving since Magna Carta. The English gentry were used to running their own local, county affairs and equally used to sitting in Parliament, where, together with the sovereign, they enacted the laws of the country.

With the Parliament at Westminster much too far away to be practical, giving Virginia its own legislature for local affairs was considered nothing more than giving Virginians the "rights of Englishmen."

(Sir Edwin Sandys never visited Virginia, but his brother George, a distinguished poet, lived there for ten years, from 1621 to 1631, serving as treasurer of the colony. It was in Jamestown that he wrote his translation of Ovid's *Metamorphoses,* the first English poetry composed in the New World.)

Three weeks after the House of Burgesses met for the first time, a Dutch ship sailed into the Chesapeake, its captain intent on selling its cargo of human beings to the planters who were avid for laborers to work the ever-expanding tobacco fields. There was nothing unusual about the arrival of the ship except for one thing: the men had not been loaded on board of their own free will in one of the English ports. They came, instead, from Africa, where they had been sold to the captain.

Still, they were not quite slaves. The planters, including the governor, who bought most of them for his own plantation, purchased the indentures, not the men. When they had finished their term of servitude, they would be free, just as English indentured servants were. Indeed, many blacks transported to Virginia in the early years of the colony became exactly that and ended up owners of substantial property themselves and even of their own slaves. Given the short life expectancy of immigrants to Virginia, slaves, who cost more, were not an economic proposition compared with indentured servants.

In 1650 there were only about three hundred slaves in Virginia, less than 2 percent of the population. It would be the 1660s before black slavery was even formally recognized in Virginia law, and as late as the 1680s, indentured servants still far outnumbered slaves. But as economic conditions improved in England, lessening the pressure to immigrate, and life expectancy improved in Virginia as the colony expanded and developed, slaves began to overtake indentured servants as the prime source of labor. The number of slaves doubled in the 1680s and doubled again in the next decade.

At the end of the seventeenth century, an indentured servant cost about 15 pounds to purchase four years of his labor; a slave cost 25 to 30 pounds but was bound for life and even beyond in the form of his children. Black slaves began to surge as a percentage of the population, constituting almost 14 percent by 1710.

It is hard for us, who are the beneficiaries of so much hindsight, to understand, but people in the seventeenth century did not regard slavery as a moral issue. It would be the middle of the eighteenth century before the idea that slavery was inherently and ineluctably immoral took hold. Once born, that idea then spread very quickly throughout both Europe and America. At least it did among the nonslaveholding parts of society, for economic self-interest is always a severe impediment to clear thinking on the moral and political aspects of an issue.

In the seventeenth century, when most people felt one's "station in life" was determined by God, slavery was regarded as nothing more than a personal misfortune, not the abomination we see it as today. Nor were slavery and race, at least at first, intertwined. In the middle of the seventeenth century a black man named Anthony Johnson owned a 250-acre tobacco plantation on Virginia's Eastern Shore and at least one slave. He regarded himself, and indeed he was largely regarded by his neighbors, as an equal. He had no hesitation asking the court to enforce his rights when his slave ran away, and enforced they were.

But as the number of black slaves increased steadily both absolutely and as a percentage of the population, while tobacco began to be less profitable per unit of labor as the market reached saturation, attitudes changed. Strictures on the activities of slaves, and of free blacks as well, increased as the fear of rebellion and the economic necessity to get more work out of the slaves increased. By the beginning of the eighteenth century blacks could not assemble in groups of more than four and needed written permission to leave their home plantations. Patrols enforced the new strictures. Discipline increased as well. One "unhappy effect of owning many Negroes," planter William Byrd wrote, "is the necessity of being severe. Numbers make them insolent, and then foul means must do what fair will not."

As the social chasm between blacks and whites deepened and the treatment of slaves grew harsher, the justifications for their social status and economic condition slowly congealed into a virulent racism. That racism proved to be a cancer in the body politic of the United States that would cost much blood and much treasure to excise. Indeed, only three hundred years later is it at long last—and now rapidly—vanishing.

That black slavery and the racism it engendered came about through an attempt to alleviate a chronic American economic problem, a shortage of labor, is of course no excuse. But it is at least an explanation. The most grievous of our self-inflicted wounds and our greatest moral failing as a people was acquired innocently and without forethought.

With the success of tobacco as a reliable export crop, Virginia's economic viability was soon assured. The English-speaking people were in the New World to stay.

Chapter Two

IN THE NAME OF GOD
AND PROFIT

T HE GOOD THING ABOUT an economic learning curve is that a society usually has to climb it only once. None of the other colonies that would one day form the United States had nearly as difficult a time as Virginia in reaching economic viability.

Maryland, next door to Virginia, was formed in 1632, when King Charles I granted his friend Cecilius Calvert, second Lord Baltimore, about twelve million acres north of the Potomac River and south of the fortieth parallel of latitude. Calvert, in gratitude, named the new colony for Charles's queen, Henrietta Maria. Maryland was thus the first proprietary colony, its government in the hands not of a corporation but of a private individual. Baltimore hoped both to establish a colony where his fellow Catholics could find a refuge from the disabilities they suffered in England and, of course, to derive a handsome income from his vast landholding, nearly a fifth as large as all England. The first two ships bearing colonists, the *Arc* and the *Dove,* arrived in 1634.

Baltimore sent his younger brother, Leonard Calvert, to serve as governor, presiding over a government similar to Virginia's. And although Catholics never constituted a majority in Maryland, the Toleration Act passed by the colonial assembly in 1649 guaranteed the rights of all Christians (the colony's few Jews, although unprotected by the Toleration Act, were never persecuted). It was the first such legislation in American history.

Maryland flourished quickly. Its original Indian inhabitants did not prove troublesome, and its rich soil and warm, humid climate were perfectly suited to growing tobacco, which, thanks to Virginia, was already a proven moneymaker. Further, Maryland's long frontage on Chesapeake Bay, with its highly indented shoreline, gave much of its land easy access to cheap freight transportation.

Calvert granted generous head rights to immigrants to encourage immigration to the colony, although charging a quitrent of 4 shillings per hundred acres. The quitrents were, in legal theory, feudal dues. In practice, they were a land tax. As in Virginia, a landed gentry, owners of often thousands of acres, soon emerged, along with a far larger number of "middling planters." Freemen, without land, and indentured servants made up the rest of the early population, with slavery, as in Virginia, slowly increasing as the plantation economy developed.

The other seventeenth-century southern colony, Carolina, was also named for royalty, in this case King Charles II, who granted it to a group of eight favorites, known as the Lords Proprietor. But the impetus for forming the new colony came not from England but from the British West Indies, specifically Barbados.

By 1670 sugar had come to dominate the economies of the chain of relatively small islands that ran north from South America before swinging westward toward the far larger Spanish island of Puerto Rico. The islands had been largely ignored by the Spanish, who thought them too small to bother with. This allowed British, French, and Dutch pirates to seize control of them and use them as bases from which to attack Spanish shipping. By the middle third of the seventeenth century, however, the islands were beginning to settle down as permanent residents took up agriculture, principally tobacco, and the various countries began asserting political control.

At first the West Indies attracted far more settlers than did North America. In 1650 more English emigrants lived in the Lesser Antilles than in New England and the Chesapeake combined. But the Chesapeake outdid the Antilles in producing tobacco, and the islands needed a new crop. Sugar was the answer, for Europe had developed an insatiable appetite for the product that came from sugarcane. Cane is a tropical plant that originated in Polynesia, but by the sixteenth century was widely grown in areas around the Mediterranean. As Europeans began exploring the Atlantic, Portugal and Spain introduced sugar production to their new island possessions such as Madeira and the Canaries.

The Portuguese then introduced sugar to Brazil and the Spanish to the Greater Antilles. They also introduced black slavery to the Americas, for sugar is a crop that requires backbreaking labor in the tropical sun. It also requires more capital than most plantation crops to buy the equipment and buildings necessary to process the cane juice into sugar.

European consumption of sugar was growing at a rate of about 5 percent a year, thus doubling demand every fourteen years on average. Sugar, properly financed and with enough production to use economies of scale, became one of the great cash crops in world history, making the big sugar planters rich beyond dreams.

As a result, when sugar production came to Barbados, only 166 square miles in size and already densely populated with European settlers, a Darwinian struggle among the landholders developed to acquire enough land to make sugar production as profitable as possible. Between 1643 and 1670, the number of landholders with more than a hundred acres fell by two-thirds.

The Lords Proprietor of Carolina (the colony would not be formally divided into North and South until 1712) wanted to populate its new colony on the mainland of North America with the men who could no longer make a living in Barbados. In 1670 they enticed two hundred with generous offers of land, 150 acres to each member of a family with only a modest quitrent that wouldn't be collected until 1689. And they granted 150 acres additional for every slave imported.

They also offered a formal constitution that had been written for Carolina by one of the Lords Proprietor, Lord Ashley, soon the Earl of Shaftesbury. Ashley, to be sure, had a good deal of help from his personal secretary, the political philosopher John Locke—a man whose writings would have an immense influence on the Founding Fathers a century hence. The constitution granted religious liberty to all believers and an assembly with control of local taxes. The constitution even provided for two degrees of peerage, entitled landgraves and caciques (pronounced ka-SEEKS) who were to be given vast land grants (forty-eight thousand and twenty-four thousand acres respectively).

Not surprisingly, this "extraordinary scheme of forming an aristocratic government in a colony of adventurers in the wild woods, among savages and wild beasts," as the South Carolina historian Edward McCrady described it, did not long survive intact its encounter with reality. But South Carolina would be the most aristocratic of the American colonies. As its first governor, the Lords Proprietor appointed the Barbadian planter Sir John Yeamans, who had ruthlessly clawed his way to the top of Barbados's social and economic structure, murdering one rival and shortly afterward marrying the man's widow. One of the Lords Proprietor said of him, "If to convert all things to his present private profit be the mark of able parts, Sir John is without doubt a very judicious man."

A thriving community was soon established at Charles Town (the name was shortened to Charleston in 1783), where the Ashley and Cooper rivers (named for Lord Shaftesbury, whose family name was Ashley-Cooper) came together in Charleston harbor. By 1700 the population of the colony had reached about sixty-six hundred (thirty-eight hundred whites and twenty-eight hundred black slaves).

At first the economy of the colony, too far north to grow sugar, depended on trading with the Indians of the interior. The Appalachian Mountains, a formidable barrier farther north, trailed off into modest hills in what is now northern Georgia, making it easy for traders to reach deep into the interior. As early as 1707, the governor of Carolina boasted that "Charles-Town Traded near 1000 Miles into the Continent."

The area was too far south to produce good furs, but in exchange for blankets, cloth, metal goods, guns and ammunition, rum, beads, and other European manufactures, the Indians supplied deerskins, which were shipped to Europe to be made into bookbindings, belts, gloves, and whatever else required soft, pliable leather. This was no small trade. Deer were very plentiful, and with firearms the Indian hunters could kill far more than previously, while the animal's wondrous fecundity kept the population stable. Between 1699 and 1715, Carolina exported an average fifty-three thousand deerskins a year to England, worth some 30,000 pounds.

Lumber from the colony's abundant pine forests was much in demand in the West Indies, and naval stores, such as tar, came from the same source. In 1717 Carolina exported forty-four thousand barrels of tar.

Livestock also became a major export, especially to the West Indies, where every inch of land was given over to the immensely profitable sugar crop. Cattle and pigs thrived in the lush, marshy lowlands and forests. Allowed to run free in the unsettled parts of the colony, the herds were often tended by black slaves, who had learned the tricks of the trade herding animals in Africa. Many of the techniques of the cattle industry associated with the West, such as branding, annual roundups, and cattle drives, were actually first developed in early South Carolina. Even the storied American word *cowboy* was first applied to the black slaves who herded cattle in colonial Carolina.

But what made the Carolina economy was rice. It was supposedly introduced in the 1690s by one of the early landgraves, Thomas Smith.

But the demanding techniques of its cultivation were learned from slaves who had cultivated rice in West Africa. The extensive tidal marshes of the Carolina low country, once diked to control the water flow, were ideal for rice growing, and the West Indies, the northern colonies, and England provided ample markets. The result was a bonanza. Carolina exported four hundred thousand pounds of rice in 1700 and forty-three million pounds forty years later. The so-called Rice Kings, families with names such as Middleton and Ball, became the richest British subjects in North America as Carolina rice added about a million pounds sterling a year to the GDP of the British Empire.

Later indigo, a plant that yielded a blue dye that was much in demand by the now burgeoning British cloth industry, which gave rise to the Industrial Revolution in the middle third of the eighteenth century, became a second highly profitable staple crop. Indigo, rice, lumber, and livestock made Charles Town the busiest port in the colonial South and its largest town, richly adorned by the houses of the wealthy merchants and planters, and the churches and institutions they built, such as the Charleston Library, the third oldest public library in the United States.

Because Carolina and the Chesapeake colonies found crops that had large export markets and could be produced more cheaply than elsewhere, they developed to a large extent as plantation economies. Because of economies of scale, these export crops came to be produced more and more on large landholdings, worked by large numbers of black slaves. This, in turn, produced highly stratified societies with a small, rich class of great planters who dominated the middling planters, the landless freemen, and, as a result, the politics of each colony.

Plantation economies, dependent on one or two cash crops, usually fail to develop other parts of a well-rounded economy. Charleston was the only place in the South that merited the term *city,* even by colonial standards. Local manufacturing languished as economic resources were devoted to the crops that were so profitable, and the southern colonies were thus dependent on importing needed goods and even foodstuffs from England and New England, which, lacking a cash crop, had developed very differently.

———

NEW ENGLAND WAS NOT FOUNDED by men bent, first of all, on adventure and profit. Instead, the most important reason for settling there was to build a "city on a hill," a place where saints—those destined to be saved—could live, unmolested by corruption, according to God's commandments.

But that city, to be sure, is a project still under construction after nearly four hundred years. And even saints, in the short term, have to eat, buy necessities, and pay for the costs of crossing an ocean to establish a New Jerusalem in what one Puritan called a "howling wilderness." Nor were the Puritans in the least averse to prosperity in this world as long as the worship of God came first. Indeed, they regarded it as a sign of God's grace, a sign that the individual was indeed saved. Sixteenth- and seventeenth-century merchants, many of them Puritans, would often write at the head of their ledgers, "in the name of God and profit."

They gave their own chances of prosperity a considerable boost as well by firmly believing that idleness was the devil's workshop and behaving accordingly. Puritans were always *doing*. The Jacobean playwright Ben Jonson gave a comic Puritan character in *Bartholomew Fair* the all too apt name of Zeal-of-the-Land Busy.

New England was not like the southern colonies in many ways besides the first impetus behind its founding. The climate was much cooler (but also much healthier), with a short growing season. And the soil was stony and thin, the landscape having been largely scraped clean to bedrock by the last glaciation of the ice age. (The soil pushed off New England by the glacier became Long Island, Martha's Vineyard, Nantucket, and Cape Cod.)

Nor were the early immigrants similar, other than being English. The immigrants to Virginia and Maryland came largely from southern England and were often desperately poor, while a few came from wealthy, landowning families. The Puritan emigration centered on East Anglia, the flat, agriculturally rich land that bulges out into the North Sea northeast of London as well as from the Home Counties, the most commercially developed part of the country. More, many of the Puritans were what today would be called middle class: small landowners or tenant farmers, shopkeepers, and skilled craftsmen, along with a good number of professionals such as doctors, lawyers, and especially clergymen.

Because many of its adherents were middle class and believed firmly in reading the Bible, early New England had perhaps the highest literacy rate in the Western world in the seventeenth century. When new towns were founded—as they were at an astonishing rate—a school was built almost as soon as a church. The Massachusetts Bay colony founded a college, Harvard, only six years after settlers arrived in Boston, and more than half a century before Virginia had a college of its own, William and Mary. A printing press was in operation in Boston as early as 1640.

Most of New England's immigrants were able to pay their own way and bring their families with them when they came. But they brought relatively few indentured servants and imported even fewer slaves, although slavery violated no tenet of their strict religion. While the ratio of men to women in early Virginia was about four to one, in New England it was only six to four. With a more nearly normal balance between men and women, New England society did not go through a "wild west" phase before settling down. Further, the deep religious commitment of most of its early settlers, and firm leadership of such men as John Winthrop and William Bradford, ensured that that society would be far more law-abiding than what developed in the southern colonies.

And because families could be more easily established and the healthy climate allowed more children to reach maturity, New England's population grew very quickly. Only about twenty-one thousand people immigrated to New England in the seventeenth century, but by the end of that century the population was ninety-one thousand, more than the white population in the Chesapeake, which had received many more immigrants.

Both Plymouth and Massachusetts Bay colonies were founded by joint-stock companies. The members of these corporations who came to New England were known as planters. Those who remained in England and invested money in the enterprise were called adventurers, a word that is still echoed today in the term *venture capitalist*.

These adventurers, while as anxious as any to build a New Jerusalem, were also hoping for a return on their investment as soon as possible. When the *Mayflower* returned to England in the spring of 1621 in ballast, the company directors wrote Governor Bradford a stinging letter admonishing him for not sending back a load of sellable goods.

The stony soil of New England made a highly profitable cash crop such as tobacco unlikely, and New England agriculture would never produce export crops in large quantity, although New England was an exporter of cattle to the West Indies. But if the soil was unpromising, the surrounding seas were rich with possibilities.

Captain John Smith had explored the coast of New England in 1614 and gave the area its name. He had, as always, been hoping to find gold. When he didn't, he set his men to fishing for cod.

Cod, a fish that can grow up to two hundred pounds, had been a staple of the European diet for centuries. Dried and salted, it would keep for months and was one of the main sources of animal protein. The traditional European codding grounds had been in the North Sea and westward toward Iceland. Then, in the fifteenth century, Basque fishermen had discovered far richer areas off the east coast of North America. Cod prefer fairly shallow waters, and a series of banks off New England and Atlantic Canada provide large areas of exactly that. Further, the south-flowing, cold Labrador Current and the north-flowing, warm Gulf

Stream meet over these banks, roiling the waters and stirring up large quantities of nutrients. The result is cod heaven, the richest fishing grounds in the world.

There were English fishermen living in New England even before the Puritan migration began in 1620, in villages in Maine, New Hampshire, and Massachusetts north of Boston, especially the towns of Marblehead and Gloucester, which are fishing ports to this day. Most of these fishermen were not Puritans, but a rowdier, less educated, and far less pious group. Marblehead, although well endowed with taverns, would not have a church of its own until 1684. But the cod they brought in proved the mainstay of the early New England economy. In 1641, as the English Civil War severely disrupted the British economy, New England shipped six hundred thousand pounds of dried cod to Europe and the West Indies. Thirty years later, six million pounds of cod were shipped out of New England.

An activity in one area often has unanticipated side effects in quite another area of the intricate ecosystem that is an economy. Sometimes these side effects are good, sometimes not. In this case they were wholly good. The vast quantities of cod waste—the skin, bones, heads, and guts of the fish that were left over from processing—were plowed into New England fields as fertilizer, greatly increasing the productivity of the soil.

But cod never dominated the New England economy the way tobacco did Virginia's. It was the largest export but not the only one, for New England's economy became by far the most diverse in British North America. Besides cod, New England exported lumber, ships' masts, soap, butter, cheese, and whatever small surpluses its farmers produced of wheat, peas, oats, and other crops.

Lumber, in fact, became the great cash crop of New England and its first great industry. As early as 1655 there were more than twenty sawmills on the Piscataqua River in New Hampshire. By 1705 there were seventy. Other New England rivers had as many. By the end of the colonial era, wood would be one of the largest of North American exports.

Between 1771 and 1773 New England exported just to the British West Indies seventy-seven million board feet of lumber, sixty million shingles, and fifty-eight million barrel staves.

One of the most important uses of New England lumber was the building of the ships that carried the other products. By the end of the seventeenth century, New England had become one of the great ship-building regions of the world, with great consequences for the New England economy as a whole.

Shipbuilding is a very complex enterprise, requiring large numbers of workers, many of them highly skilled in a number of crafts, such as iron work, sail making, rope production, timber cutting, and barrel making. A good-sized ship by seventeenth-century standards might require the coordinated efforts of two hundred workers or more.

And while New England had higher labor costs than England, because there was more competition for the labor of its artisans, it had far lower raw materials costs, especially for the major component of sailing ships—wood. As a result, New England could build a ship for about half the cost of building one in England. In the forty years between 1674 and 1714, Boston alone averaged forty ships a year, producing more than the rest of the North American colonies combined. Indeed, it was, after London, the greatest center of shipbuilding in the British Empire, with fifteen shipyards in operation by 1700.

And New Englanders were not just shipbuilders, they were soon major ship owners as well. By 1700 only the ports of London and Bristol within the British Empire outstripped Boston in shipping. The carrying trade that New England developed extended throughout the North Atlantic, the Mediterranean, the Caribbean, and beyond. And it carried far more than just New England products and imports.

New England was well named economically speaking, because its economy was the most like England's of all the British North American colonies. With a large fishing fleet and shipbuilding industry of its own, England had small need of New England's main products. But New England had an ever-expanding need for Britain's manufactured goods. In response, several "triangle trades" developed. New England lumber, fish, and meat was taken to the West Indies and traded for sugar and salt. Those went to Britain, where they were traded for manufactured goods, principally textiles and hardware, which were then sold in New England. West Indian molasses was distilled in New England into rum and then sent to Africa and traded for slaves. The slaves were sold in the West Indies. New England fish was traded in Spain and Portugal for wine and fruit, which was sold in Britain and paid for manufactures.

New England ships also carried much of the produce of the Carolinas and the Chesapeake to Europe and the West Indies. Going wherever opportunity beckoned, New England merchants, like all merchants, sought to buy cheap and sell dear. And they earned a well-deserved reputation for doing exactly that. Their success can still be seen in the increasingly grand houses that were erected in many New England seaports by the prosperous merchants.

ALTHOUGH NEW ENGLAND developed the most diverse economy to be found in North America, it still needed to import most manufactured goods. Britain was not yet the "workshop of the world," but it was on its way as the first stirrings of the Industrial Revolution began.

Among New England's greatest needs were iron goods, which were indispensable to nearly all economic activities. Iron is one of the most abundant elements on earth, but copper was the first metal used on a regular basis. The reason is that copper is much easier to work with. Copper has a melting point easily achieved in an ordinary fire, and some copper ores, when heated, produce pure copper with no further processing needed.

But iron is never found in a pure state (except, very rarely, in meteorites) because it is chemically highly reactive, which is why it rusts. Iron's melting point is so high that special techniques, such as bellows, are needed to reach it, and, when smelted, massive hammering techniques are needed to rid the metal of impurities such as carbon. So while artisan-ready copper can be produced by techniques available to a Boy Scout troop, iron requires an industrial enterprise.

Once the necessary technology was developed, however, by around 1400 BC, iron's superiority over copper and the harder bronze (copper alloyed with tin) was so apparent that it quickly became indispensable to civilization.

The early settlers in North America had no choice but to import the nails, horseshoes, pots and pans, plows, weapons, and a hundred other artifacts of seventeenth-century life that they needed. And because the nearest iron foundry was two months' upwind sailing away, iron goods were very expensive and thus a limiting factor in the development of the colonial economies.

But only eleven years after the fleet under the command of John Winthrop sailed into Massachusetts Bay and established the colony there, Winthrop's son, also John, was on his way back to England to arrange for the creation of an ironworks in his adopted land.

The Winthrops had been a gentry family in Suffolk, and John Winthrop the elder had practiced as a lawyer when he agreed to become governor of the projected colony. It was he who coined the enduring American phrase, "a city upon a hill." His son was also a lawyer and deeply interested in both science and commerce. In 1664 he would be elected a fellow of the newly founded Royal Society, the first American so honored.

The younger Winthrop, who had served as deputy governor of Massachusetts and governor of Connecticut, had set up a saltworks to supply the new colonies with another indispensable commodity that was very expensive to import. But to establish an ironworks he needed an economic necessity that America as yet wholly lacked: capital. The one place he could hope to get it was England, which would be the source of much American capital for more than the next two centuries.

One would think it would have been a tough sell to get capitalists to invest in a major industrial interprise located three thousand miles away in the middle of a wilderness. But a business plan, a few salient facts, and a persuasive salesman is often all it takes to attract capital to an untried idea. Winthrop could point to one great American comparative advantage: wood. Charcoal, which is wood heated in the absence of air until it is reduced to pure carbon, is as necessary to iron production as the ore itself. England's forests were being rapidly cut down, and coal was only beginning to be exploited for industrial uses. America had an unlimited supply of wood that was free for the taking.

Winthrop argued that by using America's cheap raw materials, Massachusetts could manufacture iron goods that could be sold profitably not only in New England and the Chesapeake, but even in England itself. Winthrop must have been persuasive, as he raised 1,000 pounds from various investors, including Lionel Copley, one of England's most prominent ironmongers. Eventually a total of 15,000 pounds would be invested in the project. To give some idea of what that sum represented in Massachusetts in the 1640s, consider that the highest annual salary in the colony in 1648 was 90 pounds, paid to the Reverend Zechariah Symmes of Charlestown. (It says much, of course, about early Massachusetts that the highest salary in the colony would be paid to a clergyman.)

Winthrop, back in Massachusetts, scouted locations for his ironworks and got the government to grant his "Company of Undertakers" a monopoly on iron production in the colony for twenty-one years and an exemption from taxes. Unfortunately, Winthrop chose badly for the site of the first furnace, in Braintree, south of Boston, where both the iron ore supplies and water to power the mill proved inadequate.

With Winthrop busy with many projects, and increasingly interested in Connecticut, where he would be governor for the last thirty years of his life, the company decided to import expertise, another commodity for which this country would often depend on Britain until long after independence. They hired Richard Leader, thoroughly familiar with running an iron business, and arranged as well to bring in several skilled ironworkers.

Leader built an ironworks in Lynn, north of Boston, on a site that is now in the town of Saugus, on the Saugus River. By 1646, only sixteen years after John Winthrop the elder had first stepped ashore, a major industrial enterprise was in operation in Massachusetts. But it got off to a rocky start, with frequent accidents as the local workmen painfully and sometimes fatally sought to learn the demanding techniques of iron production. Meanwhile, the imported workers were causing troubles of their own. Selected for their skills, not their piety, like the fishermen, they fit uneasily into Massachusetts society. The early court records of Lynn are filled with cases involving ironworkers hauled in for drunkenness, adultery, nonattendance at church, and other sins against the Puritan world order.

Dr. Robert Child, one of the English investors who happened to be visiting Boston, wrote to Winthrop in 1647, grumbling that "our Iron works as yet bring in noe considerable profit." But another investor noted that "Every new undertaking hath its difficulties." Still, by the end of the summer of 1648, the five-hundred-pound hammer imported from England and attached to the waterwheel powered by the Saugus River was pounding steadily away, driving the impurities out of the smelted iron, and John Winthrop the elder could write his son that "the furnace runnes 8 tun per weeke, and their barre Iron is as good as Spanish."

Heavy industry had come to North America. The Saugus Iron Works by that time was also displaying another aspect of the aborning American economy. In 1646 a blacksmith at the works, Joseph Jenks, received a patent for a device described as "engines for mills to goe with water," for manufacturing edged tools, such as scythes. This is perhaps the very first instance of the "Yankee ingenuity" that has so characterized the American economy, and often astonished the world, ever since.

But while the Saugus Iron Works was turning out increasing quantities of both pig iron and products made from it, it was still not producing a profit. The stockholders began pressing for changes in production and management that they hoped would make the works profitable. But matters only continued to deteriorate financially, and in 1653 the troubles exploded in a blizzard of lawsuits that reached even the attention of the man then ruling England, Oliver Cromwell, the Lord Protector.

While the lawyers argued, the works crumbled and many of the workers drifted off to other opportunities, including the new ironworks erected by Winthrop the younger in Connecticut. By 1676 the first ironworks in America was in ruins. The town of Lynn petitioned to have the dam removed so that alewives could once more run up the river to spawn.

But if the first industrial-scale enterprise in North America was a failure, the idea behind it—that there was money to be made producing iron in the American colonies—was sound. By the end of the colonial era, a hundred years hence, the colonies were producing one-seventh of the world supply of pig iron.

Chapter Three

THE ATLANTIC EMPIRE

L IKE THE NEW ENGLAND COLONIES and Virginia, New York was founded by a profit-seeking corporation. But the Dutch West India Company was a far larger concern than the startups that founded the other colonies. It had been created in 1621 and given a monopoly of trade in an area that stretched from West Africa to Newfoundland. The colony of New Netherland was established by the company at a cost of 20,000 guilders to exploit the fact that the Hudson River (called the North River by the Dutch) gave an easy entry to the source of the furs so much in demand in Europe. In the first year, the company shipped to Europe 45,000 guilders worth of furs, easily recouping the cost of establishing the colony.

In the early seventeenth century, the Dutch had the most advanced and most market-oriented economy in Europe. They invented or developed to new levels of sophistication stock and commodity exchanges, insurance, and corporate governance. They also had the most religiously tolerant government in Europe. Both the Dutch capitalist spirit and religious freedom were soon implanted in their new colony in North America. When the governor, Peter Stuyvesant, a sincere member of the

Calvinist Dutch Reformed Church, tried to expel Quakers and Jews from Nieuw Amsterdam, they appealed to the Dutch West India Company in the Netherlands in a document known as the Flushing Remonstrance. The company quickly wrote Peter Stuyvesant, telling him in no uncertain terms to mind his own business so that the Quakers and the Jews could mind theirs.

As early as the 1640s, while its population was still under a thousand, the little city at the tip of Manhattan Island was the most cosmopolitan in North America. A French priest counted no fewer than eighteen languages being spoken on the street in that decade. Almost all of those thousand citizens were there to make money. The Dutch didn't even get around to building a proper church for seventeen years. Indeed, the Dutch purpose for being in the New World could hardly have been clearer. The seal of New Netherlands was a beaver encircled with wampum, the form of money used by the Indians.

While the Dutch ruled on the Hudson for only forty years, they left a deep impress on the city they founded. It is still, nearly 350 years after the English seized it in 1664, the most commercially minded great city in the world.

———

LIKE THE PURITANS, the Quakers came to North America to escape religious persecution. But, also like the Puritans, they regarded prosperity as a sign of God's approbation.

William Penn was the son of Admiral Sir William Penn, who owned estates in both England and Ireland, and to whom the crown owed the large sum of 16,000 pounds. He had become a Quaker in his youth but stayed on good terms with both King Charles II and his brother James, Duke of York, owing to his connections and considerable income. In 1681 Charles, in exchange for canceling the debt, granted Penn a vast area of land west of the Delaware River, amounting to more than forty-five thousand square miles. These nearly thirty million acres made Penn the largest private landowner in history.

But, of course, it was just wilderness when he acquired it, and Penn wanted both to establish a "Holy Experiment" in America and to prosper in the process. "Though I desire to extend Religious freedom," he explained, "yet I want some recompense for my trouble." He certainly achieved the first part of his ambition, granting the colony nearly complete toleration and no established church, not even the Friends. He got fellow Quakers to purchase 750,000 acres, raising 9,000 pounds to finance the colony.

Pennsylvania (named for Penn's father, not himself) developed with extraordinary speed. Virtually uninhabited in 1680, only six years later it had a population of more than eight thousand. Forty-three ships, bringing three thousand settlers, arrived in the first two years. As with New England, they were largely intact families, and they multiplied rapidly in the new colony's temperate climate. By 1700 Pennsylvania's population was more than eighteen thousand, and Philadelphia was rapidly becoming the largest city in British North America. By 1776 Philadelphia would be second only to London among the cities of the British Empire, with a population, including its suburbs, of about forty thousand.

Pennsylvania's soil was far better than that of New England, and the colony had a longer growing season. Its ever growing number of farmers produced ever growing surpluses that could be traded. But the crops, such as wheat, that grew well in Pennsylvania (and in New York) were also those that grew in England. So there was a limited market in the mother country for the agricultural produce of the middle colonies. They, like the New Englanders, needed to develop other markets. The surplus wheat was turned into flour and sold to the West Indies, both French and British, in exchange for sugar and molasses. The export of flour became so important to New York's economy that the seal adopted by the English in 1686 shows two flour barrels between the blades of a windmill.

Because the middle colonies, like New England, could not rely on a cash crop that was in demand in the mother country, they did not develop economically as typical colonies of that time did. The planters in colonies such as Virginia depended on so-called factors in England to market their tobacco for them, and to function as bankers and purchasing agents, shipping back to Virginia goods that were not obtainable there. Pennsylvania and New York, like New England, developed their own merchant class, every bit as sophisticated as that in the mother country and with contacts spread just as widely over the globe.

The area in which Britain traded greatly expanded in the seventeenth century as its American empire expanded as well. In 1600 the overwhelming bulk of English trade was with its neighbors in northwestern Europe. A hundred years later Britain had overtaken the Netherlands as Europe's greatest trading nation. English ships reached as far as India, and trade with Asia and America accounted for 40 percent of the British merchant marine.

Not surprisingly, the government in London wanted to regulate this trade, and for two reasons. The first reason, of course, was to be better able to tax it. The second was to make it conform with the principles that then dominated economic thought. Those principles are now known as the "mercantile system," a term coined, as so often happens, by one of its enemies, Adam Smith. (It would be Karl Marx who would coin the word *capitalism*.)

Mercantilism held that the best form of wealth was precious metals: gold and silver. If a country lacked mines of the precious metals, it should export as much as possible and import as little as possible, so as to run a favorable trade balance and thus accumulate gold and silver. This idea was never universally accepted. Sir Dudley North (1641–91), for instance, demonstrated that the idea that one nation could grow rich only at the expense of another was a fallacy and that the more trade, both exports and imports, the better. Adam Smith would draw heavily on North and others in writing *The Wealth of Nations*.

But mercantilism coincided with powerful economic self-interests among merchants and manufacturers who wanted protection from foreign competition, and it held intellectual sway until Adam Smith blew it away with one of the most powerfully argued and influential books in Western history.

In 1651 England began passing a series of Navigation Acts to regulate the trade of its American colonies. These acts restricted the colonies to using ships built, owned, and manned by British subjects. The Dutch, far more efficient merchant mariners than the English in the mid-seventeenth century, were able to profitably ship the tobacco of the Chesapeake to Europe for as much as a third less than English ships could. But as the English merchant marine grew and as New England became a major shipping center in its own right, shipping costs declined even without Dutch competition.

The Navigation Acts also required that certain commodities exported by the American colonies could be shipped only to England. Many of these commodities—tobacco, rice, sugar, indigo, furs, copper, and naval stores (tar, pitch, and turpentine)—were reexported to continental Europe. This assured both that these commodities passed through English customs and were taxed, and that English merchants handled the trade with Europe. Other colonial exports, such as flour from the middle colonies and pig iron, could be exported by the colonies directly to wherever markets could be found.

Third, the Navigation Acts required that European goods imported to America had to pass first through England and, of course, English customs, except for certain products of southern Europe that England didn't produce in the first place, such as wine from Spain, Madeira, and the Azores. The main purpose of this legislation was to protect the American market for British manufactures. But as Britain quickly became the most efficient producer of these goods in Europe, British manufacturers almost always offered better prices anyway.

As the American colonies continued to develop in the eighteenth century, Britain placed increasing restrictions on American manufactures to protect its burgeoning domestic industries. No products were expressly forbidden, but the size of markets was limited and the building of new factories and mills for producing certain goods was forbidden.

Had these acts been scrupulously enforced, they would have severely impacted the developing economies of the North American colonies. Ships and cargoes that contravened the Navigation Acts were liable to seizure. But they were not enforced scrupulously. At times some of them were hardly enforced at all, and at all times a well-placed bribe could usually produce a studied lack of attention on the part of those charged with their enforcement. The office of collector of the port in the various American harbors was a particularly prized one, because of its profit potential. Throughout the colonial period, smuggling was rampant.

———

WHILE THE AMERICAN ECONOMY grew ever more productive and complex during the colonial era, it was never a complete economy, and the colonies remained dependent on the mother country for certain goods and services they could not provide for themselves. One of these services was banking.

British law effectively forbade the establishment of banks in the colonies and also forbade the export of British coinage from Britain, to preserve its own money supply. This left the colonies to create a money supply as best they could.

Money is a commodity, no different from pork bellies, legal services, or computer keyboards, except in one vital respect. Money, by definition, is a commodity universally acceptable in exchange for every other commodity. Money is one of the seminal inventions of *Homo economicus*. In a barter economy, someone wanting, say, to sell oranges and buy apples, needs to find another trader who has apples and wants oranges. Economists, with their usual talent for a clunky phrase, call this a "double coincidence of want."

But in an economy that uses money, the first trader can sell his oranges for money to anyone who wants oranges and use the money to buy apples from anyone who has apples for sale. This enormously increases the number of transactions that can take place in an economy. Thus money functions economically in much the same way that a catalyst does in chemistry: it speeds up reactions while remaining itself unchanged.

Money serves two other functions besides acting as a medium of exchange. It is a unit of account; that is, the value of all other commodities is expressed in terms of money. And money acts as a store of value, a place to hold wealth temporarily between productive investments.

Many commodities have functioned in some respects as money. Cattle were often used, and indeed still are in some cultures. (The English word *pecuniary,* in fact, comes from the Latin *Pecus,* meaning *ox.*) When metals came into use, they were pressed into service as quasi money and had many advantages over cattle. A bar of copper, for one thing, can be divided to make smaller units. When pieces of metal that had traded by weight were stamped with a certain value and traded by count, they became coins. Coins are real money, a commodity with no other function but to act as money.

Gold, silver, and copper, being more valuable than iron, were the usual metals for coins. And because they were elements, they could not be manufactured, only dug, expensively, out of the ground. But if they could not be manufactured, they could be debased, alloyed with cheaper metals while carrying the same face value, or short-weighted. Rulers frequently resorted to these expedients when short of funds, as rulers usually are. The long-term result, of course, was always the same. The value of the debased coins fell relative to other commodities as people in the economy adjusted their notion of the value of the coinage to take the debasement into account. Because money is a special commodity, we have a special economic term for a fall in its price: inflation.

But inflation can be caused by other things than governments trying to pay their bills with shoddy goods. As we have seen, the vast influx of gold and silver to Spain from the New World in the sixteenth century caused a great inflation throughout the European economy. The reason was simply the inexorable operation of the law of supply and demand. As the supply of money (gold and silver) rose relative to other goods, the price of money fell.

With the English embargo on exporting coins, the new English colonies in America had to solve the problem of getting a money supply another way. In 1652 Massachusetts began minting its own coins, despite strict laws forbidding anyone other than the royal mint to do so. The pine tree shilling, the first coin minted in North America, was scrupulously produced. People had to bring in their own silver and have it assayed before coins with a silver content equal to three-quarters of an English shilling were minted from it. The pine tree shilling was such an asset to the Massachusetts economy that the British government did not suppress its production for more than thirty years. Only when the original Massachusetts charter was revoked in 1684 was the mint ordered closed.

Other colonists turned to the nearest thing to an international monetary standard then in existence, the Spanish dollar. The Spanish dollar accounted for perhaps half the coinage in circulation in the North American colonies, the rest being a hodgepodge of British coins brought by travelers, French coins, and so on. But because of the drain on American specie owing to the fact that the colonies all ran persistent trade deficits with Britain, it wasn't nearly enough to meet the demand for money.

As with any superior technology, the English settlers of North America, used to using money in their economic exchanges, wanted to continue enjoying the benefits a money economy. They looked for substitutes for "real money."

In New Netherlands and elsewhere, the fur-trading Indians used wampum as a medium of exchange, and so too did their Dutch- and English-speaking customers. Wampum is beads made from the shells of the freshwater clams that abound in the local lakes and rivers. They were sewn onto leather belts in elaborate patterns. Steel drills made it much easier to drill the holes in each bead and, therefore, greatly increased the amount that could be produced with a given amount of labor. This caused the value of wampum to decline significantly, but it continued in use as money until the latter part of the eighteenth century. In 1760, however, J. C. Campbell of New Jersey opened a factory for making counterfeit wampum, destroying the value of the genuine article.

Maryland and Virginia resorted to what economists call "commodity money," using tobacco. The trouble with commodity money, however, whether tobacco or cattle, is that the commodity can be difficult and expensive to transport, vary in quality, and fluctuate in real value. When the price of tobacco collapsed in the 1680s as production in the Chesapeake caught up with and then temporarily surpassed world demand, the colonies were economically devastated. "Tobacco, our money, is worth nothing," one Marylander complained, " . . . and [there is] not a shirt to be had for tobacco this year in all our country."

Legislation setting standards for minimum quality revived the price of tobacco, and it again was used universally as money in the tobacco-growing areas. In 1696 Virginia clergymen were paid sixteen thousand pounds of tobacco a year in salary. By the turn of the eighteenth century, the legislatures had established tobacco as a legal tender for paying taxes and public debts.

In 1730 Virginia set up an inspection system, requiring planters to bring their tobacco to public warehouses where it would be inspected and warehouse receipts issued for its value. These warehouse receipts functioned in the same way as banknotes, although they fluctuated in purchasing power far more, being tied to a volatile commodity, tobacco, instead of gold and silver. Maryland quickly followed Virginia's lead.

Banknotes had been issued by English banks since the turn of the eighteenth century. Redeemable in gold on deposit in the banks, they were negotiable instruments, the gold represented by them belonging absolutely to the holder in good faith. Banknotes had many advantages over other forms of money. They were cheap to produce (originally they were handwritten) and much easier to carry around than the gold or silver that backed them in the vaults of the issuing banks.

With no banks, American colonies could not use banknotes. That did not stop them from issuing paper money. In 1690 Massachusetts raised troops to fight against the French in King William's War (called the Nine Years' War in Europe). To pay them, the colony issued bills of credit—promises to pay in the future. Issued in denominations of 5, 10, and 20 shillings, they read, "This indented Bill . . . due from the Massachusetts Colony to the Possessor shall be in value equal to money and shall be accordingly accepted by the Treasurer and Receiver Subordinates to him in all Public payments and for any stock at any time in the Treasury-New England, February the third, 1690. By order of the General Court."

Because they were legal tender in payment of taxes and other government obligations, they circulated as money (although often at a discount from face value). These instruments were not only the first paper money issued in North America, they were the first paper money issued in the Western world.

The idea worked so well that it soon spread to other colonies in New England and to Pennsylvania, which issued its first paper money in 1723. Benjamin Franklin, in 1729 when he was only twenty-three, published a pamphlet entitled "A Modest Enquiry into the Nature and Necessity of Paper Currency." He was soon rewarded with a contract to print future issues of Pennsylvania Δbills of credit and, typically, devised several means of foiling counterfeiters, some of them still in use to this day.

Franklin, however, minimized the fatal flaw inherent in what economists call "fiat money," money that is money only because the government says it is money rather than being made of or backed by a valuable commodity. Since the earliest days of civilization, politicians have faced tough choices between raising unpopular taxes and controlling popular spending. When they possessed the power to pay their bills with cheaper money, such as by debasing or short-weighting the coinage, all too often they did so.

But coins must be made of some metal and are expensive to produce. Paper money costs almost nothing to produce. The temptation to use it as a short-term solution to fiscal problems has proved irresistible to politicians. The politicians in British North America were no exception. The Massachusetts government, issuing more and more paper money, soon drove gold and silver coins out of circulation, owing to the operation of Gresham's law ("Bad money drives out good"). People passed the paper money but kept the specie in the mattress because they regarded it as a superior store of value, which it was.

Inflation soon caused the paper money to shrink in value. In 1716 Massachusetts abolished paper money and imported Spanish dollars, but soon was back to printing bills of credit once more. Every province but Virginia eventually issued various forms of paper money, but it never replaced other forms of money. In North Carolina in the 1730s, there were no fewer than seventeen different forms of legal tender.

The one unifying factor was that the pound sterling was used as the universal unit of account, even though British coins made up only a small portion of the specie in circulation and British banknotes hardly circulated at all. The costs of evaluating and converting as needed the various forms of money was a very considerable cost on the aborning American economy.

———

DESPITE THE MONETARY CACOPHONY that characterized its economy, by the middle of the eighteenth century British North America was prospering as few other places on earth. Besides being a major, sometimes dominant, exporter of agricultural products and raw materials, builder of ships, and trader, it was increasingly supplying its own needs in manufactured items.

Nearly every town and village had a blacksmith, cooper, wheelwright, cobbler, carpenter, tanner, and other artisans able to supply local demand. Flour and sawmills grew in number and size of their operations. In the larger towns, many artisans were expanding their businesses into proto-industrial enterprises. William Johnson (1741–1808) had been born in New York City but came to Charleston, South Carolina, as a young man and established himself as a blacksmith. He was soon much more, running a substantial business with employees, apprentices, and slaves turning out iron products in large numbers. Johnson was successful enough to become a major landowner and join Charleston's elite as a longtime member of the legislature and a vestryman of St. Philip's Church. His son, also William Johnson, would serve on the U.S. Supreme Court for thirty years.

Boston, New York, Philadelphia, and the other American cities all had an increasing number of William Johnsons. By this time American rum distilleries were supplying 60 percent of the American market, and a growing number of sugar refineries were converting the West Indian brown sugar into the white sugar increasingly preferred by American consumers. The trade in rum, refined sugar, foodstuffs, and manufactured goods between the colonies was increasing rapidly. In 1770 some 20 percent of the tonnage clearing the port of New York was bound for other colonial ports, not Europe or the West Indies.

The British North American economy was ceasing to be colonial in nature and was becoming more like that of the mother country, a diversified and developed one. A clear sign of that is the growing production of luxury goods in the colonies. In 1774 Philadelphia had more than three hundred workers engaged in the production of carriages. Cabinetmakers such as Thomas Elfe in Charleston, Thomas Affleck in Philadelphia, John Townsend and John Goddard in Newport, and

John Cogswell in Boston were producing masterpieces of furniture design that were every bit the artistic equal of those produced in Britain at the same time. Painters such as John Trumbull were doing a brisk business in portraits.

And that economy was not only developing rapidly, it was growing rapidly as well. The population of the thirteen colonies nearly doubled from 1750 to 1770, from 1,176,000, to 2,131,000, thanks both to immigration and to very large families, most of whose children reached maturity. The settled area of the colonies was also growing quickly and by the eve of the Revolution was about 180,000 square miles, half again as large as Britain and rapidly approaching the size of France, the largest country in western Europe.

This prosperity was widely shared among the population. Although in the 1770s the top 20 percent of the population owned about two-thirds of the wealth, while the bottom 20 percent owned only 1 percent, that raw datum gives a distorted picture because it does not take time into account. (Modern statistics do exactly the same thing, now usually for tendentious, political reasons.) The population of British North America was a very young one, and children usually do not possess significant wealth. As people get older they tend to get richer, and that was certainly true in the thirteen colonies. One economic historian has calculated that of the colonial population in their forties, only about 8 percent would have been considered poor by the standards of the day, and even fewer in their fifties.

The reason was, simply enough, that colonial America before the Revolution was a land of opportunity such as the world had not yet seen. The economy of the Western world in the mid-eighteenth century was beginning to change as the first effects of the Industrial Revolution stirring in the English midlands were felt. But it was still dominated by agriculture, and land remained the basis of wealth. Europe and the West Indies had little or no undeveloped land suitable for agriculture.

Canada, while rich in land, had a very short growing season, limiting the number of crops that could be grown and the return from them.

But the thirteen colonies had millions of acres of potentially rich land for the taking. If the family farm did not have enough land for all the children, the frontier, where there was land aplenty, was seldom more than a day or two away by horseback. Moving on was soon an American characteristic, and America to this day is the most mobile society on earth.

And it wasn't just the native-born children who felt the siren song of financial independence being sung in America. There was a steady flow of immigrants, although the number varied from year to year, climbing steadily after 1750, as word of America's prosperity and opportunity spread. To raise the cost of their passage, these people were willing to accept a limited term of slavery as indentured servants.

In 1767 Sir Henry Moore, royal governor of New York, explained that "as soon as the time stipulated in their indentures is expired, they immediately quit their masters and get a small tract of land, in settling which for the first three or four years they lead miserable lives, and in the most abject poverty. But all this is patiently borne and submitted to with the greatest cheerfulness, the satisfaction of being land holders smooths every difficulty and makes them prefer this manner of living to that comfortable subsistence which they could procure for themselves and their families by working at the trades in which they were brought up."

This willingness to accept present discomfort and risk for the hope of future riches that so characterized these immigrants, and the millions who would follow over the next two centuries, has had a profound, if unmeasurable, effect on the history of the American economy. Just as those who saw no conflict between worshipping God and seeking earthly success in the seventeenth century, those who sought economic independence in the eighteenth had a powerful impact on the emerging American culture.

And while there was a growing elite, the British colonies also had a larger percentage of their population in the middle class (not that the term was known in the eighteenth century, which called them the "middling sort" instead) than any other area of the Western world. The prosperity was very widely shared. The native-born American soldiers in the Revolution, for instance, averaged a full two inches taller than their British counterparts, who were overwhelmingly of the same genetic stock and, indeed, were often closely related. That can only be attributed to a far better childhood diet.

Of course, none of this prosperity applied to the slaves, who, by definition, had neither wealth, nor income, nor prospects, only endless unrequited toil for the benefit of others. But if the slaves were not free, at least their plight was increasingly being recognized and condemned. The first stirrings of the antislavery movement began in the late seventeenth century in England. George Fox (1624–1691), the founder of the Society of Friends, denounced slavery (but William Penn owned several slaves). Quakers would long be in the forefront of the abolition movement when it arose a century later. Many non-Quakers decried the horrors of the slave trade and the treatment of the slaves on the plantations in the West Indies, but not the institution of slavery itself. By the mid-eighteenth century, however, the institution itself was coming under attack.

The changing attitude of Benjamin Franklin, whose life spanned the eighteenth century, mirrors the evolving attitude of society at large. In his youth he regularly ran advertisements in the *Pennsylvania Gazette* for slaves he was selling. And he owned two, George and King, who worked in his household. By 1750 he regarded slavery as injurious to the welfare of a country because it bred a contempt for labor, and he thought that slavery was economically inefficient at best. By the last decades of his life, however, he was an abolitionist. The first abolitionist society was founded in Philadelphia in 1775, and Franklin accepted its presidency in

1787. By that time even major slaveholders such as George Washington and Thomas Jefferson thought that slavery was immoral, but neither knew how to rid the country, or even their own plantations, of it.

In 1772 Lord Mansfield, the lord chief justice, ruled that slavery was contrary to the common law and that slaves became free as soon as they set foot in the United Kingdom. But that, of course, did the slaves in the colonies no good. Their bondage continued even as the opprobrium regarding it spread ever deeper through society.

———

BY THE LATTER PART of the eighteenth century, Americans were quickly ceasing to think of themselves as colonial dependents of a mother country. The very word "American" is evidence of this. It was first used to denote a person of European descent living in British North America only in 1765, but quickly gained currency thereafter. Instead Americans increasingly thought of themselves as being both loyal subjects of the British crown and also fully the equal of other British subjects living in other parts of a great and growing Atlantic empire, including the mother country.

And like all British subjects, they felt themselves the heirs of a struggle for liberty, a struggle that was already more than five hundred years old.

The government in London had largely ignored the American colonies during much of the seventeenth and eighteenth centuries, other than to use them as a convenient dumping ground for convicts and other undesirables, and to protect the economic interests of both the crown and those with influence in Parliament, such as West Indian sugar planters and British merchants.

But the geopolitical position of Great Britain had been rapidly changing during that time. In the reign of Charles II Britain had been, at best, a middling European power. Charles II had even accepted a secret subsidy from Louis XIV, in return, of course, for acquiescing in French ambitions. But when the Glorious Revolution of 1688 replaced the openly Catholic brother of Charles, James II, with the Protestant and virulently anti-French William III, Britain embarked on a seemingly endless series of wars with France that had the result of transforming Britain into a Great Power.

Britain's secret weapon in these wars was its advanced system of taxation and its ability to finance its military through a modern national debt. At a time when the other major European powers were still using tax farmers (men who promised a certain revenue to the government in exchange for the right to collect taxes in a given area, keeping any surplus for themselves), Britain turned its tax collectors into bureaucrats. As a result, a much higher percentage of taxes raised actually made it into the British treasury.

Before the eighteenth century, government debts were usually the personal debts of the sovereign, individually negotiated with the lenders. But in 1694 the British government gave a corporate charter to the Bank of England, which evolved in the next few decades into the country's central bank. The Bank of England soon was arranging government loans, issuing bonds that could be bought and sold in the marketplace. The effect was to greatly liquefy the nation's available wealth. Instead of keeping surplus capital in the form of gold and silver, investors could keep it in bonds that had a ready market and provided a steady income. These bonds could, in turn, serve as the collateral for personal loans to provide working capital for new enterprises. As a result, the British economy grew rapidly in the eighteenth century as more and more of its capital could be mobilized.

And a true national debt allowed Britain to wage war successfully with countries that were far larger in population and more richly endowed with natural resources. As the Roman statesman Cicero had explained two thousand years earlier, "the sinews of war are infinite money." Because of its national debt, Britain became the linchpin of European power politics.

But Britain's new status as a Great Power didn't come cheap. Its national debt had stood at 16.3 million pounds in 1700, at the end of the Nine Years' War. By 1748, at the end of the War of the Austrian Succession, it stood at 76 million pounds. Fifteen years later, at the end of the Seven Years' War (called the French and Indian War in North America) it was 131 million pounds, an almost incomprehensible sum to a society where a family could live in reasonable comfort on 100 pounds a year and where an annual income of 1,000 pounds made one very rich.

The French and Indian War, fought across the globe from Fort Duquesne in western Pennsylvania to India, had ended in British triumph. France was forced to cede its North American empire to Britain, ending the menacing French presence at the back door of the British colonies. With the Royal Navy commanding the Atlantic, and ten thousand British soldiers stationed on the frontier to keep the peace with the Indians, the colonies were, for the first time in their short history, secure from foreign attack.

The British government, saddled with servicing a huge national debt—60 percent of the government's budget in these years went to paying interest on it—while continuing to fund a large military, sought new sources of revenue. With the British Empire in North America rapidly becoming a major economic power in its own right, it is not surprising that Britain looked there. The colonists, after all, had greatly benefited from the outcome of the French and Indian War, and they were far more lightly taxed than were British subjects living in the mother country. The average Briton paid 26 shillings a year in taxes, the average American only one. It was only fair, the government in London quite reasonably thought, that the colonies should contribute more to the costs of empire.

Further, the closer attention that the French and Indian War had brought to the colonies revealed things that did not please the government in London one bit. The customs officials in the British colonies were so corrupt and inefficient that they brought in a tariff revenue to the state that was only one-fourth the cost of administering the tariff. And the parts of the Navigation Acts that the colonial merchants found inconvenient, they simply ignored. In 1733, at the behest of British sugar interests, Parliament had required that all molasses imported to the North American colonies come from British sugar islands. But molasses was much cheaper in the French West Indies, and the American merchants kept right on buying it there, even during periods when Britain was at war with France.

People with an economic advantage, however "unfair" that advantage may be, will always fight politically as hard as they can to maintain it. Whether the advantage is the right to benefit from another's labor, or an unneeded tariff protection, or an exemption from taxation makes no difference. And because the advantage for the few is specific and considerable, while the cost to the disadvantaged many is often hidden and small, the few regularly prevail over the many in such political contests.

Certainly the colonists furiously resisted the very light taxes Britain sought to impose and the tightening of the trade regulations. They argued that British subjects could only be taxed by their representatives in Parliament, and they were represented in their colonial assemblies, not in the Parliament at Westminster. Ergo Parliament had no power to tax them. And, as British subjects were all equally entitled to liberty, what right had Parliament to pass laws advantageous to British merchants at the expense of colonial ones?

As so often happens in a family quarrel, as the argument deepened, neither side made much of an attempt to understand the other's point of view, while more and more grievances were aired. The colonists frequently stated that knuckling under to British demands would reduce them from free men to slaves, a condition of servitude with which they were all too familiar. And, realizing that they no longer needed British military power for their own defense, they could not see its necessity for the empire as a whole.

The British commentary in the crisis, on the other hand, almost invariably uses words like *plantations,* and *children,* in reference to the colonies and their inhabitants. They were subordinates and needed to be treated as such. Further, most of the British establishment had no doubt that if push came to shove, their superb military could deal easily with any resistance the colonies might mount.

But they were wrong. William Pitt, Earl of Chatham, who as prime minister had been responsible for the triumph of the French and Indian War, knew better. "You cannot conquer America," he told the British Parliament. But the members were in no mood to listen.

PART II

A COUNTRY THAT COULD
MAKE ITSELF AS IT PLEASED

———

Men make their own history, but they do not make it as they please;
they do not make it under circumstances chosen by themselves,
but under circumstances directly encountered and transmitted from
the past.

— Karl Marx
*The Eighteenth Brumaire
of Louis Bonaparte*, 1852

THE AMERICAN REVOLUTION

T HE COUNTRY THAT DECLARED its independence on July 4, 1776, had many advantages in the military struggle with Britain that was already under way. Finances, however, was not one of them.

The United States was fighting on its home ground and could react quickly. Britain had to fight from a distance of three thousand miles and with a communications time lag of at least three months, often four. The American military commanders and politicians were intimately familiar with that ground; their British counterparts were often profoundly ignorant. Most of all, the United States had only to avoid losing the war until the British government and people tired sufficiently of the struggle and its mounting costs. Britain had to defeat and pacify a vast country awash in rebellion.

But Britain had virtually unlimited financial resources; the Americans had hardly any. Because of those resources it could deploy the largest and best navy in the world (although it had been allowed to decay considerably since the end of the Seven Years' War). The British army was second to none in training and equipment, and could be easily augmented with hired foreign troops. The Americans had to scratch together what forces they could, using state militias and privateers as much as if not more than the Continental Army and Navy.

They also had to scratch together a means of paying for the war. That was not easy, especially as there was, in a very real sense, no national government. The thirteen states, having thrown off British control, were not willing to cede much of their new-gained sovereignty to a central government. The Second Continental Congress had no power to tax. Instead it had to estimate its revenue needs and then call on the states to provide the money. With their own war efforts to fund, few did, and only about 6 percent of total revenues came from taxes.

The rest had to come from borrowing, some from wealthy Americans committed to the cause, but mostly from France and Holland, who were both, of course, far more interested in humbling Britain than in helping the Americans. Along with money, they also supplied about 60 percent of the gunpowder used by American forces, as they did most of the uniforms and firearms. Even the British, quite unintentionally, provided much military matériel for the American forces. During the course of the war, American privateers seized some two thousand British vessels, worth, together with their cargoes, some 18 million pounds.

Beyond borrowing, the only source of revenue was the printing press. Beginning in 1775 the Continental Congress issued negotiable bills of credit, called continentals. By the end of 1779 it had issued bills with a face value of no less than $225 million, a huge sum relative to the size of the American economy at that time. This ballooning of the money supply (made still worse by states and even counties doing the same) inevitably caused a huge inflation. Prices doubled in 1776 and doubled again the next year and the next. From early 1779 to early 1781, prices rose nearly tenfold. Congress tried to stem the spiral by revaluing the continentals already in circulation at 2.5 percent of their face value. They quickly depreciated into near worthlessness. The phrase "not worth a continental" would be part of the American lexicon for more than a hundred years.

Many farmers had no choice but to accept quartermaster and commissary certificates—which also circulated as money—at whatever value the requisitioning officers chose to place on them at the time they forcibly purchased supplies. Fortunately, the British sometimes acted even more arbitrarily, seizing livestock and grain as spoils of war.

And because the Continental Congress had no experience in administering a large bureaucracy (the quartermaster's department of the Continental Congress had more than three thousand employees at one point), chaos, corruption, and inefficiency reigned. Only when Robert Morris, a highly successful Philadelphia merchant, took charge in 1781 did some semblance of order come to government procurement and finances.

Most important, Morris was able to arrange financing to allow Washington to move the Continental Army from New York State to Yorktown, Virginia. There, with the French fleet acting as the stopper in the entrance to the Chesapeake Bay, cutting off relief, Lord Cornwallis was forced to surrender the main British army in North America.

If the British war effort was to continue, London would have to raise, equip, and transport a new army. With the national debt increasing rapidly (it was already well over 200 million pounds at this point), there was little political support for doing so. The British began negotiating a peace treaty that resulted in formal recognition by Great Britain of American independence in 1783.

The United States had won by not losing.

BUT IT HAD PAID A FEARFUL PRICE. Much of the Carolinas and parts of Virginia had been devastated by British troops destroying farms and plantations. Large numbers of slaves were taken by the British as well. The British blockade had seriously disrupted commerce, as had British occupation of some of the major ports.

New York was occupied by British forces from the fall of 1776 to

November 25, 1783 (celebrated as Evacuation Day in New York for a hundred years). That is the longest period of time in which a city in the Western world has been held by an occupying power in modern times. During the occupation, two fires had broken out that destroyed half the buildings in Manhattan. The city's population had fallen by half in these years. Many of the city's merchant elite, hopelessly compromised by dealing with the British, evacuated with them.

And the coming of peace also brought British commercial retaliation. The British West Indies, previously a huge market for American food-stuffs and lumber, were closed to American ships. Favorable tariff treatment, such as for indigo, ended.

But Britain remained by far the largest customer for American goods and the largest exporter to the United States. British merchants, anxious to reestablish their position in the lucrative American market, offered generous terms. And with the end of the war, the American economy soon began to recover, if patchily. South Carolina remained largely mired in depression, while the mercantile economies of the Middle Atlantic states rebounded. And if markets such as the British West Indies were lost (temporarily, it turned out), new ones opened up. Northern Europe, closed by the British Navigation Acts, now welcomed American products. Foreign goods that previously had to come through Britain could now come directly at lower cost. And the Far East, once the zeal-ously guarded monopoly of the British East India Company, opened up to American merchants. In 1784 the *Empress of China* cleared New York harbor, bound for the Orient, the first ship in what would be a growing armada. It carried a cargo of furs and ginseng root—much prized in China as a cure-all—that was bartered for tea, silks, china and porcelain, exotic plants and birds, and other luxury goods. When it returned fifteen months later, it disposed of its cargo at a profit of between $30,000 and $40,000.

New York City itself recovered from its devastation at the hands of the British with astonishing speed. By the end of the decade its population had not only recovered but reached new highs, thirty-three thousand in the census of 1790.

———

BUT IF THE COUNTRY'S ECONOMY was slowly recovering, its finances were not. The states had finally agreed to a basic frame of government, the Articles of Confederation, in 1781, to replace the ad hoc administration of the Second Continental Congress. But it proved woefully inadequate to the job at hand. Most power was still vested in Congress, whose members were appointed by the state governments and served at their pleasure. More, it lacked the power to tax, and instead had to requisition funds from the various states, many of which did not pay on time and some of which simply did not pay at all. Under the Articles of Confederation, the government of the United States much more closely resembled the present-day United Nations than a real government.

An attempt to give the national government a secure income through a 5 percent impost on imports failed when New York imposed so many conditions that Congress declined to accept them. Without unanimous consent from the states, the measure failed.

As a result, the national government was unable to fulfill its obligations. The navy was disbanded and the army reduced to a nullity, with only eighty privates. In 1785 the government stopped paying interest on its debts to France. Two years later it stopped paying on the principal as well.

Knowing the powerlessness of the United States, foreign powers treated it with increasing contempt. Britain encouraged separatist movements in the Northwest and Vermont, and refused to evacuate its forts in what was now United States territory. Spain refused to recognize United States sovereignty west of the mountains and south of the Ohio River, and closed the Mississippi, whose mouth it controlled, to American commerce. With settlers pouring over the mountains into the fertile lands that became Kentucky and Tennessee, this was a major problem. These people needed to export their increasing agricultural output to survive, and the Mississippi was the only outlet to the sea. Spain hoped to earn their allegiance, and it might well have gotten it. In 1784 George Washington described these people as standing on a pivot. "The touch of a feather would turn them any way."

Domestic creditors as well as foreign ones were unable to collect either interest or principal. The great mass of bills of credit and commissary certificates continued to circulate at rates that were only a tiny fraction of their face value.

The biggest problems with the Articles of Confederation—the national government's lack of power to fund its operations through taxes, and to regulate commerce among the states—were obvious. How to fix the problems, especially given the reluctance of the states to cede aspects of sovereignty, was not. But, pushed by James Madison, the Virginia legislature invited the other states to meet at a convention to consider "how far a uniform system in their commercial regulations may be necessary to their common interest and their permanent harmony."

This resulted in the Annapolis Convention in September 1786. But only five states showed up, and all they accomplished was to call for another meeting in May 1787, "to take into consideration the situation of the United States, to devise such further provisions as shall appear to them necessary to render the constitution of the Federal Government adequate to the exigencies of the Union."

The nationalists, as those who favored a stronger central government were called, then got a stroke of luck. Just as the Annapolis Convention was meeting, Shays's Rebellion broke out in Massachusetts. In the western part of the state, many farmers were in extremis with debts and had no way to pay them. But the Massachusetts legislature, thoroughly under the thumb of the Boston merchant establishment, adjourned without acting on petitions to issue paper money and to provide debt relief by staying foreclosure proceedings.

By November Daniel Shays, an army captain during the Revolution but now a destitute farmer, was leading a force of twelve hundred men. The Massachusetts government, thoroughly alarmed, sent General William Shepherd with six hundred men to guard the arsenal at Springfield and authorized General Benjamin Lincoln to raise a force of forty-four hundred men. Shepherd, armed with artillery against muskets and pitchforks, made short work of the men who attacked Springfield on January 24, 1787, and Lincoln finished off the rebellion by an attack on its forces at Petersham on February 4. Shays fled to Vermont.

Massachusetts pardoned everyone involved (even including, the next year, Daniel Shays himself). Sympathizers with Shays's Rebellion won the legislative elections that spring and quickly passed legislation exempting such things as household goods, clothes, and the tools of one's trade from foreclosure proceedings.

Although it collapsed quickly, Shays's Rebellion added greatly to the public perception that things were seriously amiss with how the country was being governed and that fundamental change was needed. This prepared the way for the Constitutional Convention that met in Philadelphia late that spring to quickly decide to throw out the Articles of Confederation and start afresh. Referring to the changes in the debtors' law brought about by Shays's Rebellion, a Boston newspaper snootily noted in May 1787 that "sedition itself will sometimes make laws." In a very real sense, Shays's Rebellion helped make a constitution.

———

MASTERPIECES CREATED BY A COMMITTEE are notably few in number, but the United States Constitution is certainly one of them. Amended only twenty-seven times in 215 years, it came into being just as the world was about to undergo the most profound—and continuing—period of economic change the human race has known. The locus of power in the American economy has shifted from sector to sector as that economy has developed. Whole sections of the country have risen and fallen in economic importance. New methods of doing business and economic institutions undreamed of by the Founding Fathers have come into existence in that time, while others have vanished. Fortunes beyond the imagination of anyone living in the preindustrial world have been built and been destroyed. And yet the Constitution endures, and the country continues to flourish under it.

By no means the least of the lucky breaks that the United States has had in its history was the time at which it came into existence and established its fundamental laws. In one of history's great coincidences, Adam Smith published *The Wealth of Nations* in 1776. It destroyed the intellectual underpinnings of the mercantilism on which the economic policies of Western nations had been based for two hundred years.

It showed in example after example, each more powerfully argued than the next, that unfettered trade, both within and without the country, and a government that did not take sides as individuals competed in the marketplace resulted in greater prosperity for all and thus greater power for the country as a whole. Many of the Founding Fathers had read Smith, and all knew the thrust of his arguments.

Because the United States was new, it did not have long-established monopolies and systems of privilege to be dismantled. It had no immensely wealthy British East India Company or entrenched aristocracy that dominated the country's politics. It had no ancient royal grants, such as the rights to collect local tariffs that abounded in prerevolutionary France. Thus it was much easier for the United States to inculcate the ideas of Adam Smith into its economic system and politics than it was for the other major Western nations. This gave it immense advantages in the new economic world that was being born as the Founding Fathers met in Philadelphia.

In *The Eighteenth Brumaire of Louis Bonaparte,* Karl Marx wrote that "Men make their own history, but they do not make it as they please; they do not make it under circumstances chosen by themselves, but under circumstances directly encountered and transmitted from the past." That is very true, almost tautological. Marx, however, never visited the United States. (For that matter he never visited a factory—all Karl Marx knew of the proletariat he claimed to champion was what he read in books written by his fellow intellectuals.) Had Marx ever ventured to the New World, he would have seen a country that, because of circumstances, did make its history as it pleased far more than any other Great Power.

To be sure, the United States did not create a purely Smithian economy. People in government will always try to help those who are powerful at the expense of those who might become so. What-is can always wield influence that what-might-be cannot match, regardless of any campaign finance laws that may be in place. The power of what-is made the abolition of slavery—by the 1780s widely seen as immoral and inefficient—politically impossible. Indeed, what-is was able to force a provision in the new Constitution that counted the disenfranchised and powerless slaves at three-fifths their actual numbers for determining the distribution of seats in Congress, greatly increasing the political power of the states with large slave populations.

But the United States has consistently come closer to the Smithian ideal over a longer period of time than any other major nation. The results can be seen wherever one looks in the United States and, indeed, around the world.

Chapter Four

THE HAMILTONIAN CREATION

T HE IMPORTANCE THAT THE Washington administration, which took office on April 30, 1789, placed on dealing with the financial situation confronting the government under the new Constitution can be judged by the numbers. While the newly created State Department had five employees, the Treasury had forty.

The tasks before the Treasury were monumental. A tax system had to be created out of whole cloth and put in place. The debt left over from the Revolution had to be rationalized and funded. The customs had to be organized to collect the duties that would be the government's main source of revenue for more than a century. The public credit had to be established so that the federal government could borrow when necessary. A monetary system had to be implemented.

The last already existed, at least in theory, established by Congress under the Articles of Confederation. In what was to be his only positive contribution to the financial system of the United States, it had been devised by Thomas Jefferson.

As we have seen, before the Revolution, the merchants of the various colonies had kept their books in pounds, shillings, and pence, but the money in actual circulation was almost everything *but* pounds, shillings, and pence. The question of what new unit of account to adopt was nearly as complex, because the inhabitants of the various colonies "thought" in terms of so many different, often incommensurate units.

Robert Morris, who had done so much to keep the Revolution financially afloat, tried to bridge the differences by finding the lowest common divisor of the most often encountered monetary unit of each state. He calculated this to be 1,440th of a Spanish dollar. Jefferson thought this far too infinitesimal to be practical, and Morris agreed. He proposed that his unit be multiplied by one thousand and made equal to 25/36ths of a dollar. Jefferson argued instead for just using the dollar, already familiar throughout the United States, as the new monetary unit.

The origin of the word *dollar* lies in the German word for valley, *Thal*. In the fifteenth century major silver deposits had been discovered in Bohemia, in what is now the Czech Republic. In 1519 the owner of mines near the town of Joachimsthal, the Graf zu Passaun und Weisskirchen, began minting silver coins that weighed a Saxon ounce and were called thalers, literally "from the valley." These coins, new and pure, met with great acceptance from merchants, and other rulers in the Holy Roman Empire began to imitate them with their own coinage.

The Holy Roman Emperor, the Hapsburg Charles V, also adopted the thaler as the standard for his own coinage in both his Austrian and Spanish lands and his new-won, silver-rich empire in the New World. The staggering amounts of gold and silver mined in Spanish America in the sixteenth and seventeenth centuries (just between 1580 and 1626, more than eleven thousand tons of gold and silver were exported to Spain from the New World) made the thaler the standard unit of international trade for centuries. *Thaler* became *dollar* in the English language, much as *Thal*, centuries earlier, had become *dale* and *dell*. It also became the most common major coin in the British North American colonies.

Jefferson, in his "Notes on the Establishment of a Money Unit, and of a Coinage for the United States," advocated not only using the dollar but making smaller units decimal fractions of the dollar. Today this seems obvious. After all, every country in the world now has a decimal monetary system, and as Jefferson himself explained, "in all cases where we are free to choose between easy and difficult modes of operation, it is most rational to choose the easy." But Thomas Jefferson was the first to advocate such a system, and the United States, in 1786, was the first country in the world to adopt one.

Spanish dollars had often been clipped into halves, quarters, and eighths, called bits, to make small change (which is why they were often called "pieces of eight"). But Jefferson advocated coinage of a half dollar, a fifth, a tenth (for which he coined the word *dime*), a twentieth, and a hundredth of a dollar (for which he borrowed the word *cent* from Robert Morris's scheme). In 1785 Congress declared that the "monetary unit of the United States of America be one dollar." But the next year Congress, while adopting the cent, five-cent, dime, and fifty-cent coins advocated by Jefferson, decided to authorize a quarter-dollar coin rather than a twenty-cent piece.

The quarter is with us yet, now the last, distant echo of the old octal monetary system of colonial days. But other echoes held on for decades. The New York Stock Exchange still gave prices in eighths of a dollar as late as 1999. And the term *shilling* long remained in common use to mean twelve and a half cents, an eighth of a dollar, although there has never been a United States coin in that denomination. The east side of Broadway, in New York, where the less fashionable stores were located, was still called the "shilling side" as late as the 1850s, while the west side of the street was called the "dollar side."

One reason the term *shilling* held on so long, of course, was that American coinage was not adequate to the ever-growing demand for it, and the old hodgepodge of foreign coins thus held on as well. The first United States coin, a copper cent bearing the brisk motto "Mind Your Business," was privately minted. The Philadelphia mint was established in 1792 but minted few coins in the early years for lack of metal with which to do so.

———

ROBERT MORRIS, bent on making money, turned down Washington's offer to name him as secretary of the treasury in the new government (it was a bad decision—he ended up in debtors' prison). The president then turned to one of his aides-de-camp during the Revolution, Alexander Hamilton, only in his early thirties.

Hamilton was the only one of the Founding Fathers not to be born in what is now the United States. He was born in Nevis, one of Britain's less important Leeward Island possessions. He was also the only one—besides Benjamin Franklin, who had made a large fortune on his own and an even larger reputation—not to be born to affluence. Indeed, he grew up in poverty after his feckless father—who had never married his mother—deserted the family when Hamilton was only a boy.

Living in St. Croix, now part of the U.S. Virgin Islands but then a possession of Denmark, Hamilton went to work at a trading house owned by the New York merchants Nicholas Cruger and David Beekman, when he was eleven years old. Extraordinarily competent and ferociously ambitious, Hamilton was managing the place by the time he was in his mid-teens, quite literally growing up in a counting house. Thus, of all the Founding Fathers, only Franklin had so urban and commercial a background. Even John Adams, a lawyer by profession, considered his family farm in Braintree (now Quincy), Massachusetts, to be home, not Boston.

Cruger, recognizing Hamilton's talents, helped him come to New York in 1772 and to attend King's College, now Columbia University. After the Revolution he studied law and began practicing in New York City, where he married Elizabeth Schuyler, from one of New York's most prominent families. After the Revolution he wrote a series of newspaper articles and pamphlets outlining his ideas of what was needed to create an effective federal government. In 1784 he founded the Bank of New York, the first bank in that city and the second in the country.

He attended the Constitutional Convention in Philadelphia and worked tirelessly to get the document ratified, writing two-thirds of *The Federalist Papers*. When Robert Morris took himself out of consideration, Hamilton, whom Morris called "damned sharp," was more than happy to take the job of secretary of the treasury.

He was also one of the very few competent to do so. While Americans had already distinguished themselves in many fields of endeavor, they "were not well acquainted with the most abstruse science in the world [public finance], which they never had any necessity to study."

Hamilton, a deep student of economics, understood public finance thoroughly, a fact that he would make dazzlingly clear in the next few years. But like so many of the Founding Fathers, he was also a deep student of human nature and knew that there was no more powerful motivator in the human universe than self-interest. He sought to establish a system that would both channel the individual pursuit of self-interest into developing the American economy and protect that economy from the follies that untrammeled self-interest always leads to.

Even before the Treasury Department was created on September 2, 1789, and Hamilton was confirmed by the Senate as its first secretary on September 11, Congress had passed a tax bill to give the new government the funds it needed to pay its bills. There was no argument that the main source of income was to be the tariff, but there was lengthy debate over what imports should be taxed and at what rate. Pennsylvania had had a high tariff under the old Articles to protect its nascent iron industry and wanted it maintained. The southern states, importers of iron products such as nails and hinges, wanted a low tariff on iron goods or none at all. New England rum distillers wanted a low tariff on its imports of molasses. Whiskey manufacturers in Pennsylvania and elsewhere wanted a high tariff on molasses, to stifle their main competition.

Congress finally passed the Tariff and Tonnage Acts (the latter imposed a duty of 6 cents a ton on American ships entering U.S. ports and 50 cents a ton on foreign vessels) in the summer of 1789. But, second only to slavery, the tariff would be the most contentious issue in Congress for the next hundred years. Pierce Butler of South Carolina even issued the first secession threat before the Tariff Act of 1789 made it through Congress.

With funding in place, Hamilton's most pressing problem was to deal with the federal debt. The Constitution commanded that the new federal government should assume the debts of the old one, but how that should be done was a fiercely debated question. Much of the debt had fallen into the hands of speculators who had bought it for as little as 10 percent of its face value.

On January 14, 1790, Hamilton submitted to Congress his first "Report on the Public Credit." It called for redeeming the old debt on generous terms and issuing new bonds to pay for it, backed by the revenue from the tariff. The report became public knowledge in New York City, the temporary capital, immediately, but news of it spread only slowly to other parts of the country, and New York speculators were able to snap up large quantities of the old debt at prices far below what Hamilton proposed redeeming it for.

Many were outraged that speculators should profit while those who had taken the debt at far higher prices during the Revolution should not see their money again. James Madison argued that only the original holders should have their paper redeemed at the full price and the speculators get only what they had paid for it. But this was hopelessly impractical. For one thing, determining who was the original holder would have often been impossible.

Even more important, such a move would have greatly impaired the credit of the government in the future. If the government could decide to whom along the chain of holders it owed past debts, people would be more reluctant to take future debt, and the price in terms of the interest rate demanded, therefore, would be higher. And Hamilton was anxious to establish a secure and well-funded national debt, modeled on that of Great Britain and for precisely the purposes that Great Britain had used its debt.

Many of those in the new government, unversed in public finance, did not grasp the power of a national debt, properly funded and serviced, to add to a nation's prosperity. But Hamilton grasped it fully. One of the greatest problems facing the American economy at the start of the 1790s was the lack of liquid capital, capital available for investment. Hamilton wanted to use the national debt to create a larger and more flexible money supply. Banks holding government bonds could issue banknotes backed by them. And government bonds could serve as collateral for bank loans, multiplying the available capital. He also knew they would attract still more capital from Europe.

Hamilton's program eventually passed Congress, although not without a great deal of rhetoric. Hamilton's father-in-law, a senator from New York in the new Congress, was a holder of $60,000 worth of government securities he hoped would be redeemed by Hamilton's program. It was said the opposition to the program made his hair stand "on end as if the Indians had fired at him."

Hamilton also wanted the federal government to assume the debts that had been incurred by the various states in fighting the Revolution. His main reason for doing so was to help cement the Union. Most of the state debt was held by wealthy citizens of those states. If they had a large part of their assets in federal bonds, instead of state bonds, they would be that much more interested in seeing that the Union as a whole prospered.

Those states, mostly northern, that still had substantial debt were, of course, all for Hamilton's proposal. Those that had paid off their debts were just as naturally against it. Jefferson and Madison—Virginia had paid off its debts—were adamantly opposed and had enough votes to defeat the measure. Hamilton offered a deal.

If enough votes were switched to pass his assumption bill, he would see that the new capital was located in the South. To assure Pennsylvania's cooperation, the capital would be moved from New York to Philadelphia for ten years while the new one was built. Jefferson and Madison agreed. Hamilton's program passed and was signed into law by President Washington, who was delighted at the prospect of the new capital being located on his beloved Potomac River.

The program was an immediate success, and the new bonds sold out within a few weeks. When it was clear that the revenue stream from the tariff was more than adequate to service the new debt, the bonds became sought after in Europe. In 1789 the United States had been a financial basket case, its obligations unsalable, it ability to borrow nil. By 1794 it had the highest credit rating in Europe, and some of its bonds were selling at 10 percent over par.

Talleyrand, the future French foreign minister, then in the United States to escape the Terror, explained why. The bonds, he said, were "safe and free from reverses. They have been funded in such a sound manner and the prosperity of this country is growing so rapidly that there can be no doubt of their solvency."

Talleyrand might have added that the willingness of the new federal government to take on the debt of the old, rather than repudiate it for short-term fiscal reasons or political advantage, also helped powerfully to gain the trust of investors. The ability of the federal government to borrow huge sums at affordable rates in times of emergency—such as during the Civil War and the Great Depression—has been an immense national asset. In large measure, we owe that ability to Alexander Hamilton's policies that were put in place at the dawn of the Republic. It is no small legacy.

To be sure, Hamilton, and the United States, had the good fortune to have a major European war break out in 1793, after Louis XVI was guillotined. This proved a bonanza for American foreign trade and for American shipping, which was protected from privateers by the country's neutrality. European demand for American foodstuffs and raw materials greatly increased, and the federal government's tariff revenues increased proportionately. In 1790 the United States exported $19,666,000 worth of goods, while imports not reexported amounted to $22,461,000. By 1807 exports were $48,700,000 and imports $78,856,000. Government revenues that year were well over five times what they had been seventeen years earlier.

———

THE OTHER MAJOR PART of Hamilton's fiscal policy was the establishment of a central bank, to be called the Bank of the United States and modeled on the Bank of England.

Hamilton expected a central bank to carry out three functions. First, it would act as a depository for government funds and facilitate the transfer of them from one part of the country to another. This was a major consideration in the primitive conditions of the young United States. Second, it would be a source of loans to the federal government and to other banks. And third, it would regulate the money supply by disciplining state-chartered banks.

The money supply was a critical problem at the time. Specie—gold and silver coins—was in very short supply. In 1790 there were only three state-chartered banks empowered to issue paper money, including Hamilton's Bank of New York, but these notes had only local circulation. Hamilton reasoned that if the Bank of the United States accepted these local notes at par, other banks would too, greatly increasing the area in which they would circulate. And if the BUS refused the notes of a particular bank, because of irregularities or excess money creation, other banks would refuse them as well, helping to keep the state banks on the straight and narrow.

Hamilton had learned not to like the idea of the government itself issuing paper money, knowing that in times of need the government would be unable to resist the temptation to solve its money problems by simply printing it. Certainly the Continental Congress had shown no restraint during the Revolution, but at least it had had the excuse of no alternative. And the history of paper money since Hamilton's day has shown him to be correct. Without exception, wherever politicians have possessed the power to print money, they have abused it, at great cost to the economic health of the country in question.

Hamilton proposed a bank with a capitalization of $10 million. That was a very large sum when one considers that the three state banks in existence had a combined capitalization of only $2 million. The government would hold 20 percent of the stock of the bank and have 20 percent of the seats on the board. The secretary of the treasury would have the right to inspect its books at any time. But the rest of the bank's stock would be privately held.

"To attach full confidence to an institution of this nature," Hamilton wrote in his "Report on a National Bank," delivered to Congress on December 14, 1790, "it appears to be an essential ingredient in its structure, that it shall be under a *private* not a *public* direction—under the guidance of *individual interest,* not of *public policy;* which would be supposed to be, and, in certain emergencies, under a feeble or too sanguine administration, would really be, liable to being too much influenced by *public necessity.*"

The bill passed Congress with little trouble, both houses splitting along sectional lines. Only one congressman from states north of Maryland voted against it and only three congressmen from states south of Maryland voted for it. Hamilton thought the deal was done.

But he had not counted on Thomas Jefferson, by now secretary of state, and James Madison, who then sat in the House of Representatives. Although Jefferson had personally enjoyed to the hilt the manifold pleasures of Paris while he had served as minister to Louis XVI under the old Articles of Confederation, nonetheless he had a deep political aversion to cities and to the commerce that thrives in them.

Nothing symbolized the vulgar, urban moneygrubbing he so despised as banks. "I have ever been the enemy of banks . . . " he wrote to John

Adams in old age. "My zeal against those institutions was so warm and open at the establishment of the Bank of the U.S. that I was derided as a Maniac by the tribe of bank-mongers, who were seeking to filch from the public their swindling, and barren gains."

Jefferson, born one of the richest men in the American colonies—on his father's death he inherited more than five thousand acres of land and three hundred slaves—spent money all his life with a lordly disdain for whether he actually had any to spend. He died, as a result, deeply in debt, bankrupt in all but name. And regardless of his own aristocratic lifestyle, his vision of the future of America was a land of self-sufficient yeoman farmers, a rural utopia that had never really existed and would be utterly at odds with the American economy as it actually developed in the industrial age then just coming into being.

Jefferson and his allies Madison and Edmund Randolph, the attorney general, fought Hamilton's bank tooth and nail. They wrote opinions for President Washington saying that the bank was unconstitutional. Their arguments revolved around the so-called necessary and proper clause of the Constitution, giving Congress the power to pass laws "necessary and proper for carrying into Execution the foregoing Powers."

As the Constitution nowhere explicitly grants Congress the power to establish a bank, they argued, only if one were absolutely necessary could Congress do so. This "strict construction" of the Constitution has been part of the warp and woof of American politics ever since, although even Jefferson admitted that it appealed mostly to those out of power. The fact that the Constitution nowhere mentions the acquisition of land from a foreign state did not stop Jefferson, as president, from snatching the Louisiana Purchase when the opportunity presented itself.

Hamilton countered with a doctrine of "implied powers." He argued that if the federal government were to deal successfully with its enumerated duties, it must be supreme in deciding how to do so. "Little less than a prohibitory clause," he wrote to Washington, "can destroy the strong presumptions which result from the general aspect of the government.

Nothing but demonstration should exclude the idea that the power exists." Further, he asserted that Congress had the right to decide what means were necessary and proper. "The national government like every other," he wrote, "must judge in the first instance of the proper exercise of its powers." Washington, his doubts quieted, signed the bill.

The sale of stock was a resounding success, as investors expected that the bank would prove very profitable, which it was. It also functioned exactly as Hamilton thought it would. The three state banks in existence in 1790 became twenty-nine by the turn of the century, and the United States enjoyed a more reliable money supply than most nations in Europe.

With the success of the Bank of the United States stock offering, the nascent securities markets in New York and Philadelphia had their first bull markets, in bank stocks. Philadelphia, the leading financial market in the country at that time, thanks to the location there of the headquarters of the Bank of the United States, established a real stock exchange in 1792. In New York a group of twenty-one individual brokers and three firms signed an agreement—called the Buttonwood Agreement because it was, at least according to tradition, signed beneath a buttonwood tree (today more commonly called a sycamore) outside 68 Wall Street. In it they pledged "ourselves to each other, that we will not buy or sell from this day for any person whatsoever any kind of Public Stock, at a less rate than one quarter per cent Commission on the specie value, and that we will give preference to each other in our negotiations."

The new group formed by the brokers was far more a combination in restraint of trade and price-fixing scheme than a formal organization, but it proved to be a precursor of what today is called the New York Stock Exchange.

A speculative bubble arose in New York, centered on the stock of the Bank of New York. Rumors abounded that it would be bought by the new Bank of the United States and converted to its New York branch. Numerous other banks were announced and their stock, or, often, rights to buy the stock when offered, was snapped up. The Tammany Bank announced a stock offering of 4,000 shares and received subscriptions for no fewer than 21,740 shares.

An unscrupulous speculator named William Duer was at the center of this frenzy in bank stocks. He had worked, briefly, for the Treasury, but had resigned rather than obey the rule Hamilton had put in place forbidding Treasury officials from speculating in Treasury securities. Hamilton was appalled by what was happening on Wall Street. "'Tis time," he wrote on March 2, 1792, "there should be a line of separation between honest Men & knaves, between respectable Stockholders and dealers in the funds, and mere unprincipled Gamblers."

It didn't take long for Duer's complex schemes to fall apart, and he was clapped into debtors prison, from which he would not emerge alive. Panic swept Wall Street for the first time, and the next day twenty-five failures were reported in New York's still tiny financial community, including one of the mighty Livingston clan.

Jefferson was delighted with this turn of events. "At length," he wrote a friend, "our paper bubble is burst. The failure of Duer in New York soon brought on others, and these still more, like nine pins knocking down one another." Jefferson, who loved to calculate things, estimated the total losses at $5 million, which he thought was about the total value of all New York real estate at the time. Thus, Jefferson gleefully wrote, the panic was the same as though some natural calamity had destroyed the city.

In fact, the situation was not nearly that dire, especially as Hamilton moved swiftly to stabilize the market and ensure that the panic did not bring down basically sound institutions. He ordered the Treasury to buy its own securities to support the market, and he added further liquidity by allowing customs duties—ordinarily payable only in specie or Bank of the United States banknotes—to be paid with notes maturing in forty-five days.

The system Hamilton had envisioned and put in place over increasing opposition from Thomas Jefferson and his political allies worked exactly as Hamilton had intended. Several speculators were wiped out, but they had been playing the game with their eyes open and had no one to blame but themselves. The nascent financial institutions, however, survived. "No calamity truly *public* can happen," Hamilton wrote, "while these institutions remain sound." The panic soon passed and most brokers were able to get back on their feet quickly, thanks to Hamilton's swift action.

Unfortunately, Thomas Jefferson was a better politician than Hamilton, and a far better hater. The success of the Bank of the United States and its obvious institutional utility for both the economy and the smooth running of the government did not cause him to change his mind at all about banks. He loathed them all. The party forming around Thomas Jefferson would seize the reins of power in the election of 1800 and would not lose them for more than a generation. In that time, they would destroy Hamilton's financial regulatory system and would replace it with nothing.

As a result, the American economy, while it would grow at an astonishing rate, would be the most volatile in the Western world, subject to an unending cycle of boom and bust whose amplitude far exceeded the normal ups and downs of the business cycle. American monetary authorities would not—indeed could not—intervene decisively to abort a market panic before it spiraled out of control for another 195 years.

Thomas Jefferson, one of the most brilliant men who has ever lived, was psychologically unable to incorporate the need for a mechanism to regulate the emerging banking system or, indeed, banks at all, into his political philosophy. His legion of admirers, most of them far less intelligent than he, followed his philosophy for generations as the country and the world changed beyond recognition. As a direct result, economic disaster would be visited on the United States roughly every twenty years for more than a century.

Chapter Five

A TERRIBLE SYNERGY

N OWHERE IS JOHN DONNE'S DICTUM that we are all a part of the continent more true than in the world economy. It is built, by definition, upon an endless exchange of goods between individuals, industries, and nations, the most intricate net in the human universe. When a change happens anywhere in that economy, that change ripples through the whole. And when two separate developments happen to interact in a major way, it can produce an economic synergy that is both great and terrible.

No better example of this can be found than in what happened when a young New Englander's simple, bright idea to help depressed southern agriculture interacted with the aborning Industrial Revolution in the Midlands of faraway England. Besides producing the most successful cash crop in American history, it also revived the rapidly decaying institution of bondage labor, and nearly destroyed the United States.

———

AMONG THE MAJOR economic casualties of the American Revolution was the indigo trade of South Carolina and Georgia. *Indigofera tinctoria,* a plant native to central Asia and north Africa, produces a blue dye that was in wide demand in the British cloth industry. The best indigo came from Spanish and French sources, but imperial preference gave the British market, by far the world's largest, to Georgia and South Carolina. At the end of the colonial era, about 10 percent of the slave labor in the American colonies was used in producing indigo.

With American independence, however, Britain turned to India for its indigo supply and, without the British market, the indigo industry in South Carolina and Georgia quickly collapsed. Rice, the mainstay of Carolina plantation agriculture, continued to be profitable but was now dependent for any growth on the small American market. If the Deep South was to flourish in a postimperial economy, it needed a new cash crop.

One possibility was cotton, then grown mainly on the Sea Islands that line that part of the Atlantic coast. But this long-staple cotton (today it is often called Egyptian cotton) could not be grown inland any distance because it requires an exceptionally long growing season and sandy soil. Indeed, so little was grown that when a bale was exported to England in 1784, the first instance of cotton as an American export, it ran afoul of British navigation laws. These laws required that raw products arrive in British ports either in British ships or in ships of the country of origin. Customs officials simply refused to believe that there was such a thing as American cotton, and the bale was left to rot on the docks of Liverpool.

Short-staple cotton, or upland cotton, required only a two-hundred-day growing season and was not fussy about soil conditions, although it preferred a rich loam. It could be grown easily in upland South Carolina and Georgia. But there was a big problem. Unlike Sea Island cotton, the seeds of this upland cotton are sticky and cling tenaciously to the fibers that surround them. Separating the seeds from the lint, as the cotton fibers are called, was an immensely time-consuming task. While a field hand could pick as much as fifty pounds of cotton bolls in a day, it took twenty-five man-days to remove the seeds from that much cotton by hand, a process called ginning.

As so often happens in the history of the American economy, Yankee ingenuity solved the problem. Eli Whitney was born in Westboro, Massachusetts, in 1765, the son of a farmer who had a small manufacturing enterprise on the side, making things other farmers needed but weren't handy enough to supply themselves. Eli soon proved to have a marked aptitude for mechanics and a keen sense of what the market was looking for. While still a boy during the Revolution, when nails were in short supply, he talked his father into building a forge so he could manufacture them. The enterprise did well enough that he was soon talking about hiring an assistant.

After graduating from Yale in 1793, Whitney accepted a post as a tutor in South Carolina. En route, he visited a friend, Phineas Miller, who managed a plantation in Georgia, owned by the widow of General Nathaniel Greene. There Whitney saw cotton growing for the first time. He soon wrote his father that "I have . . . heard much said of the extreme difficulty of ginning cotton. . . . There were a number of respectable Gentlemen at Mrs. Greene's who all agreed that if a machine could be invented which would clean the cotton with expedition, it would be a great thing both to the country and the inventor. . . . I involuntarily happened to be thinking on the subject and struck out a plan for a machine in my mind."

The machine was simplicity itself. Whitney studded a roller with nails, set half an inch apart. When the roller was turned, the nails passed through a grid, pulling the cotton lint from below through the grid, but leaving the seeds behind. A rotating brush swept the lint off the nails into a compartment while the seeds fell into a separate compartment.

With Whitney's cotton gin, a laborer could do in a single day what had taken twenty-five laborers to do in that amount of time before. Its economic utility was so obvious that Whitney's first model was stolen, but he received a patent on a new, improved model the following year and, in partnership with Phineas Miller, set up a factory to manufacture cotton gins near New Haven, Connecticut.

Unfortunately for Whitney and Miller, the concept of the cotton gin is so simple that any competent carpenter could make one in an afternoon, and it proved impossible to enforce the patent as the idea spread throughout the parts of the South where cotton could be grown, and production began to soar. Whitney spent years fighting expensive patent-infringement cases, and it was only in 1807 that his rights were secured, not long before the patent itself expired.

Altogether Whitney would realize only about $100,000—to be sure, that was a good-size fortune by the standards of the early nineteenth century—from an invention that quite literally changed the world. Most of that came not from royalties, however, but from grateful state governments. South Carolina voted Whitney $50,000 from the state treasury to cover patent infringements, and North Carolina put a tax on cotton for five years to compensate him, yielding him about $30,000. Tennessee chipped in $10,000.

The effect of the cotton gin on the economy of the South and thus of the United States as a whole was enormous. In 1793 the United States produced about five million pounds of cotton, almost all of it of the Sea Island variety. That was less than 1 percent of the world's total cotton crop, most of which was then grown in India. By the first decade of the nineteenth century, thanks to the cotton gin, American production had increased eightfold to forty million pounds. And annual American exports of cotton to Britain had grown to about fifty thousand bales.

Each decade thereafter, American cotton production doubled on average, reaching two billion pounds by 1860. By 1830 the United States was producing about half the world's cotton. Two decades later the percentage had risen to nearly 70 percent, three-quarters of which was exported.

The Deep South turned out to be the best place in the world to grow cotton. While it could be grown very profitably in the Piedmont region of South Carolina and Georgia (the setting for *Gone With the Wind*), the Cotton Kingdom really came into its own in the rich soil of Alabama's black belt and the deep, alluvial soils of the Mississippi Delta, where cotton thrived as nowhere else on earth.

As these new areas opened up, the locus of American cotton production moved ever westward. South Carolina was the leading cotton state until the 1820s, when Georgia took over, only to be replaced in turn by Mississippi and Alabama a decade later. Louisiana also soon became a major cotton producer, and these five states were producing three-fourths of American cotton by the time of the Civil War. It is by no means a coincidence that they, along with the cotton-producing states of Florida and Texas, were the first states to secede from the Union after the election of 1860.

Although the price of cotton, thanks to the cotton gin, had fallen low enough to reach a mass market, it remained a labor-intensive crop. It required about 70 percent more labor per acre to produce a crop than did corn. One reason for this is that cotton is very susceptible to weed infestation and must be hoed regularly. Chopping cotton, as it is called, was backbreaking toil in the fierce heat of a southern summer. But there was a large pool of involuntary labor available to do it: the slaves. As we have seen, slavery was on the decline in the late eighteenth century, as world opinion shifted ever more firmly against it.

Shortly after the Revolution began, so too did the abolition of slavery. Not surprisingly, the areas with the least economic stake in slavery were the first to abolish it. Vermont, in declaring its own independence from Britain in 1777, also abolished slavery, becoming the first place in the Western Hemisphere to make the practice illegal. Other northern states soon followed. Even New York, with nineteen thousand slaves in 1790, about 5.5 percent of the population, began gradual emancipation in 1799 and had freed all its slaves by 1827. The Northwest Ordinance of 1787 forbade slavery north of the Ohio River.

In the South, manumission became fashionable, and many planters, George Washington among them, freed their slaves upon their own deaths. The Constitutional Convention, in 1787, found it necessary, in order to reach agreement, not only to forbid outlawing the slave trade prior to 1808, but to make that clause unamendable. But by 1808 public opinion, even in the South, had swung so against the trade that Congress abolished it as soon as it legally could do so, on January 1 of that year. (Abolishing the slave trade and actually suppressing it, of course, were two different matters. It would be the primary task of the navy for most of the next fifty years.) The price of slaves had been declining through most of the late eighteenth century.

Cotton changed all that. After 1793 the price of a slave ratcheted upward. A slave who would have sold for $300 before the cotton gin was selling for $2,000 and more by 1860. The slave holders, possessed of an increasingly valuable asset, were less and less inclined to part with what became, in the early decades of the nineteenth century, an enormous capital investment. Even the parts of the South outside the cotton-growing area became deeply involved economically with continuing what came to be called the South's "peculiar institution." As tobacco became less and less important to the economy of such states as Virginia and Maryland, they began selling their surplus slaves to the new cotton states. Between 1790 and 1860, some 835,000 slaves were "sold south."

While slavery was found throughout the South, it was not widely spread among the population. In 1860, while the white population of the South was more than eight million, there were only 383,637 slaveowners among them. And only 2,292 held more than a hundred slaves. But slavery had become bound up with the very identity of the South and its way of life. The South remained largely an exporter of raw materials, while the North's economy grew ever more diversified. Even the South's cotton was largely brokered through New York. And because of slavery, skilled immigrants gravitated to the North, where opportunities were much greater, while the North's native-born, and free, human capital acquired skills needed in the new industrial world that was coming into being in the advanced countries. The South remained a largely agricultural economy and an exporter of raw materials to ever more advanced economies elsewhere.

———

COTTON PRODUCTION in the Deep South could grow so quickly only because demand for cotton was growing so quickly across the Atlantic Ocean. The weaving of cloth, a home task since time immemorial, became in the late eighteenth century the first great industry of the Industrial Revolution. In the early eighteenth century cotton had been a luxury fabric because producing it had been very labor-intensive. Cotton fiber has a natural twist that makes it much more difficult to spin into thread by hand than wool, linen, or even silk. It required as much as twenty days for a woman (those who spun thread were almost always women—it's the origin of the word *spinster*) to spin a pound of cotton into thread.

This spinning was done at home under the putting-out system. An entrepreneur would give workers the necessary materials and then pay them by the piece for completed work. It took, on average, four spinners to supply one loom. But John Kay's invention of the flying shuttle in 1733 upset the balance because it made weaving much faster. Either more spinners or a faster way of doing the spinning was needed.

In 1764 James Hargreaves introduced the spinning jenny, which could spin eight threads at a time, and five years later James Arkwright improved on it with the water frame, so called because it was powered by a waterwheel. This mechanization made thread abundant, and, now to speed up the weaving, the Reverend Edmund Cartwright developed the power loom in 1785. The new machinery required a shift from home production by the putting-out method to factory production by labor paid hourly wages.

Cotton cloth production increased dramatically. In 1765 about five hundred thousand pounds of cotton was spun into thread in Britain, almost all of it at home. Twenty years later sixteen million pounds were spun, mostly in factories. And the price of cotton cloth had dropped precipitously, causing an upsurge in demand for it. This caused the price of raw cotton to surge until Whitney's invention solved the problem of ginning and made upland cotton cheap to produce. Now American cotton production and British cloth production could surge together.

By the end of the antebellum period, cotton dominated both the British and southern economies, and the two economies were, it seemed, inextricably linked. "Let any great social or physical convulsion visit the United States," wrote the British magazine *The Economist* in 1855, "and England would feel the shock from Lands End to John O'Groats. The lives of nearly two millions of our countrymen are dependent upon the cotton crops of America; their destiny may be said, without any kind of hyperbole, to hang upon a thread. Should any dire calamity befall the land of cotton, a thousand of our merchant ships would rot idly in dock; ten thousand mills must stop their busy looms; two thousand thousand mouths would starve, for lack of food to feed them."

The South meanwhile was equally dependent on the British market but increasingly felt itself to be in a position of strength, especially in the 1850s, when cotton prices were rising. "What would happen if no cotton was furnished for three years?" asked James Henry Hammond, senator from South Carolina. "I will not stop to depict what every one can imagine, but this is certain: England would topple headlong and carry the whole civilized world with her, save the South. No, you dare not make war on cotton. No power on earth dares to make war upon it. Cotton is king. Until lately the Bank of England was king; but she tried to put her screws as usual, the fall before last, upon the cotton crop, and was utterly vanquished. The last power has been conquered. Who can doubt, that has looked at recent events, that cotton is supreme?"

———

THE IMPORTANCE OF COTTON EXPORTS to the southern economy and the need to import most manufactured goods made the South the natural opponent of high tariffs, the principal source of federal revenue in the nineteenth century. But precisely because of cotton, the North, especially New England, became the natural exponent of a high tariff. The tariff would prove to be the thin edge of the wedge that eventually drove the two sections of the union apart.

When George Washington was inaugurated as president on April 30, 1789, he wore a simple brown suit with silver buttons, white stockings, and shoes with silver buckles. But he had taken care that the cloth was American made, woven in Hartford, Connecticut. Conscious of symbolism, as politicians always are, Washington wanted to encourage American manufactures, as industrial goods were then called, and virtually all cloth of good quality at that time had to be imported from England.

But the industrialization of the British cloth industry in recent decades had given Britain an insuperable competitive advantage, and Britain was determined to keep it. The export of textile machinery was strictly forbidden. If the United States was to develop a textile industry of its own, therefore, it had only two choices. It could reinvent what was then high technology on its own, or it could steal it. The former was not very likely as the United States then lacked people familiar with the intricacies of textile manufacture.

So it had to be stolen. Although British newspapers were forbidden to print them, clandestine advertisements—a sort of early capitalist samizdat—circulated by hand in the textile areas of the Midlands, offering big rewards to anyone willing and able to set up working textile machinery in the United States. One person who undoubtedly was aware of these enticements was Samuel Slater, who had been born in Belper, Derbyshire, in 1768, in the heart of the burgeoning textile manufacturing area. Slater was apprenticed in 1782 to Jedidiah Strutts, the owner of a textile mill in Belper and one of the first to make a fortune out of the new technology.

Slater was fascinated by machinery and would often spend his Sundays at the factory studying the machines and figuring out how they worked. While still a teenager, he invented a means of winding the newly spun yarn evenly on the spindles, and Strutts rewarded him with a guinea, the better part of a year's wages for an apprentice. Slater also showed a talent for management and was soon overseeing the mill and repairing and building machinery.

When his apprenticeship ended in 1789, Slater wanted to emulate his former master and become a mill owner in his own right. But lacking capital, he knew that he would have a much better chance of achieving his goal in the United States, where his skills would be in very high demand, than in his native country. His self-interest commanded him to head across the Atlantic Ocean. The problem was how to get there. Besides forbidding the export of textile machinery or even drawings of it, Britain also did not allow those with expertise in the textile industry to emigrate.

But Slater laid his plans carefully and, before leaving Strutts's employ, carefully memorized every detail of the machinery he had been supervising. Aware of the vigilance of the British customs, he kept his intentions so secret that he did not even tell his mother what he was doing until he mailed her a letter from London, a few hours before he sailed for New York, listed on the manifest as a farm laborer.

He arrived in the New World on November 11, 1789, and soon heard that Moses Brown, a Quaker of Providence, Rhode Island (Brown University was named after his family in 1804), had some spinning machines that he could not get to work. He wrote Brown offering his services. Brown was only too happy to accept. "We are destitute of a person acquainted with water frame spinning. . . . If thy present situation does not come up to what thou wishest . . . come and work [with] ours and have the credit as well as the advantage of perfecting the first watermill in America." He offered Slater all the profits over and above interest on the capital and depreciation, a deal Slater could never have made in Derbyshire.

When Slater arrived in Providence, he was disappointed to find that

Brown's spinning machines were beyond hope. He offered, however, to build new ones from scratch. Over the next twelve months, Slater slowly built the machinery needed for a spinning mill. With American carpenters and mechanics wholly unfamiliar with textile equipment, it was a struggle, and at one point Slater almost gave up when a carding machine refused to work. But on December 20, 1790, the first cotton spinning mill in the United States opened for business, and the American Industrial Revolution began.

Brown wrote to Secretary of the Treasury Alexander Hamilton, who would submit his "Report on Manufactures" to Congress a year later, that "Mills and machines may be erected in different places, in one year, to make all the cotton yarn that may be wanted in the United States." This was, of course, a rather considerable exaggeration. But cotton mills quickly began to spread in New England, whose many clear, swift-flowing rivers provided ready power. Brown and Slater (and Brown's cousin William Almy) became partners when the first building designed as a spinning mill was erected in 1793, and Slater became a wealthy man. He also married William Almy's daughter, Hannah, who became the first woman to receive a U.S. patent, for an improved means of spinning thread. By the time of Slater's death in 1835, New England was the center of the second largest cloth industry in the world.

Two years earlier, Slater had been visited by President Andrew Jackson on a tour of New England, and Jackson bestowed on Slater the title "Father of American Manufactures."

———

NEW ENGLAND, besides having many of the swift-flowing rivers needed to turn the waterwheels that powered the new mills, had one other comparative advantage at the turn of the nineteenth century: a source of ready, cheap, and willing labor. New England had always been marginal farming country, with thin and often sandy soil and an unforgiving climate. But as long as most food had to be produced locally, because of transportation costs, local farming was a necessity.

Increasingly, however, sons of New England farmers were emigrating west, into New York State and then beyond. Once the Erie Canal was built, they would begin to pour into the lush farmlands of the old Northwest in what is known as the New England diaspora. But the unmarried sisters of these emigrants were stuck at home, with ever decreasing prospects for marriage as they got older. Thus the mills that began springing up along New England rivers at this time were a welcome means of escape from the dreariness and loneliness of New England farm life that would be so brilliantly evoked by Edith Wharton in her novel *Ethan Frome*.

The mills hired these young, single women in increasingly large numbers, housing them in dormitories under the strict supervision of matrons. Church was compulsory and classes in various subjects were offered as well. Many of the women who became teachers, librarians, and social reformers in nineteenth-century New England began their formal education as mill workers. And of course, despite the ever-vigilant matrons, many more found husbands and went on to establish families.

Samuel Slater's factory on the Pawtucket River in Rhode Island turned cotton fiber into thread. The thread was then turned over to home weavers to be made into cloth. Even in England, spinning and weaving were done in separate factories. But in 1814 Francis Cabot Lowell went one step further and built a factory in Waltham, Massachusetts, that began with bales of cotton and turned them into finished cloth, with even dyeing being done in the same building. It was the first fully integrated cloth manufacturing enterprise in the world.

It was not, however, entirely without precedent. Oliver Evans, one of the country's first great inventors, had developed an integrated flour mill, a stunningly new concept. Flour mills had used water power since the Middle Ages. But only the grinding was powered by the waterwheel. Everything else, the spreading, bolting, and packing, was done by human effort. Evans designed a series of conveyances, using bucket elevators and screw conveyors, all powered by the waterwheel, that moved the grain, meal, and flour, from one process to another. In effect, Evans turned the flour mill into a machine: grain was poured in one end and barrels of flour came out the other. Except for adjusting, maintaining, and monitoring, the whole process needed hardly any human labor.

While requiring far more human labor, the cotton mill built by Francis Cabot Lowell was in the same mold. Born in Newburyport, Massachusetts, to one of the state's well-established families, Lowell entered Harvard at the age of fourteen, and after graduation, like so many New Englanders of his day, he went into the trading business.

Trade had been the source of most early American great fortunes since the founding of the colonies, and this remained true in the 1790s. Elias Derby of Salem, Massachusetts, a major ship owner, became the country's first millionaire in that decade, trading as far away as China. John Jacob Astor also began trading with China—a rich market for furs—often making a profit of $50,000 on a single voyage.

As noted, the outbreak of the Wars of the French Revolution in Europe in 1793 had proved a bonanza for American foreign trade. But as the European war deepened, each side attempted to disrupt the trade of the other, placed increasing restrictions on neutral shipping, and seized an increasing number of ships deemed in violation. Between 1803 and 1807 Britain seized 528 American ships and France seized another 389. In addition, the Royal Navy, ever in need of trained seamen, frequently boarded American vessels and impressed any sailors declared to be British subjects.

In hopes of forcing France and Britain to respect neutral rights, President Jefferson rammed through Congress the Embargo Act, which he signed on December 22, 1807. It was one of the most remarkable acts of statecraft in American history. Indeed it is nearly without precedent in the history of any country. The Embargo Act forbade American ships from dealing in foreign commerce, and the American navy was deployed to enforce it. In effect, to put pressure on Britain and France, the United States went to war with itself and blockaded its own shipping.

The Embargo Act devastated New England, still so heavily dependent on maritime commerce. Legal exports fell from $48 million in 1807 to $9 million in 1808, while an epidemic of smuggling erupted along the Canadian border. In fact, it became so rife on Lake Champlain, which straddles the border, that President Jefferson declared the area to be in a state of insurrection.

The reaction against the Embargo Act in all the seaboard cities was so intense that it lasted only fourteen months, but the Nonintercourse Act, which replaced it, forbade commerce with both Britain and France, our largest trading partners, and American foreign commerce stayed in deep depression.

With his trading business in a shambles, Francis Cabot Lowell went to England in 1810 and visited cotton mills in Lancashire. He was deeply impressed. He committed to memory as many details of layout and design as he could, planning to build a mill on his return to the United States. Like Samuel Slater before him, Lowell was engaging in what today is called industrial espionage, and like Slater, he smuggled the fruits of it out of England. But he also improved the machinery with the aide of a mechanic named Paul Moody.

By the time he did return to the United States, in 1813, the War of 1812 had broken out and had ended what remained of American foreign commerce and New England was economically prostrate. Lowell established the Boston Manufacturing Company, capitalized at a then huge $300,000, a sum that was soon doubled.

The disruption of trade, while throwing New England into deeper depression, also created a national shortage of cloth in the United States, and Lowell's company was soon operating very profitably. But with the return of peace in 1815 came the return of British goods, especially cheap cotton cloth from the vast mills of Lancashire. The New England textile industry had expanded rapidly in the years before and during the War of 1812, when the Embargo Act and the British blockade had served the same function as a protective tariff. Now, confronted with renewed British competition (and, in fact, British dumping of cheap cotton cloth on the American market), the textile manufacturers, led by Francis Cabot Lowell, went to Washington for relief.

What they wanted was an actual protective tariff. They were the first of what would become an endless parade, one that continues to this day, of American manufacturers seeking protection from foreign competitors who, due to their comparative advantages, can sell in the American market for less than domestic manufacturers and still make a profit.

A protective tariff always has a surface plausibility that makes it attractive to politicians: a protective tariff saves jobs and protects profits in the short term, which is what always most concerns politicians—who are usually up for reelection in the near future. Even Alexander Hamilton recommended one in his "Report on Manufactures." But the arguments against it, laid out at length in *The Wealth of Nations,* are economically persuasive.

Most important, a tariff is not paid by foreign producers; it is paid by domestic consumers, to whom the cost is passed along. And consumers not only pay higher prices for foreign goods, they also pay higher prices for domestic goods as well, for local producers will invariably take advantage of any opportunity to raise their own prices. And a protective tariff partially insulates domestic producers from competition, the engine of innovation and cost cutting in a capitalist economy.

Opposing the idea of a protective tariff were the New England shipping interests and the South. But Lowell and the other textile manufacturers succeeded in getting Congress to place a 25-cent-a-yard duty on cotton cloth, the first protective tariff in American history, in 1816.

By this point the shipping interests were losing their clout in Congress, as New England manufacturing began surpassing them in influence with local legislators. The South, however, was united behind the idea of low tariffs. Ever more dependent on the export of cotton to Britain, and, increasingly, France, for its prosperity, the South feared retaliatory tariffs. As a large importer of manufactured goods, it regarded a high tariff as little more than a means for northern industrialists to gouge southern customers.

But northern pressure kept the tariff rising until in 1828 Congress passed what the South—always a major exporter of catchy political phrases—called the Tariff of Abominations. This led directly to the Nullification Crisis of 1832, when South Carolina declared that states had the power to rule federal laws unconstitutional. President Andrew Jackson made it clear that he would use force to see them upheld. The threat of secession was ended when a new tariff bill, calling for gradually lowered rates, was passed.

———

FRANCIS CABOT LOWELL, long in precarious health, died in 1817, but his enterprise flourished, and in 1823 it moved to a site on the Merrimack River north of Boston, where more water power was available. When the area was incorporated as a town in 1826, it was named for Lowell. But Lowell's company was hardly the only American manufacturer to prosper at this time. By 1824 there were two million people employed in manufacturing. That was ten times the number only five years earlier (and fully two-thirds of the total population of the United State at the time of the American Revolution fifty years earlier).

The United States was becoming an industrial power second only to Britain.

Chapter Six

LABOR IMPROBUS OMNIA VINCIT

I N 1791 THE FEDERAL GOVERNMENT enacted an excise tax on distilled spirits. This was, of course, unpopular with the numerous producers of rum and whiskey, although they could, and did, pass the tax on to their customers. But it was a very serious matter for farmers in western Pennsylvania, beyond the crest of the Appalachian Mountains. Because of the lack of good roads, the only way the farmers could profitably ship their grain to the markets in the East was to first distill it into whiskey, giving it a much lower weight-to-value ratio.

In July 1794 opposition to the tax flared into insurrection, and five hundred armed men burned the house of General John Neville, who was the regional inspector of the excise. On August 4 President Washington issued a proclamation ordering the rebels to disperse and militia to muster. When negotiations failed, Washington ordered thirteen thousand troops into western Pennsylvania under the command of General Henry Lee and accompanied the troops himself as far as Bedford, Pennsylvania, before returning to Philadelphia. Before such overwhelming force, the rebellion melted away. Two leaders were captured and convicted of treason, but Washington pardoned them.

Today the so-called Whiskey Rebellion is remembered mostly for being the only time in American history that the commander in chief has taken the field with his troops. But to contemporaries, it made it plain that the country faced no greater economic problem, now that Hamilton's program was in place and working, than that of transportation.

A vast country—four times the size of France, ten times the size of England—the United States had few roads worthy of the name. And it was already a restless country, indeed the most restless on earth, with a population that had been pushing westward since settlement began. One of the causes of the Revolution had been the Quebec Act of 1764, which had extended Canada's boundaries to the Ohio River and forbidden white settlement west of the Appalachian Mountains.

Reaching the area west of the mountains before the Revolution had been accomplished largely by Indian trails that had existed for centuries, such as the Wilderness Road that Daniel Boone had taken into the Kentucky country and the road through the Cumberland Gap that led to Tennessee. Usage slowly enlarged these trails into roads as travelers cut down trees and brush along the way to accommodate wagons. In swampy areas, travelers would cut trees and lay them across the pathway to make what were called corduroy roads.

But the early roads, even in settled areas, were rutted and dusty in the summer and often morasses of mud in the spring and fall. Wagons and stagecoaches, when they were able to negotiate the roads at all, could take hours to go only a few miles. Travel was often easiest in winter, when the ground was frozen hard. Traveling in western New York State in 1822, the Englishman Henry Addington reported seeing "the wreck of a coach or wagon, sticking in picturesque attitudes in some hole in the low road. . . . How any vehicle can get through such fearful breakers as those routes present sometimes for four or five miles continuously it is difficult for those who have experienced their ferocity to comprehend."

But it was commerce, not migration, let alone the comfort of adventurous tourists like Henry Addington, that spurred road development in the years after the Revolution. Philadelphia wanted to funnel the output of the rich farmland in Lancaster County to its markets and port, rather than see it travel down the Susquehanna River and the Chesapeake Bay, to the benefit of Baltimore. So in 1790 it authorized the formation of a private company to build a toll road.

Unlike earlier American roads, which had been largely created by the feet that trod them, the Philadelphia-Lancaster Turnpike was built to precise specifications, with a uniform width, and layers of crushed stone and gravel to provide an even surface. A rounded, or "cambered," surface allowed water to drain off quickly.

A Scottish engineer named John McAdam, in the early part of the nineteenth century, would perfect the technology of road building using layers of stone and gravel and give his name (slightly misspelled as *macadam*) to the process, which was widely used in both the United States and Britain. Toward the end of the nineteenth century, when engineers began mixing the top layer of gravel with tar to provide a relatively waterproof surface, the roads were said to be tarmacadamized, or tarmacked.

The Philadelphia-Lancaster Turnpike was an immediate financial success for the company that built it, and this resulted in many turnpike projects getting under way in the New England and mid-Atlantic states in the next few decades. Governments also noticed that where turnpikes ran, economic development soon followed, as inns, taverns, livery stables, and such sprang up to meet the needs of travelers.

In 1802 the act of Congress that created the state of Ohio set aside funds from the sale of public lands for road construction. In 1811 a road from Cumberland, Maryland, on the Potomac, to Wheeling, in what is now West Virginia, on the Ohio River was authorized. The Cumberland Road would eventually extend all the way to Vandalia, Illinois, a distance of nearly five hundred miles.

By the 1840s southern New England, the lower Hudson Valley in New York, New Jersey, and southeastern Pennsylvania were well served by a network of improved roads, and these greatly speeded travel. In the 1780s a stagecoach had taken from four to six days to travel from Boston to New York, depending on the weather. By 1830 the trip took a day and a half. In the South, however, better served by navigable rivers, the road system remained primitive.

And the vehicles had greatly improved as well. The early stage-coaches had been little more than farm wagons, fitted with benches and lacking springs. The passengers bounced around in them like dice in a cup. By a few decades later, however, they had evolved into the far more comfortable stagecoach familiar to anyone who has watched a Holly-wood Western.

Traveling for pleasure rather than business had been nearly unknown in the colonies, but by the 1830s it was common for affluent couples to go on "nuptial journeys" (and Niagara Falls was already a popular destination for them). Doctors at that time began to prescribe travel for the health of their wealthier patients.

But while individual travel increased, it was still commerce that crowded the expanding road net. In 1836 Ralph Waldo Emerson noticed an "endless procession" of wagons, passing his house in Concord, Massachusetts, headed "to all the towns of New Hampshire and Vermont." By 1840 nearly fifteen thousand men made their living as full-time teamsters in the United States, hauling produce and freight into and out of the growing cities. Many more did so on a part-time basis, especially in the winter when farming was at a standstill and the roads were at their best.

But there was a strict limit to what a horse-drawn wagon could haul, regardless of how good the roads became. Commodities with a high weight-to-value ratio had to move by water if they were to be moved any great distance and still be sold at a profit. Where no natural waterways existed, the preindustrial world had only one solution to this problem: the artificial rivers called canals.

———

THE IDEA OF THE CANAL dated to antiquity. Both China and Mesopotamia built them, and the Persian king Darius the Great ordered the construction (or perhaps reconstruction) of a canal linking the Nile and the Red Sea in 510 BC.

The invention of the lock in the mid–fifteenth century transformed the possibilities for canals, allowing them to cross far more uneven terrain. In the middle of the seventeenth century a canal connected the Loire and Seine rivers, and by the end of that century the Languedoc Canal in the south of France ran an astonishing 142 miles, connecting the Mediterranean with the Gironde River, which flowed past Bordeaux into the Atlantic. By the end of the eighteenth century Britain was laced with a network of canals, many of them built by the Duke of Bridgwater, that greatly facilitated the Industrial Revolution.

The need for canals to help develop the country and reduce the cost of many commodities in the new United States was obvious. By 1790, the Constitution only a year in effect, no fewer than thirty canal companies had been chartered in eight of the thirteen states. Many of these, of course, never got beyond the planning stage, and most were quite modest in their ambitions, seeking only to extend the navigable length of rivers by providing a way around rapids and falls. George Washington was a tireless promoter of a canal to extend the navigability of the Potomac River, but could not extract the funding from Congress.

The first major canal project to reach the construction stage was in Massachusetts, when the Middlesex Canal Company was chartered by the state to build a twenty-seven-mile canal between Boston and the town of Chelmsford on the Merrimac River. It was hoped that it would funnel the products of New Hampshire—lumber, granite, naval stores, and firewood—to Boston.

Deciding to build a canal is one thing, of course; actually constructing it quite another, especially as the United States was almost totally lacking in trained engineers at this time. The Middlesex Canal Company hired Loammi Baldwin to do the job. But while he had read books on canal building, he had never actually seen a canal lock, and he soon convinced the company to hire an Englishman, William Weston, who had practical experience in building canals.

The company issued shares of stock in 1794, and they were eagerly taken up by investors who were enamored with the economic promise of the new technology and ignorant of the practical difficulties involved in building a canal and operating it profitably. In the first instance—but by no means the last—of a technological investment bubble in the United States, the shares in the Middlesex Canal Company, issued at $225, rose as high as $475, thanks to speculators, before the canal actually opened for business ten years later.

Once in operation, while the canal greatly helped the economy of Boston and the area served by it, it never made money. By the time the company was dissolved in 1860, it had returned to its investors only about 75 percent of the money they had put into the enterprise.

The high capital cost of canals and the lack of engineering expertise in the country impeded construction in the early years. Speculators, once burned, were reluctant to invest in new canal ventures. Then New York State decided to undertake a canal project that was not only the largest yet undertaken in the United States, but was more than twice as large as any canal yet built in the world, at a projected cost that rivaled the annual budget of the federal government. The Erie Canal would prove to be the first of the long, and continuing, list of megaprojects— the Atlantic cable, the transcontinental railroad, the Brooklyn Bridge, the Panama Canal, the Hoover Dam, the interstate highway system, the Apollo project—that would become so much a part of the American experience.

And it was a titanic roll of the economic dice. Failure might have crippled the New York economy for decades. But success would ensure that New York, already the most populous state by 1810, would outstrip all its rivals as the American economy developed.

Thanks to a now largely forgotten giant of American politics, DeWitt Clinton, the project succeeded beyond anyone's wildest dreams, even his own. The Erie Canal would prove the most consequential public works project in American history and make New York, both state and city, the linchpins of the American economy for more than a century.

————

THE PATHWAY THAT LED from the Hudson River just north of Albany, westward through a gap in the Appalachian Mountains, between the Adirondacks and the Catskills, to the Great Lakes had been known from early colonial times. The Indians and fur traders had used it constantly. The headwaters of the Mohawk River, which tumbled into the Hudson through a sharp series of rapids, reached almost to the headwaters of Wolf Creek, which ran westward to join Lake Oneida. From the other end of the twenty-mile-long lake, the Oswego River flowed to Lake Ontario.

As early as 1724 the Irish-born Cadwallader Colden, a New York merchant and gifted amateur scientist (and, later, politician: as acting colonial governor he would be nearly lynched during the Stamp Act crisis) proposed improving this route to increase its commercial possibilities. Later it was proposed that the canal, instead of going to Lake Ontario, should run to Lake Erie. There were two reasons for this. For one, it would obviate the need to portage around Niagara Falls to reach the Great Lakes beyond Lake Ontario. And, once in Lake Ontario, it was thought, traffic would likely prefer traveling down the St. Lawrence River to Montreal and the Atlantic Ocean, at least during the warm months, rather than using the canal, threatening its economic viability.

The argument against the Lake Erie route, of course, was that it would greatly lengthen the canal and greatly increase the engineering difficulties that had to be overcome. Lake Erie is only 563 feet above the Hudson River, which is at sea level at Albany (the Hudson is actually an estuary, not a river at all). But west of Lake Oneida, the canal would have to cross the Irondequoit and Genesee rivers, which flow into Lake Ontario; pass a considerable swamp; and break through a ridge of rock that ran north and south just east of Lake Erie.

After the Revolution, Philip Schuyler, Alexander Hamilton's father-in-law, and others founded the Western Inland Lock Navigation Company to improve navigation on the Mohawk and began to back the idea of a canal. The wealthy and influential Gouverneur Morris also backed the idea of a canal, but he was afraid that "our minds are not yet enlarged to the size of so great an object."

DeWitt Clinton's mind quickly became so enlarged. Born in 1769 to wealth and position (his uncle, George Clinton, would be governor of New York and vice president under James Madison), Clinton graduated from Columbia at the age of just seventeen (delivering an address in Latin at the commencement). Soon elected to the state senate, he was appointed U.S. senator in 1802. But he resigned the next year to become mayor of New York City, a post he would hold for most of the next twelve years. In 1810 the state legislature appointed him to the newly formed Canal Commission. He almost immediately became the driving force behind the idea.

The final design was an awesome project. It would run 363 miles through what was still a semiwilderness, and would require no fewer than eighty-three locks. At forty feet wide and four feet deep, the canal would require digging and removing no less than 11.4 million cubic yards of earth and rock—well over three times the volume of the Great Pyramid of Egypt—entirely by hand. The canal project was budgeted at $6 million, a sum equal to three-quarters of the federal budget in 1810.

It was hoped that the federal government would pay a considerable portion of the cost of what would be the biggest public works project in the Western world *since* the Great Pyramid. But Clinton got little encouragement. Thomas Jefferson, no mean engineer himself and friendly to what were then called "internal improvements" and westward expansion, thought the whole idea ridiculous. He pointed out to Clinton that George Washington had been unable to get Congress to appropriate $200,000 to build a thirty-mile canal through the well-settled Virginia countryside, "and you talk of making a canal *three hundred and fifty miles through the wilderness!* It is a splendid project, and may be executed a century hence. It is little short of madness to think of it at this day."

The War of 1812 brought any hope of getting the project under way to a halt, but after the war the drive to build the canal revived. By 1817, although Clinton's opponents were still calling it "Clinton's Ditch," public opinion in the state was moving strongly in favor of it, and one hundred thousand people signed a petition to the legislature to have it built. As to the technical problems that were bound to be encountered, the collective attitude of the state can best be summed up in the airy, Latin words of Gouverneur Morris, *"Labor improbus omnia vincit"* ("Enormous labor conquers all").

On April 15, 1817, the legislature acted. It passed the Canal Act, declaring that the canal would "promote agriculture, manufactures, and commerce, mitigate the calamities of war, enhance the blessings of peace, consolidate the Union, advance the prosperity and elevate the character of the United States."

It was still hoped that the federal government and the other states that stood to benefit directly, such as Ohio, would contribute, but New York was now willing to go it alone. The federal government and the other states, not surprisingly, decided to let it, and President James Madison vetoed a bill that would have provided funding.

Just how big an undertaking the Erie Canal was for the state of New York can be seen by the fact that the final projected cost, $7 million, was equal to more than one-third of all the banking and insurance capital in the state. New York put a tax on steamboat travel, salt, and land within twenty-five miles of the canal to service the bonds it issued, and two London insurance firms made major investments in them. But most of the money came from New Yorkers, by no means all of them people of great wealth. Out of the sixty-nine subscribers to the bond issue of 1818, fifty-one invested $2,000 or less. Twenty-seven invested less than $1,000.

On July 4, 1817, newly inaugurated as governor, DeWitt Clinton turned the first shovelful of dirt outside Rome, New York, and promised that the work would be completed within ten years.

It was done in eight.

To get revenue flowing as soon as possible, the first work began on the "long level," the sixty-nine and a half miles between Syracuse and Herkimer that required no locks at all. Clinton wanted it done in a year and it was.

Thousands of workers were employed in building the canal, as many as fifty thousand at the peak, many of them local but many of them Irish and Welsh immigrants hired right off the docks of New York City. And while it was dug entirely by hand, much Yankee ingenuity made the task easier and cheaper than it might have been. A new way of felling trees was developed, using a screw and cable to pull the trees down. With another machine invented on the fly, seven men and a team of oxen could pull thirty to forty stumps a day.

One of the many self-taught engineers (who would come to be known as the Erie School of Engineers) was Canvass White. Rather than import very expensive hydraulic cement from Europe, he prospected for, and found, a local supply of trass, the type of pumice that is the key ingredient in hydraulic cement.

By 1821 some 220 miles of the canal was complete, and traffic was building steadily and attracting attention abroad. The London money market began making serious investments; Baring Brothers alone would buy more than $300,000 worth of canal bonds.

It took four years to blast through seven miles of the rocky ridge near Lockport, just east of Buffalo, and build a series of five double locks there. Those locks, the great aqueduct over the Genesee River at what is now the city of Rochester, and the twenty-seven locks needed in the first fifteen miles of the canal at the Hudson River, were each larger engineering projects than any previously undertaken in the United States.

But by the fall of 1825 it was done. On October 26 Governor Clinton and numerous dignitaries came aboard the canal barge *Seneca Chief,* which left Buffalo and proceeded along the canal and then was towed down the Hudson to New York City, while the state of New York had what amounted to a five-hundred-mile-long party.

At the end of it, off Sandy Hook, New Jersey, the entrance to the Lower Bay of New York harbor, Governor Clinton poured a cask of Lake Erie water into the Atlantic. It was a wedding of the waters, "to commemorate," Clinton said, "the navigable communication which has been established between our Mediterranean Sea and the Atlantic Ocean." One of the orators of the day described the project as the "the longest canal, in the least time, with the least experience, for the least money and to the greatest public benefit."

The Erie Canal was an immediate and astonishing economic success. "Having taken your position at one of the numerous bridges," an early observer wrote, "it is an impressive sight to gaze up and down the canal. In either direction, as far as the eye can see, long lines of boats can be observed. By night, their flickering head lamps give the impression of swarms of fireflies."

In 1825 a total of 13,100 boats used the canal and paid tolls amounting to half a million dollars, more than enough to fund the debt incurred in the canal's construction. In less than a decade the debt was paid off and the surplus funds were used to extend feeder canals. The first year an average of forty-two barges a day were passing through Utica, and that year carried 221,000 barrels of flour, 435,000 gallons of whiskey, and 562,000 bushels of wheat. A total of 334,605 tons of freight was carried on the canal that year. The population of the counties through which the canal passed tripled in only a few years as what had been villages became towns and towns became cities.

Even more important, thousands of people used the canal to leave New England's stony fields for the lush farmland of the Middle West, and then sent back the products of their labor via the canal as the cost of transporting goods eastward collapsed. It had taken three weeks and cost $120 to send a ton of flour from Buffalo to New York City before the canal, triple the cost of the flour itself. Soon it cost only $6 and took eight days.

Transportation is what economists call a "transaction cost," one that adds to the cost of an item without adding to its intrinsic value. Advertising, sales, and packaging are all examples of transaction costs. The lower these transaction costs, obviously, the lower the final price, and thus the higher the demand. When a transaction cost is lowered as dramatically as the Erie Canal lowered the cost of the produce of the Middle West, delivered to New York City, the consequences will always be far-reaching.

In this case they transformed the city from merely the largest of American cities to the metropolis of the Western Hemisphere. As early as 1822, before the Erie Canal was even finished, the *Times* of London saw it coming, writing that year that the canal would make New York City the "London of the New World." The *Times* was right. It was the Erie Canal that gave the Empire State its commercial empire and made New York the nation's imperial city.

In 1820 the population of New York City was 123,700, five times what its population had been at the beginning of the Revolution. But Philadelphia, which had been ahead of New York in 1776, was then only slightly behind at 112,000. The Erie Canal, however, turned New York into the greatest boom town the world has ever known. Manhattan's population grew to 202,000 in 1830, 313,000 in 1840, 516,000 in 1850, and 814,000 in 1860. The population of what are now the five boroughs of New York was more than twice the population of Philadelphia by the time of the outbreak of the Civil War. In 1800 about 9 percent of the country's exports passed through the port of New York. By 1860 it was

62 percent, as the city became what the Boston poet and physician Oliver Wendell Holmes (the father of the Supreme Court justice) rather grumpily described as "that tongue that is licking up the cream of commerce and finance of a continent."

The city rushed pell-mell up Manhattan Island, adding on average about ten miles of newly developed street front per year in the quarter century after the opening of the canal. No wonder John Jacob Astor (a major investor in the canal) is supposed to have said on his deathbed in 1848 that his only regret in life was not having bought all of Manhattan. But he had bought enough of it early enough to make himself by far the richest man in the country and his family legendarily rich for generations thereafter.

———

THE YEAR 1817 had been an *annum mirabilis* for New York City. Not only had construction begun on the Erie Canal, but two other far-reaching developments also occurred that year. One was the establishment of the Black Ball Line of sailing packets. Before 1817, someone wanting to travel to Europe booked passage on a ship that was essentially designed for cargo and then waited for the ship to be ready to sail, which happened when the captain felt he had enough cargo to make the trip profitable, and the wind and weather favored. It was not uncommon for there to be a wait of one or even two weeks after the announced sailing date before the ship finally departed.

New York textile importers and cotton brokers had to travel often to Britain on business, and many of them were British by birth. They regretted the time wasted waiting for ships to sail, and one of them, Jeremiah Thompson, had a revolutionary idea for how to waste less of it. He proposed the establishment of a "line" of ships under common management that would keep to a regular schedule, always sailing on the advertised day.

He formed a syndicate with four other textile businessman and put an ad in a New York paper in the fall of 1817 announcing that "the subscribers have undertaken to establish a line of vessels between NEW-YORK and LIVERPOOL, to sail from each place on a certain day in every month throughout the year."

The first vessel left New York on January 6, 1818, a large black ball painted on its fore-topsail for identification. The concept soon proved a commercial success, so much so, indeed, that it quickly spread not only to other ports but to other forms of transportation as they developed, such as railroads. Today regularly scheduled passenger transportation seems so obvious that it is hard to imagine a world without it. But it was regarded as a wonder at the time. "Such steadiness and despatch is truly astonishing," wrote *Niles' Weekly Register* in 1820, "and, in a former age, would have been incredible." The Black Ball Line did much to cement New York's reputation as the country's leading port, for passengers as well as freight.

The other significant event in New York history in 1817 was the formal establishment of a stock exchange. Wall Street had been growing quickly ever since the city recovered from the Revolution, but it still lacked a real stock exchange such as Philadelphia had had since 1792. With the near tripling of the national debt due to the War of 1812, business on the nation's exchanges had increased markedly, but much of this business went to Philadelphia, thanks to its large banks.

The New York brokers who still operated under the old Buttonwood Agreement sent a broker named William Lamb to Philadelphia to study that stock exchange. On February 25, 1817, in the offices of Samuel Beebe, several leading New York brokers met and drew up a constitution that was, in fact, nearly identical with the constitution of the Philadelphia exchange. There were just twenty-eight original members, belonging to only seven firms, in what they named the Board of Brokers, a name soon changed to the New York Stock and Exchange Board. They rented the second floor of the Bank of New York headquarters at 40 Wall Street for $200 a year, a price that included heat. It was a modest beginning to an organization that would in 1863 change its name again, this time to the New York Stock Exchange.

After a sharp but short-lived contraction in 1819, the Wall Street securities market prospered along with the country. Brokers who had once dealt in insurance, cotton, and even lottery tickets, as well as stocks and bonds, now began to specialize in securities exclusively.

But the New York Stock and Exchange Board remained only one of several exchanges in major cities around the country, such as Boston, Philadelphia, and Cincinnati, and no more important than they. Indeed, the New York Stock and Exchange Board was not even the only action on Wall Street. Much stock trading took place on the street itself, the various lampposts serving as places where particular securities were traded, quite unregulated.

And business was often less than brisk, with volume sometimes under a hundred shares a day. On March 16, 1830, the New York Stock and Exchange Board traded a mere thirty-one shares in both the morning and afternoon auctions, still the all-time, low-volume record.

That would soon change. Roads and canals, for all the improvement for life and commerce that they provided, were still the technology of the past, known to the Romans. The new technology of steam, once applied to transportation, would far surpass them in importance and create the modern world. Financing steam transportation would make Wall Street, a mere convenience in 1830, indispensable to the American economy.

Chapter Seven

THE JEFFERSONIAN DESTRUCTION

T HE CHARTER OF THE BANK OF THE UNITED STATES had a
term of twenty years and was due to run out in 1811. By that time
there would be more than a hundred state banks in operation, each pursuing profit in the very peculiar business that is banking. For banks are in
the money business: they safeguard it, loan it, and most importantly,
they create it.

Money was in the form of coinage for the first two thousand years of
its existence. And minting coins was usually a jealously guarded government monopoly. But the great influx into Europe of gold and silver from
the New World in the sixteenth century not only caused a very considerable inflation, it brought about a new form of money. People often left
precious metals on deposit with goldsmiths for safekeeping, taking
receipts for the value. Before long, people started using the receipts for
trading purposes. As long as the goldsmith was trusted, his receipt was
just as good as the precious metal in his vault that it represented, and a
lot more convenient to carry around. Once they were made negotiable
by law (in 1704 in Britain) and thus belonged, absolutely, to the holder in
due course, banknotes became money.

The goldsmiths had long been in the business of making loans and now began issuing receipts when doing so, rather than giving the borrower the metal itself. Soon, seeking increased profit, they began issuing more receipts than there was metal in the vaults to cover them. As long as the goldsmith had a good reputation, there was no problem with this, as it was highly unlikely that everyone holding receipts would try to redeem them at the same time.

By issuing more receipts than there was gold to cover them, the goldsmiths—who by this point had become bankers—created money out of thin air. Many highly intelligent and educated people failed to understand what banking was all about and thought that this was inherently dishonest. John Adams, for one, wrote that "Every dollar of a bank bill that is issued beyond the quantity of gold and silver in the vaults represents nothing and therefore is a cheat upon somebody."

But Adams was wrong. It was not dishonest in the least. The loans were, of course, collateralized with things of value, such as land and buildings. If the borrower defaulted, the lender would seize the collateral, sell it, and pay himself back. In fact, what a banker does is turn the dead capital of fixed assets into living—that is to say, liquid—capital that can be invested in new assets, multiplying the wealth of the community and creating a far more dynamic economy. Thus the assets behind the money created to make the loan are, essentially, the borrower's not the lender's. The precious metal in the vaults was simply the bank's capital, the wealth the banker puts at risk as a vouchsafe of his honesty and business sense.

But it all hinges on people not deciding to redeem the receipts en masse, as the banker could never call in his loans fast enough to meet a flood of demand for the precious metal the receipts represented. He would be forced to suspend business, and by the time everything was sorted out he would likely be bankrupt. And panic—which is essentially a psychological term, not an economic one—is highly contagious. It can spread from one depositor to all of them, and then to depositors in other banks with frightening speed.

Thus the most precious asset a banker can have is his reputation for maintaining a sound bank. As the great British political scientist of the nineteenth century, Walter Bagehot, explained, "Every banker knows that if he has to *prove* that he is worthy of credit, however good may be his arguments, in fact his credit is gone."

But despite the overwhelming need to maintain an impeccable reputation, bankers are human. They are sometimes too agreeable to their friends, sometimes too optimistic, sometimes too greedy, sometimes dishonest. Just as with politicians, once he is possessed of the magic power to create money, the banker's temptation to create too much is always strong. This is precisely why Alexander Hamilton thought the country needed a central bank: to discipline the state banks and prevent excess money creation. And this is also, of course, why many of the country's new commercial bankers resented the Bank of the United States that constrained their activities.

At the turn of the nineteenth century, obtaining a bank charter required an act of the state legislature. This of course injected a powerful element of politics into the process and invited what today would be called corruption but then was regarded as business as usual. Hamilton's political enemy—and eventual murderer—Aaron Burr was able to create a bank by sneaking a clause into a charter for a company, called the Manhattan Company, to provide clean water to New York City. The innocuous-looking clause allowed the company to invest surplus capital in any lawful enterprise. Within six months of the company's creation, and long before it had laid a single section of water pipe, the company opened a bank, the Bank of the Manhattan Company. Still in existence, it is today J. P. Morgan Chase, the second largest bank in the United States.

The Bank of the United States, headquartered in Philadelphia, had opened branches in New York, Boston, Baltimore, and Charleston, the major American ports, within a year of its creation. By 1810 it also had branches in Washington, New Orleans, Savannah, and Norfolk as well. Its interstate branches and its monopoly on deposits of the federal government made it by far the most powerful bank in the country and the only one whose notes were accepted at par everywhere.

This power, of course, attracted many enemies, especially the "Old Republicans," who still looked to Thomas Jefferson, now retired, for political leadership, and the owners of the state banks. The bankers both chafed under the discipline of the BUS and wanted their share of federal deposits, which would allow them to expand their loans. Their political connections in state legislatures—which at that time appointed United States senators—made them a powerful lobby. This was especially true in the new western states, where banking laws were often lax and where resentment against eastern financial power was greatest and Anglophobia most intense. Westerners seldom missed an opportunity to point out that much of the bank's stock was now held by British shareholders. Henry Clay, then a senator from Kentucky, led the opposition to rechartering the bank.

In favor of rechartering it were the merchants and traders in eastern cities, as close as the country then came to a financial establishment. Even James Madison, who had opposed its creation and was now president, recognized the bank's utility both as agent for the federal government and as a provider of a uniform national currency. His secretary of the treasury, Albert Gallatin, also pushed hard to have the bank's charter renewed.

It was due to expire on March 4, 1811, and the Madison administration submitted a bill to renew it for twenty years on January 24. Unfortunately Madison, while richly deserving of his place in the American pantheon as the father of the Constitution, was a largely ineffective president. He did not push hard enough to get the bill through or even to keep members of his own administration in line. When his vice president, George Clinton of New York, broke a tie vote in the Senate against the bank bill, the measure died. It was the most significant independent political act—nearly the only one—in the history of the vice presidency, and it would have disastrous consequences.

Most of the branches were recharted by the states where they were located, and the headquarters was sold—building, furniture, and all—to Stephen Girard, a Philadelphia merchant who had been born in France. Girard paid $115,000 for it, then a very large sum indeed to have in liquid assets. It turned out that he had far more where that came from. Girard, who traded around the globe, had been repatriating his money from abroad as the relations with both Britain and France deteriorated in the first decade of the nineteenth century.

Having bought a turnkey operation, Girard quickly opened a private bank of his own, backing the venture with negotiable securities amounting to $1.2 million and with cash in the staggering sum of $71,000. At a time when the country's millionaires could be counted on the fingers of one hand, Stephen Girard had shown himself to be a multimillionaire.

The diplomatic situation continued to deteriorate and many in the West agitated for war, hoping to conqueror Canada in the process. In June 1812 President Madison sent a "warning message," to the Senate, laying out the country's grievances against Great Britain in detail. He did not specifically ask for a declaration of war. Instead, constitutionally scrupulous to a fault, Madison called that "a solemn question which the Constitution wisely confides to the legislative department of the Government."

The West and South were very much in favor of war, the Northeast, with its great merchant marine and international trade that would be directly at risk, opposed it. The House voted 79–49 in favor of war, with only eight southern and western votes opposed. The Senate vote was even closer, 19–13. On June 19 Madison proclaimed a state of war to exist with Great Britain.

Having declared war on the only Great Power capable of attacking the United States, Congress voted to increase the army's pay for privates from $5 to $8 a month and to provide generous enlistment bonuses of $31 and 160 acres of land, later increased to $124 and 320 acres. These were no small sums at the time, when an unskilled laborer might be lucky to earn $2.50 a week. But Congress, having decided to take on the most expensive of all public policies, war, and having raised the pay of the army substantially, took no action to pay for any of this and adjourned the following month. It was, without a doubt, the most feckless act in the history of the United States Congress, a title for which there has been much competition over the years.

Even worse, because the year before Congress had shut down the Bank of the United States, it had terminated the only institutional mechanism the government had for borrowing money to pay for a war. The secretary of the treasury would have no choice but to rely on public subscription to sell bonds, and most of the people with the liquid assets to buy them lived in the Northeast, where opposition to "Mr. Madison's war" was most intense.

Military success in the early days of the war might have helped its popularity and therefore its funding. But while there were several stirring single-ship triumphs over the Royal Navy, they affected the outcome not a bit and the British blockade of American ports tightened inexorably, causing revenue from the tariff and tonnage duties to fall precipitously. On land, all was disaster. General William Hull, ordered to invade Upper Canada (today called Ontario) from Detroit, got no more than a few miles into Canada before retreating hastily to Detroit, which he then surrendered without firing a shot. Hull was court-martialed and sentenced to die for cowardice and neglect of duty, but Madison, in view of his service in the Revolution (Hull had served with Washington at the Battle of Trenton), pardoned him and only dismissed him from the army. Two other thrusts into Canada, at Niagara and Lake Champlain, were no more successful, if less ignominious.

By the early spring of 1813 the United States government was, financially at least, in extremis. On March 5 the secretary of the treasury had written President Madison that "We have hardly enough money left to last to the end of the month."

On February 8, 1813, Congress had authorized the Treasury to borrow $16 million, by far the largest loan in the country's short history. Gallatin did his best to structure the loan in attractive ways. People could subscribe for as little as $100 and pay for their investment in eight monthly installments. The loan was scheduled to close on March 13 but could be kept open until the end of the month if not enough subscriptions had come in. The entire $16 million had to be subscribed to or the loan would not go through.

Less than $4 million was subscribed by March 13, and Gallatin extended the subscription period. It did not help much, and while not all subscriptions had come in because of the slowness of communications, Gallatin knew the loan had failed. Later he told Madison that "The Treasury was so far exhausted on the first day of this month that the small unexpended balance, dispersed in more than thirty banks, could not have afforded any further resources." The government was broke and the war effort was likely to sputter out not because of military defeat, but because of financial collapse.

Gallatin had allowed himself an extra five days to find a lender of last resort if necessary. On April 5 he went to see the one man in the country who could help, Stephen Girard.

Gallatin told him that subscriptions had totaled $5,838,200 and that a New York syndicate led by John Jacob Astor was willing to undertake $2,056,000 if the loan went through. That left $8,105,800 that needed to be subscribed to *that day* if utter disaster was to be averted. The sum was beyond staggering. Total government revenues from all sources in 1812 had been $9,801,000.

Rich as he was, $8 million was more than even Stephen Girard's net worth. And yet he immediately said yes, subscribing, in writing, "to the residue of the said loan." He simply asked that the residue be deposited in his bank until drawn down and that he receive a commission of one-quarter of 1 percent on the loan to cover his costs in selling portions to other subscribers.

It was not quite as daring a move as it seems on its face, although it was certainly daring enough. Girard had an unimpeachable credit rating, far better than the government's, and he expected to be able sell considerable portions of it. And he had eight months in which to come up with the entire amount.

His credit rating was so much better than the government's, in fact, that within ten days, once the news spread that Girard was a major participant, he was able to sell $4,672,800 to other investors.

With money once more in the Treasury, the military situation improved quickly. William Henry Harrison recaptured Detroit, and the American army won a considerable victory at York (now Toronto) and decisively defeated a British fleet on Lake Erie. The British, having an inkling of the U.S. government's financial difficulties, had spurned an offer—accepted by Madison—from the Russian government to mediate. Now, wanting only to be done with this annoying little war on the margins of its global concerns, Britain proposed peace talks in Ghent, in what is now Belgium. The United States was able to get a draw out of what had very nearly been a catastrophe. Even so, it suffered the ignominy of having its capital burned by the enemy.

———

WITH THE EXTINCTION of the Bank of the United States, state banks proliferated, more than doubling in the two years after 1811. Most of them issued banknotes. Adam Smith had estimated that a bank could safely issue banknotes in the amount of five times capital, and some states restricted banknote issues to three times capital. But other states placed no such limits. In 1809 a bank in Rhode Island, capitalized at a mere $45, had issued banknotes amounting to $800,000, more than seventeen thousand times the capital. Of course, this was far more a fraud than a bank, but it represents the first of many thousands of bank failures in the United States; for bank failure, thanks in large part to Thomas Jefferson and his political heirs, was to become as American as apple pie.

Some states did not even limit the issuance of paper money to banks. Although technically called "scrip," and intended to relieve the shortage of small change, paper currency was issued by municipalities, trade associations, factories, and even individuals. Something called "A General Assortment of Groceries in Philadelphia" issued scrip, including a note for six and a quarter cents (a sixteenth of a dollar, or half a "shilling") that was redeemable "on demand, in groceries, or Philadelphia banknotes."

The Madison administration, its mind wonderfully concentrated by the near disaster of 1813, wanted a new central bank. Madison vetoed a bill for one in 1815 on technical grounds, but signed a bill sent to him in 1816, chartering a Second Bank of the United States for twenty years. There had been a curious reversal in the politics of a central bank. In 1811 Henry Clay of Kentucky had opposed rechartering the first bank. Now he was in favor of it, as was John C. Calhoun of South Carolina. Daniel Webster of Massachusetts, once a staunch supporter, now opposed the measure.

The areas of the country represented by Calhoun and Clay were suffering from a chronic lack of specie. The reason was that New England, its manufactures growing rapidly as was its foreign trade, thanks to its ownership of much of the nation's merchant marine, was running a large trade surplus with the rest of the country, causing specie to drain away from the West and South and toward the Northeast. Calhoun felt that only a national bank could solve this problem. Webster, who did not want the problem solved, now fought for the status quo.

But while the Second Bank of the United States, headquartered in Philadelphia like the first one, would, after a shaky start, be a stabilizing influence on the American monetary system, it would never have the power to control it that the first bank had had. The country was growing too rapidly and the state banks had proliferated too much in the five-year hiatus between the two banks.

Some states had sound banking laws. Missouri, admitted to the Union in 1821, and Indiana, admitted in 1816, had single, state-owned, many-branched central banks of their own, a system that worked well. Louisiana (1812) closely regulated its commercial banks and had a wide reputation for sound banking. Illinois (1818), on the other hand, had an equally wide reputation for flimflam, fraud, and failure among its banks. But despite the states with well-regulated banking systems, bank failure became endemic in this era. Fully half the banks founded between 1810 and 1820 had failed by 1825.

Hundreds more sprang into existence, however, many of them issuing banknotes. And many people who did not own banks at all also printed them and tried to pass them. Publishers began printing "banknote detectors" that listed, illustrated, and assessed the worthiness of banknote issues that were circulating in various areas of the country. By the mid-nineteenth century the number of paper money issues in circulation numbered in the thousands and created a monetary cacophony quite as bad as the colonial hodgepodge of bits of foreign coins, warehouse certificates, and provincial bills of credit.

———

NO ONE, not even Thomas Jefferson, had more influence on shaping the Democratic Party and its economic philosophy before Franklin Roosevelt than Andrew Jackson. Indeed, the modern Democratic Party coalesced around Jackson's extraordinary political personality. Both Jackson and Jefferson believed in pushing the locus of power down the social scale. Both had a deep-seated dislike of inherited privilege and what Jackson called "the money power," which is to say banks, especially large, well-established, and powerful ones.

But the two men were also profoundly different. Jefferson had reached his philosophy by intellectual means and was, first and foremost, a man of thought. Jackson was the quintessential man of action. It is hard to imagine Thomas Jefferson fighting a duel, but Jackson fought no fewer than three—killing his opponent in one of them—and avoided many others only because his opponents backed down. Jefferson never wore a uniform in his life; Jackson was an excellent general who achieved national fame for his smashing victory over the British at the Battle of New Orleans in 1815. (It was, however, a militarily inconsequential victory, as the peace treaty had already been signed.) Most of all, Jefferson had been born rich and was cavalier, at best, about money. Jackson had been born very poor. He had no intention whatever of dying poor, and he didn't.

Andrew Jackson represented a revolution in American politics. So much so that the great nineteenth-century historian George Bancroft thought him the last of the Founding Fathers. His, wrote Bancroft, was "the last great name, which gathers round itself all the associations that form the glory of America." Jackson was the first president who did not come from the seaboard states (indeed, he was the first president who did not come from Massachusetts or Virginia) and thus pushed the political center of the country sharply westward. He was also the first who had not been born into the colonial upper class. He was born on the South Carolina frontier in 1767 of Scots-Irish immigrant parents and orphaned while still a child, both his mother and his two brothers dying indirectly as a result of the British invasion of the Carolinas in the American Revolution. As a result, Jackson had a lifelong antipathy toward Great Britain.

Jackson did not receive much formal education, but studied law in a law office in Salisbury, North Carolina, and was admitted to the bar in

1787. He shortly thereafter moved to Nashville, Tennessee, then hardly more than a collection of log cabins. Jackson practiced law and speculated in land, the fastest if riskiest way to wealth. By the time Tennessee became a state in 1796, Jackson was wealthy by the standards of the time and place. He was elected Tennessee's first congressman and the following year served briefly as a U.S. senator.

James Parton, an early biographer of Jackson, described what he called "the secret of his prosperity." It was also, of course, the secret of most early fortunes on the frontier. "He acquired large tracts [of land] when large tracts could be bought for a horse or a cow bell, and held them till the torrent of emigration made them valuable."

Like most speculators in land, Jackson sometimes got involved in complicated deals involving credit. In 1795, wanting to establish a trading post, he sold sixty-eight thousand acres to a man named David Allison, taking the latter's promissory notes in payment. He used the notes as collateral to buy supplies for his trading post, but when Allison went bankrupt, Jackson was left holding the bag.

It would take him a decade and a half to finally settle everything in this affair, and it left him with a lifelong horror of debt and debt's various instrumentalities. To Andrew Jackson, real money was specie—gold and silver coins. Paper money and what was coming in his day to be called commercial paper—bills of exchange, promissory notes, bank checks, and such—were to Jackson, just as they had been to John Adams a generation earlier, a form of fraud.

Just how different Jackson's administration was going to be from earlier ones was clear on the very first day. Previous inaugural festivities had been invitation-only, decorous affairs for "polite society." Jackson's reception was attended by everyone who could get in the door of the White House, and several thousand dollars' worth of glass, china, and furniture was destroyed. Finally the crowd became so great that Jackson's safety was threatened and he escaped out a window while servants carried the liquor out onto the lawn to get the mob out of the White House.

Jackson's first order of financial business when he entered the White House was a simple one: pay off the national debt. The national debt, which had stood at over $80 million in 1792, had been reduced to only $45 million by 1811. The War of 1812 had then caused the debt to soar to more than $125 million by 1815. The high tariff generated large surpluses after the war, and by the time Jackson reached the presidency, it stood at $48,565,000.

Jackson had two purposes in ridding the country of debt. The first, of course, was that he thought debt was bad in and of itself. He had called it a "national curse" in his first run for the presidency in 1824. But he thought that the institutions and the people who benefited from it were a national curse as well. "My vow," he pledged, "shall be to pay the national debt, to prevent a monied aristocracy from growing up around our administration that must bend to its views, and ultimately destroy the liberty of our country."

To achieve his goal, Jackson was perfectly willing to sacrifice the "internal improvements" such as roads that were so dear to the hearts of his fellow westerners, saying that once the debt was paid off, there would be plenty of money for improvements. When a Kentucky congressman pleaded with Jackson not to veto an improvements bill, Jackson promised to study the matter thoroughly. But the congressman reported to his friends that "nothing less than a voice from Heaven would prevent the old man from vetoing the Bill, and [I doubt] whether that would!"

By the end of 1834 Jackson was able to report to Congress in the State of the Union message that the country would be debt-free on January 1, 1835, and have a balance on hand of $440,000. Jackson thought it one of his greatest achievements and so did much of the country. The *Washington Globe,* noting that paying off the debt coincided with the twentieth anniversary of the Battle of New Orleans, wrote that "New Orleans and the National Debt—the first of which paid off our scores to *our enemies,* whilst the latter paid off the last cent to *our friends.*"

Chief Justice Roger B. Taney noted that it was "the first time in the history of nations that a large public debt has been entirely extinguished."

It remains the only time to this day.

———

ALTHOUGH JACKSON'S MAIN MOTIVE for paying off the debt was the simple—indeed for sovereign governments, simplistic—notion that one should borrow only when absolutely necessary, he had another motive as well. Banks used federal bonds as backing for their issuance of banknotes, and Jackson, wanting to get rid of paper money, planned to do so by getting rid of the backing.

To Jackson, the heart and soul of the "monied aristocracy" he feared was the Second Bank of the United States and its president, the Philadelphia aristocrat Nicholas Biddle.

Biddle was everything Jackson was not: wellborn, highly educated, widely traveled, and financially sophisticated. Trained in the law, he spent three years in Europe, where he served as James Monroe's secretary while the latter was minister to Great Britain. After marrying Jane Craig, a considerable heiress, he abandoned the law and became editor of a Philadelphia literary magazine, *Portfolio*. He soon built one of the finest houses in America, Andalusia, on the shore of the Delaware River just north of Philadelphia, where the Biddle family still lives.

In 1819 President Monroe appointed Biddle to the board of directors of the Second Bank of the United States, and in 1823 he became president of the bank. The Second Bank had gotten off to a rocky start under its first president, William Jones, who had speculated in the bank's stock and engaged in other shady practices. Jones resigned after a congressional investigation and Langdon Cheves replaced him, straightening up the mess left by Jones. This, unfortunately, helped mightily to bring on the panic of 1819 and the brief business slowdown that followed.

By the time Biddle became president, most of the animosity against the bank had disappeared, thanks to economic recovery and sound policy on the part of the bank. It was not an issue in the presidential campaign of 1824—when Jackson won a plurality of the popular vote, but lost in the House of Representatives to John Quincy Adams—or in 1828, when Jackson won a smashing victory over the unpopular Adams. Biddle voted for Jackson in both elections.

But hardly had Jackson entered the White House than his visceral hatred of banks, especially large, powerful ones, manifested itself in his first message to Congress as president. He raised the question of rechartering the Second Bank of the United States a full seven years before its charter was due to expire. By 1832, when Jackson ran for reelection, it was clear that he intended to kill the bank.

Biddle fought back as best he could. Many congressmen enjoyed good relations with the bank (and many of them were beneficiaries of loans on favorable terms—again just business as usual in those days, not corruption), and Biddle pressed them to pass a bill rechartering the bank for fifteen years before Congress adjourned for the summer in 1832. He hoped that Jackson would not want to make it a campaign issue. In the bare-knuckle politics of the era, when most newspapers were openly partisan and tendentious in their coverage, the fight over the bank was a brawl.

Congress finally passed the recharter bill and prepared to adjourn. But when it was clear that Jackson would use a pocket veto to kill the bill, Henry Clay persuaded Congress to stay in session long enough to force Jackson to sign or veto the bill and state his reasons for doing so. Jackson, not exactly a man to back away from a fight, was only too happy to take up the challenge. He issued a blistering veto message that was as much a campaign speech as an act of statecraft. He argued that the bank was a monopoly and favored the rich and powerful over the ordinary citizens of the country. Further, despite Supreme Court rulings to the contrary, he declared it unconstitutional. Congress was unable to override the veto.

When Jackson won a landslide victory that November, the Second Bank of the United States, although it had four years left on its charter, was at best a dead man walking. Nor was Jackson content to simply allow the charter to expire. He began to withdraw federal deposits from the BUS and move them to what his opponents soon dubbed "pet banks." As the bank's deposits drained away, its ability to issue banknotes and exert discipline on the rapidly increasing number of state-chartered banks diminished apace. The pet banks, now flush with federal deposits, rapidly increased the amount of banknotes they had in circulation. The control of the country's money supply shifted from a single institution with a long-term and national perspective to a myriad of local ones, each concerned only with its own short-term prosperity.

Although there was a brief dip in 1834 as the Second Bank of the United States called in its loans, prosperity was widespread in the early 1830s, as high cotton prices in the South, rapidly increasing manufacturing in the North, and an improving transportation system combined to increase GDP rapidly. The number of banks in the country increased from 329 in 1829 to 788 in 1837. But the face value of banknotes in circulation more than tripled, and outstanding loans quadrupled in this time.

As always in boom times, speculation increased rapidly as well. So much did stock trading on Wall Street increase that it was at this time that the term *Wall Street* first came to be used as a metonym for the American financial community.

But nowhere was speculation more rife than in the West, where land rather than securities was the focus of attention. People who had no intention of settling bought large tracts of land from the federal government and paid for them with paper money borrowed from local banks in ever greater amounts. Land sales by the federal government, handled through the government's General Land Office, had amounted to $2.5 million in 1832. By 1836 they amounted to nearly $25 million, and in the early summer of that year were running at the astonishing rate of nearly $5 million a month. This rush to buy federal land is the origin of the phrase "to do a land office business," and it horrified Jackson, who probably never understood just how much his own policies had increased what he most abhorred: speculation and paper money.

But he understood exactly what was happening. "The receipts from the public land were nothing more than credits on the bank," he later wrote. "The banks let out their notes to speculators, they were paid to the receivers, and immediately returned to the banks to be sent out again and again, being merely instruments to transfer to the speculator the most valuable public lands. Indeed, each speculation furnished means for another."

In typical Jacksonian fashion, he resolved to do something about it. That spring he proposed to his cabinet that the land office be instructed to accept only gold or silver in payment for land, except from genuine settlers who could buy up to 320 acres and pay for the land with banknotes until December 15 that year. Most members of the cabinet, many of whom were deeply involved in the speculation themselves, opposed the plan, and it was clear that Congress, many of whose members were equally involved, would not stand for it.

So Jackson waited until Congress adjourned and then, on July 11, issued the so-called Specie Circular as an executive order. The effect, needless to say, was to bring speculation in western land to an immediate halt. But it also created a considerable increase in demand for specie in the West, draining eastern banks of gold and silver and leading to hoarding. Many western banks were soon in distress, none more so than the pet banks, thanks to another part of the Jackson financial program.

With the national debt paid off and the government running large surpluses (government revenues increased by 150 percent between 1834 and 1836, in part due to the greatly increased land sales), the question of what to do with the money was increasingly urgent. Jackson convinced

Congress to give it to the state governments, beginning January 1, 1837. Faced with having much of their government deposits withdrawn, the pet banks had begun to call in loans.

As highly illiquid borrowers defaulted on their loans, a wave of bank failure swept the West and began to roll eastward. Bankruptcies in other sectors of the economy followed as liquidity vanished. The Bank of England, trying to prevent an outflow of gold from that country, raised interest rates, and British investment in American securities declined as did British cotton purchases. Wall Street plunged. On January 2, 1837, the *New York Herald* reported that interest rates that had been 7 percent a year were now 2 or even 3 percent a month.

The American economy began to slide into deep depression. The price of cotton fell by half in March. Volume on Wall Street soared in the panic (sometimes reaching a then-staggering ten thousand shares a day). By April Philip Hone, former mayor of New York and himself badly burned, was writing in his diary that "The immense fortunes which we have heard so much about in the days of speculation, have melted like the snows before an April sun. No man can calculate to escape ruin but he who owes no money; happy is he who has a little and is free from debt." By early fall, 90 percent of the nation's factories were closed. Federal revenues fell by half in 1837.

In May the New York banks had suspended payment in specie, and were soon followed by the banks in the other major eastern cities. Worst hit were the big Philadelphia banks. Pennsylvania, with a huge state debt of $20 million and with its tax revenues falling sharply, was unable to meet payments of either principal or interest and defaulted. New York, which had a state debt of only $2 million, was able to meet its obligations, and its banks survived in far better shape. Philadelphia's days as a rival to Wall Street were over.

Andrew Jackson, with the consummate timing of a great politician, had retired from the presidency on March 4. It would be his successor, Martin Van Buren, who would suffer the political consequences of his financial policies. The country suffered as well in the longest economic depression in the nation's history. It didn't reach bottom until February 1843, fully seventy-two months after it began.

And the country's financial reputation suffered as well. The year the depression finally began to lift, 1843, was the same year that Charles Dickens published *A Christmas Carol*. In it he describes how Ebenezer Scrooge was relieved to find that the world had not ended, as he first feared it had, after the visit of the Ghost of Christmas Past. Thus a note payable to him in three days, Scrooge was relieved to know, was not as worthless as "a mere United States' security."

Chapter Eight

NEW JERSEY MUST BE FREE!

T HE STEAM ENGINE DEVELOPED by James Watt and patented in 1769 was something very new indeed under the sun: the first source of work-doing energy since the windmill had appeared in Persia in the seventh century. But Watt did not invent it; Thomas Newcomen had patented the first practical steam engine in 1712. Watt's improvements, however, made the Newcomen engine four times as fuel efficient, greatly increasing the number of possible applications. When Watt developed a rotary steam engine in 1784, which converted the reciprocal up-and-down motion of the Newcomen engine into rotary motion that could turn a shaft, the steam engine's economic potential became boundless.

Until the coming of the steam engine, only human beings, draft animals, falling water, and windmills were available to do work. They each had severe limitations. Waterwheels and windmills had to be placed where the water and wind could be found, which is why the factories of the early eighteenth century were usually located in the country, not in urban areas. Humans and draft animals have limited power and can be harnessed effectively together only to a limited extent. But a steam engine could be built large enough to produce prodigious quantities of energy and could bring that energy to bear on almost any task. Because of the steam engine, many tasks that had been difficult (and therefore expensive) became easy (and therefore cheap). Many more tasks that had been impossible were within reach. Like the printing press of the fifteenth century, the steam engine was a world-transforming technology.

With England's advanced iron industry and coal reserves providing cheap fuel, the steam engine spread quickly through the British economy. In the United States, however, with its industry at first centered in New England, where water power was abundant, steam was adopted only slowly for industrial purposes. As late as 1832 a census of 249 factories east of the Appalachians showed only 4 were steam-powered.

But almost from the moment that James Watt patented his rotary steam engine, men were at work trying to apply it to the task of moving boats through the water. The advantages of steam were more than obvious. Small vessels could be moved by oars, paddles, or sculls worked by men; large ones could move only by wind pushing on sails.

The "fuel" for sailing ships is free, as they are, in the last analysis, solar-powered. But a sailing ship can only go where—and when—the wind is willing to take it. The tubby merchant ships of the eighteenth century could barely make headway against wind forward of the beam and often had to go hundreds, even thousands of miles out of their way to find a favorable slant. The prevailing westerly winds of the North Atlantic are the principal reason that the passage from America to Europe was usually so much quicker than the reverse trip, from Europe to America.

The main problem with applying steam to navigation was how to transmit the power generated by the steam engine to the water and thus move the boat. Any number of methods was tried. James Rumsey tried a system of pumping water in at the front and expelling it out the stern, but it was too complicated to be practical. There was even an attempt to imitate the feet of a duck, likewise unsuccessful.

Another system attempted was a series of oars attached vertically to a horizontal rod. The center of the rod moved in a circle, dipping the oars into the water, moving them backward, lifting them, and bringing them forward for the next stroke. John Fitch, born in Windsor, Connecticut, was living in Bucks County, north of Philadelphia, when he became interested in the problem. He built a boat powered this way that operated on the Delaware River as early as 1787. It worked, but as with so many pioneering inventors, Fitch never paid much attention to the commercial necessity of making money, and while his boat operated for a while on a regular schedule, it did not make a profit and soon vanished.

The most promising method was the paddle wheel. The idea came from the water mill, where water pushed the wheel to turn the machinery. The paddle wheel reversed this: the machinery turned the wheel to push the water and thus move the boat. But there was one big problem. Early paddle wheels were placed low in the boat so that the bottom half of the wheel was immersed in the water. But most of the energy transmitted from the engine to the water was wasted as the paddles entered the water more or less horizontally and pushed the water down rather than backward. The opposite was true at the end of the stroke, when the paddle pushed the water upward. Only at the bottom of the stroke was useful work being done.

It was a Scotsman, William Symington, who found the answer by putting the wheel high, so that only the tips of the paddles entered the water, at a point where they could push the water efficiently. In January 1803 his *Charlotte Dundas* towed a vessel of one hundred tons from Stockingfield to Port Dundas on the Forth and Clyde Canal at the rate of three miles an hour.

Symington's vessel worked well, but not in a canal, where it was far less economical than horses towing barges. He more or less lost interest in it, but an American named Robert Fulton did not.

Fulton was born in Pennsylvania and early displayed a considerable mechanical aptitude. He became expert in gunsmithing, although he was not apprenticed to a master, the usual way to learn such skills in the eighteenth century. At the age of fourteen, he built a small boat with a hand-powered paddle wheel. Apprenticed to a Philadelphia jeweler, he soon branched out into painting miniatures and doing "hair work." He went to England in 1786 to study under the great American painter Benjamin West, but he abandoned art for engineering in the early 1790s. In 1797 he moved to France.

Fulton was not an original inventor; instead he used the ideas of others, improving them and synthesizing them into a better whole. He improved the submarine David Bushnell had designed in 1776 to attack the British fleet in New York harbor, and tried to sell it, unsuccessfully, to the French (while keeping the British fully informed of his negotiations, presumably to make a second profit).

Fulton was a gifted, if not overly scrupulous businessman, and had the knack of making friends with the powerful. His most powerful friend was Robert Livingston, appointed United States minister to France by Thomas Jefferson and a member of the great New York Livingston clan that then played a major part in the politics of the state.

Livingston had a vast estate named Clermont 110 miles up the Hudson from New York City and wanted to speed up travel between the two places. An amateur tinkerer, he had tried to develop a steamboat on his own, but without success. However, he had managed to get New York State to grant him a monopoly of steamboat navigation in New York waters, provided that he build a boat capable of traveling four miles an hour within a year. He didn't meet the deadline, but the legislature extended it in 1803, passing the bill amid gales of laughter, as most members thought the requirements beyond possibility.

When he met Fulton in Paris, Livingston decided to help fund Fulton's steamboat experiments, and the two signed an agreement to build a steamboat for use in the Hudson River back home. It boiled down to

Fulton supplying the design and Livingston supplying the money—and the monopoly that would guarantee the enterprise's profitability. Fulton had become committed to a chain-drive mechanism, a bit like a tank tread with paddles on it, for transmitting the engine's power to the water. Livingston pushed for a paddle wheel, but Fulton resisted, and Livingston went along. According to Symington's later recollection, however, Fulton visited him in Scotland, took a ride on the *Charlotte Dundas,* and was mightily impressed.

Fulton always insisted that his steamboat was wholly his own design, but Fulton is known to have hedged the truth and even lied outright about countless other matters, and it is altogether likely that he adopted vital aspects of Symington's design without acknowledgment, especially the high placement of the paddle wheel axle. Certainly he abandoned the chain drive concept on his return to France.

Returning to the United States for the first time in nearly twenty years in 1806, Fulton settled in New York City and set about building a steamboat to run on the Hudson River. When completed, it was 146 feet long and 12 feet wide, with a flat bottom and straight sides. The wrought-iron paddle wheel mechanism and the copper boiler were constructed locally, but the twenty-four-horsepower steam engine came from James Watt's firm in England.

On the morning of August 1, 1807, the *North River Boat,* as Fulton rather unimaginatively named it (only after his death would it come to be remembered as the *Clermont*), set off from the Christopher Street dock. A large crowd had assembled to see it off, many of them undoubtedly expecting the vessel—which someone likened to a sawmill placed on a raft and set afire—to sink or explode.

But it didn't. Instead it chugged steadily northward, passing sailing ships on the way, and reached Livingston's estate at Clermont by midmorning the next day. Livingston came on board, and the two partners went on to Albany, arriving in the morning of the following day. The vessel had taken thirty-two and a half hours to travel the hundred and fifty miles between New York and Albany, averaging four and a half miles an hour. The provisional monopoly of steamboat navigation in New York waters, granted in 1803, was now the property of Livingston and Fulton.

Fulton advertised for passengers for the return voyage, but only two were willing to pay the $7 charge, more than twice the usual price for a boat passage to New York. On the return, however, the shores of the Hudson were lined with people, including the cadets at West Point, cheering the boat on. Regular service was soon established between New York City and points north. By 1812 Fulton and Livingston had six steamboats plying the local waters.

———

LIVINGSTON, who had negotiated the purchase of Louisiana from Napoleon, was well aware of the potential of steamboats in the Mississippi River and its tributaries. These rivers provided no less than sixteen thousand miles of navigable waters that drained an area of more than a million square miles from western New York State to Montana. Much of it was the finest agricultural land on earth with almost limitless economic potential. Much of it was also rich in minerals.

But there was a problem before the coming of steam: these great fluvial highways were virtually one way. Flatboats, hardly more than large rafts casually knocked together, could carry thirty to forty tons of cargo apiece to be swept by the current downriver to New Orleans. The young Abraham Lincoln twice made such a voyage. Once in New Orleans, the cargo would be sold while the flatboats were broken up and sold for lumber.

But the only way upriver was by keelboats, which hugged the shore where the current was least and were poled upstream by human effort. A trip by flatboat from the Ohio Valley to New Orleans could be made in a month, with hardly any expenditure of human labor. The return trip by keelboat took three months of unremitting toil. Most didn't even try and walked home instead. The Natchez Trace, running between the Mississippi at Natchez and Nashville, Tennessee, on the Cumberland, was a major highway until the coming of steam.

In 1811 Livingston and Fulton sent the boatbuilder Nicholas Roosevelt to Pittsburgh to build a steamboat there, designed by Fulton. At the same time Livingston tried to obtain monopolies similar to what he had in New York. Most states and territories refused this scion of the eastern establishment abruptly. "Our road to market must and *will* be free," thundered the *Cincinnati Western Spy* at the very idea. "This monopolizing disposition of individuals will only arouse the citizens of the West to insist on . . . the privilege of passing and repassing, unmolested, on the *common highway* of the West."

But while he did not succeed elsewhere, he did succeed where it really mattered, the Territory of New Orleans. There the territorial governor—who by no coincidence whatever was his brother Edward, a former New York City mayor and congressman—granted him a monopoly in Louisiana waters. As New Orleans was the breakpoint between river and ocean traffic, this was nearly as good as a monopoly over the whole Mississippi Valley.

But the monopoly was often ignored or, more accurately, defied. A boat owner named Henry Shreve took the case to federal court and eventually won a verdict holding that the territory had no authority to grant such a monopoly. By this time both Livingston and Fulton were dead and the case was not appealed. The number of steamboats in the Mississippi River system exploded as a result.

Thanks largely to Henry Shreve, who had a knack for naval architecture among other things, they rapidly evolved into a new form. Multidecked, wide-beamed, shallow-drafted vessels, they were able "to float on a heavy dew," and were well adapted to the rivers, which were threaded with ever-shifting sandbars and mudflats. As these handsome, ginger-breaded vessels proliferated on the rivers of the mid-continent, they soon entered into the folk memory and became an enduring image of nine-teenth-century America, thanks to such people as Currier and Ives, Mark Twain, and Jerome Kern and Oscar Hammerstein II.

Henry Shreve also solved another major problem of navigating the Mississippi and its tributaries: snags. The river often undercut its banks, and large trees would fall into the water and make their way down-stream. Those that floated free were called "sawyers" for their resem-blance to circular saw blades as they rolled lazily in the current. Far more dangerous were the "planters," trees that had become lodged on the bot-tom. Often invisible, they could rip the bottom out of a steamboat and sink it in seconds. There would be only "a sudden wrench, the rush of sucking water, a clanging of bells, terrified screams, and the current would sweep over another tragedy."

In the 1820s, it was estimated that there were at least fifty thousand snags in the Mississippi River system, having accumulated since the ice age, and most people thought there was nothing to be done about them. Henry Shreve thought otherwise. He built two steamboat hulls, each 125 feet long, each powered by a paddle on only one side. These he con-nected together with beams that held a large wooden wedge, sheathed in iron for catching the snag, and a pulley and windlass system for lifting them from the river.

On August 19, 1829, Shreve's snag boat, the *Heliopsis,* went to work at Plum Point, Tennessee, one of the most snag-ridden places on the entire Mississippi. Shreve pointed it at a planter sticking out of the water and charged. The ram split the snag in two, and the crew winched the tree up by the block and tackle and sawed it into harmless chunks. By that evening the channel at Plum Point was clear. By the following year a newspaper was able to report that "Capt. Shreve has perfectly succeeded in rendering about 300 miles of river as harmless as a millpond."

Shreve's greatest feat of snag clearance was cutting a path through the Great Raft, a logjam 150 miles long on the Red River. It opened up all of northwest Louisiana to commerce, and today the largest city in that part of the state is called Shreveport.

———

THE WATT STEAM ENGINE, as Fulton had shown, could power a vessel. But it was very bulky and its low pressure and leisurely pace—only about twelve cycles per minute—provided little power per unit of weight. A radical new type of steam engine was needed to make the steamboat a really paying proposition. Oliver Evans in the United States and Richard Trevithick in Britain developed such an engine independently.

In the Watt engine, steam pushed the piston to the bottom of the cylinder and was then drawn off and condensed, creating a vacuum that brought the piston back up. In both the Trevithick and Evans engines, steam not only pushed the piston down, but pushed it back up as well, dispensing with the separate condenser. (Because the steam was vented twice each cycle rather than condensed, they came to be called puffer engines for their characteristic noise.)

This allowed many more cycles per minute and produced much more power per unit of weight. Evans, who had built the first steam engine of the Watt type ever constructed in the United States in 1800, built one to his new design in 1803. It had a cylinder only about six inches in diameter and eighteen inches long. It produced about five horsepower. The Watt-type engines that had been built in England, and installed shortly before in Philadelphia's Central Square as part of the city's waterworks, each had cylinders thirty-two inches in diameter and six feet long, but developed only about twelve horsepower.

Oliver Evans did not build a steamboat, but he did build the first steam-powered vehicle in the United States—and arguably the world's first automobile thereby. Commissioned to build a steam dredge for the port of Philadelphia, he produced a vessel thirty feet long and twelve feet wide, which weighed seventeen tons. He mounted a new engine in it, smaller, lighter, and even more efficient than his first model, at his shop about a mile up Market Street from the Schuykill River. He then put the whole thing on wheels and attached the engine to one axle with a chain drive. Giving the contraption the improbable name of *Orukter Amphibolos,* he set off down Market Street for the river "with a gentle motion."

When Evans reached Central Square, he circled around the water-works several times, literally as well as figuratively running rings around Watt's low-pressure engine design, before continuing on to the Schuykill, where he dropped the wheels, and the *Orukter Amphibolos* departed from history to take up its duties as a dredge.

The Watt engine, which had helped to spark the Industrial Revolution and was the most fundamentally important technological development since the printing press three hundred years earlier, was obsolete after only three decades. The pace of change had already begun the relentless acceleration that continues to this day.

Evans began manufacturing steam engines at his Mars Iron Works in Philadelphia, and later a branch, under the direction of his son, opened in Pittsburgh to supply steam engines to the ever-growing fleet of steamboats in the Mississippi Valley.

The Erie Canal and the steamboat profoundly changed the economic gravity of the upper Mississippi River basin. Where before, most of the rapidly increasing output of this area had, perforce, gone down the Mississippi to New Orleans, now it began to move east instead. By 1830 Ohio had begun trading predominantly with the East. Indiana (1835), Michigan (1836), Illinois (1838), and Wisconsin (1841) soon followed. Even Cincinnati, located right on the Ohio River, was trading predominantly with the eastern cities by 1860.

Besides contributing greatly to the growth of such cities as New York, Philadelphia, and Baltimore, this shift in economic orientation firmly attached the Upper Middle West, largely populated by immigrants from New England and upstate New York, to the Northeast, and assured its allegiance to the Union cause in the Civil War.

But New Orleans, thanks to its location at the base of this vast commercial network, still prospered beyond any other southern port. Freight exported out of New Orleans was only 65,000 tons in 1810. By 1860 it was 4,690,000 tons, increasing seventy-two fold in only fifty years.

———

THE MONOPOLY OF STEAMBOATING in New York waters lasted far longer than it did in New Orleans and would have far greater consequences.

New Jersey and Connecticut retaliated as best they could against New York by banning New York boats from their waters as theirs were banned from New York's. The monopoly was, of course, extremely unpopular with everyone except its direct beneficiaries, including New Yorkers who had to pay higher fares because of it. One man from New Jersey, Thomas Gibbons, decided to fight both in court and in the marketplace. He owned a steamboat named the *Stoudinger* (although, because it was very small, it was usually known as the *Mouse*), which he put on the New York–New Brunswick run, the first leg of the quickest route to Philadelphia. He hired as its captain a young man from Staten Island named Cornelius Vanderbilt.

Vanderbilt, still in his twenties, had already owned a small fleet of sailing ships, but he realized that the future belonged to steam and went to work for Gibbons to gain experience and build up his capital. He soon convinced Gibbons to build a larger boat, designed by Vanderbilt, which Gibbons named *Bellona,* after the Roman goddess of war. In that far more classically oriented age, the implication in the name was clear to everyone.

Flying a flag that read "New Jersey Must Be Free!" Vanderbilt would steam boldly to New York, dock wherever the New York State authorities seemed not to be, and immediately disappear into the city. The authorities didn't dare seize the boat itself, knowing that New Jersey would retaliate by seizing the first monopoly steamboat it could lay its hands on. At sailing time he would sneak back as near to the ship as possible and then make a dash for it, the crew casting off the instant he was aboard.

The authorities even tried to arrest Vanderbilt by boarding the *Bellona* in the middle of New York harbor. But all they found at the wheel was one of the female passengers—all innocence, ribbons, and bonnet—while Vanderbilt hid out in a secret compartment he had had the foresight to install belowdecks. The other passengers hooted their derision at the hapless police.

The monopoly tried to buy Vanderbilt by offering him the colossal salary of $5,000 a year, but he declined abruptly, saying, "I shall stick to Mr. Gibbons till he is through his troubles." Throughout his long career, from farm boy to the richest man in America, Vanderbilt could always be counted on to keep his end of a bargain once it was made.

While the monopoly was having limited success preventing competition in the real world, it not surprisingly kept beating Gibbons in the New York State courts. After five years the case finally reached the United States Supreme Court, and Gibbons hired two of the best lawyers in the country to represent him there—Daniel Webster, then serving as a congressman from Massachusetts, and William Wirt, who was attorney general of the United States, although here acting in a private capacity.

Webster took all day to deliver what was universally recognized as a brilliant legal oration, which he gave before a packed chamber. He claimed that the grant of power to the federal government by the Constitution to "Regulate Commerce . . . among the several States" was sweeping and exclusive. New York had no power to grant a monopoly in New York waters that excluded non–New Yorkers, according to Webster, because the federal government alone had jurisdiction in the matter.

William Wirt and the lawyers for the Livingston interests, Thomas J. Oakley and Thomas Addis Emmett, also spoke at length and, by all reports, eloquently. The decision was widely awaited, not only in New

York but elsewhere as well. "Great anxiety is manifested in this city," reported the *New York Statesman* on February 14, 1824, "to learn the decision of the great steamboat question which has lately been argued with consummate ability at Washington."

Matters were delayed when Chief Justice Marshall, returning from a visit to the White House, fell on leaving his carriage and dislocated his shoulder on February 19. Still, the opinion was read out by Marshall, in a "low, feeble voice," on March 2, only three weeks after the Court heard arguments. "Commerce undoubtedly is traffic, but it is something more, it is intercourse," Marshall wrote for a unanimous court (Justice Johnson of South Carolina wrote a concurring opinion, even more absolute in its interpretation than Marshall's), and ". . . is regulated by prescribing rules for carrying on that intercourse." Since the Constitution gave the federal government the power to "regulate interstate Commerce," the federal government and the federal government alone was empowered to prescribe the rules.

This, of course, is precisely what Webster had argued. (Webster, patting himself on the back as usual, wrote, "The opinion of the Court, as rendered by the Chief Justice, was little else than a recital of my argument.") But it was also a new and breathtaking assertion of federal power. President Monroe, in 1822, in a veto message to Congress, had written that the power granted by the Constitution to regulate interstate commerce did not extend beyond the power to lay duties on foreign commerce and to prevent duties being laid on trade between the states, which was explicitly forbidden by the Constitution.

The decision was greeted jubilantly everywhere, and many newspapers reprinted the decision in its entirety. "Some of the New Yorkers," wrote a Missouri paper, "show themselves a little restive under the late decision of the U.S. Supreme Court on the subject of the steamboat monopoly. They may rest assured that it is a decision approved of in their sister states, who can see no propriety in the claim of New York to domineer over the waters which form the means of intercourse between that State and others, and over that intercourse itself."

In fact, even most New Yorkers heartily approved. Shortly after the decision was handed down, "the steamboat *United States,* Capt. Bunker, from New Haven, entered New York in triumph, with streamers flying, and a large company of passengers exulting in the decision of the United States Supreme Court against the New York monopoly. She fired a salute which was loudly returned by huzzas from the wharves."

The huzzahs were heard all over the nation. The *Georgia Journal* reported that two steamboats arriving in Augusta were greeted with "cries of 'down with all monopolies of commerce and manufactories— one is as great an evil as the other. Give us *free trade* and sailor's rights!' " While one might well doubt that the anonymous reporter was accurately transcribing what was being yelled from a wharf, he undoubtedly captured the spirit of the moment. As a judge described it twenty years later, the decision had "released every creek and river, every lake and harbor in our country from the interference of monopolies."

The economic effects of what Charles Warren, author of the classic work *The Supreme Court in United States History,* called the "emancipation Proclamation of American Commerce," were immediate. Fares from New Haven to New York fell by 40 percent thanks to competition, and the number of steamboats operating in New York waters jumped in less than two years from six to forty-three.

But the long-term effects were even more profound. States stopped granting monopolies of any sort to rent-seeking influential citizens, as all of them were now presumptively unconstitutional. Other barriers to interstate commerce, erected for parochial benefit, fell as well. Thus, thanks to *Gibbons v. Ogden,* the United States became the world's largest truly common market, just as the power of steam to move goods cheaply over long distances—a power merely hinted at by the steamboat—was about to grow exponentially. The railroad would prove the seminal invention of the nineteenth century and create the modern economy that *Gibbons v. Ogden* had made the United States ready for.

———

LIKE SO MANY nineteenth-century inventions (and far more twentieth-century ones), the railroad was not a single invention created by a lone genius. Instead it was a system whose components were invented separately and then pieced together by people in the new profession of civil engineering (so-called because until the middle of the eighteenth century, "engineer" had been solely a military specialty).

It had been known since the sixteenth century, when it was used in mining operations, that a draft animal (or a human being) could pull a much heavier load if the wagon was set on rails. The reason is that metal flanged wheels set on metal rails have very low "rolling friction." A forty-ton locomotive accelerated to sixty miles per hour will coast about five times as far as a truck of similar weight on a level highway. This makes railroads, even today, by far the most economical means of hauling freight.

And it was not long after the coming of the steam engine that it occurred to people to marry the two technologies. Indeed, Oliver Evans foresaw the railroad nearly in its totality long before it became a practical reality. "The time will come," he wrote in 1813, fifteen years before the first commercially successful railroad, "when people will travel in stages [i.e., stagecoaches] moved by steam engines, from one city to another, almost as fast as birds fly . . . A carriage will set out from Washington in the morning, the passenger will breakfast in Baltimore, dine at Philadelphia and sup at New York on the same day. . . . To accomplish this, two sets of railways will be laid . . . to guide the carriage, so that they may pass each other in different directions and travel by night as well as by day."

Evans never adapted his engine to the concept of the railroad, but Trevithick did when he built the world's first locomotive, using the high-pressure steam engine. He tried it out on the tramway at the iron foundry of Samuel Homfray in Glamnorganshire, Wales. On February 21, 1804, the first locomotive pulled the first train along a set of tracks, and the railroad was born.

It would be another twenty-five years, however, before the multitude of problems that stood in the way of a practical railroad were solved and George Stephenson—who solved many of them—built the world's first commercially successful steam-powered railroad, the Liverpool and Manchester Railway. It opened on September 15, 1830, with the Duke of Wellington, then prime minister, present. Connecting as it did the great industrial city of Manchester with the great port city of Liverpool, it was an immediate financial success.

But railroad projects were already under way in the United States by this time. John Stevens, founder of the Stevens Institute in Hoboken, New Jersey, had been granted a charter to build a railroad between the Delaware and Raritan rivers, but it was never built. Stevens also built the first locomotive in this country, in 1825, but it only ran on a circular track at his home in Hoboken.

His son, Robert Livingston Stevens, a very gifted engineer, however, made fundamental contribution to the technology of railroads. It was the younger Stevens who developed the rail, T-shaped in cross section, that has been the basic design of railroad rails ever since. He also discovered that rails laid on wooden cross ties with gravel between them made the most satisfactory roadbed. And he invented the railroad spike to hold everything together.

The Erie Canal's success stimulated the minds of the business communities of other eastern port cities. Baltimore, which had been growing rapidly, wanted to assure further growth by tapping into the burgeoning western market that the Erie Canal was rapidly tying to New York. But the geography of the Appalachians made a canal to the West from Baltimore impossibly expensive. So it was decided to use the aborning technology of the railroad, employing horses as the power source.

On July 4, 1828, Charles Carroll of Carrollton, the last surviving signer of the Declaration of Independence, turned the first shovelful of earth for the Baltimore and Ohio Railroad. The ceremonies were a curious mixture of the old and the new. Carroll himself was ninety-one and insisted on still wearing the knee breeches of the days of his prime, although they had gone out of fashion three decades earlier. Although the B&O was to be privately financed, its beginning was publicly celebrated and the vast procession leading to the ceremony was arranged by craft and profession, like the medieval guild parades of English cities. But the technological undertaking being solemnized was as new as could be and would make a new economic world in a mere generation.

Carroll sensed it. "I consider what I have just now done," he told the crowd, estimated by the newspapers at fifty thousand, "to be among the most important acts of my life, second only to my signing the Declaration of Independence, if indeed, it be even second to that."

———

ONCE THE LIVERPOOL and Manchester Railway showed the practicality of the railroad, railroad projects were begun in many parts of this country, usually as short, local lines intended to connect a town to a part of the existing water-borne transport system. Many canal projects were converted to railroads, which had many advantages over canals. They were easier to build, could be built nearly anywhere and over most terrain, and could operate all year round.

They multiplied rapidly. In 1830 there were only 23 miles of railroad track in the country. By 1840 there were 2,818 miles; by 1850 there were 9,021. By the time of the Civil War, 30,626 miles—two-thirds of it in the North, significantly—laced the country together into what was increasingly an economically cohesive whole. This had profound consequences, for the railroad knitted what in the eighteenth century had been an infinity of local markets into an increasingly integrated national market. "Two generations ago," marveled Arthur T. Hadley in his classic work of economics, *Railroad Transportation,* published in 1886, "the expense of cartage was such that wheat had to be consumed within two hundred miles of where it was grown. Today, the wheat of Dakota, the wheat of Russia, and the wheat of India come into direct competition. The supply at Odessa is an element in determining the price in Chicago."

As the railroads made larger markets possible, they made larger industrial enterprises possible as well. But they had consequences that reached far beyond such direct economic effects. Wherever they went railroads brought economic activity into being, and cities and towns sprang up along the lines and especially at their intersections. In Europe the railroads connected existing cities. In America, in many cases, they midwifed them into existence.

And railroads were extremely capital-intensive, costing in the early days about $36,000 a mile on average at a time when $1,000 a year was a middle-class income. The first railroads were paid for by people living near the rights-of-way who were likely to benefit directly, just as the Liverpool and Manchester had been financed.

But the securities issued locally soon made their way to capital markets, especially to what was rapidly becoming the largest one, in Wall Street. And when larger railroads were envisioned, as they soon were, the securities were often floated in these markets from the beginning. In 1835 only three railroads had their securities regularly quoted in the newspapers. By 1850 there were thirty-eight. By the middle of that decade railroad stocks and bonds accounted for more than half of all the negotiable securities in the country, while volume on Wall Street rose by a factor of ten.

And railroads required enormous quantities of industrial goods: locomotives, freight and passenger cars, rails, cross ties, spikes, and bridge members, to name just a few. At first, nearly all these had to be imported from England. But as the American demand for these commodities grew, more and more American entrepreneurs began to supply them, driving the Industrial Revolution in this country more than any other single force.

In 1828, the year of the B&O groundbreaking ceremony, a budding industrialist from New York, Peter Cooper, and two partners bought three thousand acres of land in Baltimore and built the Canton Iron Works. He hoped that the B&O would be a steady source of business, as well as a means of bringing supplies such as fuel and iron ore. The B&O, however, was soon near bankruptcy. The line had discovered it could not make money using horses, but its thirteen miles of track had such sharp curves that George Stephenson, seeing the map, declared that the curves were too sharp for trains to be pulled by steam locomotives.

Cooper, a very gifted mechanic as well as a first-rate businessman, knew the great engineer was wrong. "I'll knock together an engine in six weeks," he said, "that will pull carriages ten miles an hour."

He found some old wheels that would serve and rigged them to a platform. He had a steam engine he had built for an earlier project sent down from New York and bolted it to the platform, along with a boiler. Connecting the boiler and the engine, however, was a problem. What plumbing was available in this country at the time was made of lead, which could not stand up to the pressure and temperature of a steam engine. So Cooper took a couple of old muskets, sawed off the barrels, and used them as piping.

The result was the first commercial locomotive built in America. Very small by later standards, it was later given the nickname of the Tom Thumb, after P. T. Barnum's famous midget. However small, it worked just fine, and on its first run it pulled a carriage loaded with forty people at speeds up to eighteen miles an hour, a breathtaking pace at the time. (Several of the passengers brought along paper and pencil and wrote down coherent sentences to disprove the then widely held belief that people's brains couldn't function at such speeds.)

———

POWERED BY STEAM, the Baltimore and Ohio began to prosper. The line was steadily extended, reaching Harper's Ferry on the Potomac in

1834, and the Ohio River in 1852. Peter Cooper's ironworks also prospered, as did the city of Baltimore. When Cooper sold out a few years later, he took B&O stock in payment at $45 a share, stock he later sold for $235 a share.

There could be no better illustration of the economic synergy that is so characteristic of any major new technology as its effects ripple through an economy. The railroads made it possible to travel farther and faster and cheaper than ever before. Andrew Jackson had needed a month to travel by coach from Nashville to Washington for his inauguration in 1829. Thirty years later the trip could be made easily, and far more comfortably, in three days.

But the railroads also greatly stimulated manufacturing, mining, travel, and commerce in general. And the railroads simply thrilled the people of the day, who sensed immediately that they were in a new era, one beyond the comprehension of earlier times. "It's a great sight to see a large train get underway," nineteen-year-old George Templeton Strong wrote in his diary in 1839. "I know of nothing that would more strongly impress our great-great grandfathers with an idea of their descendant's progress in science. . . . Just imagine such a concern rushing unexpectedly by a stranger to the invention on a dark night, whizzing and rattling and panting, with its fiery furnace gleaming in front, its chimney vomiting fiery smoke above, and its long train of cars rushing along behind like the body and tail of a gigantic dragon—or like the devil himself—and all darting forward at the rate of twenty miles an hour, Whew!"

But it also induced a sense of misgiving and unease, especially in the older generation. By 1844 Philip Hone, forty years older than Strong, wrote, "This world is going too fast. Improvements, politics, reform, religion—all fly. Railroads, steamers, packets, race against time and beat it hollow. . . . Oh, for the good old days of heavy post coaches and speed at the rate of six miles an hour!"

Philip Hone's use of the phrase "the good old days," is the earliest recorded. He had been born in 1781, into a world that was technologically, economically, and socially very similar to the world his parents had been born into, and even the world of his great-great-grandparents. But thanks to the steam engine and the Industrial Revolution, he had lived to see a new economic universe. Every generation since has lived through a similar experience, and it has become a commonplace to live long enough to see the technological world of one's youth slowly vanish as we age. But to Philip Hone's generation, it was a new, exciting, but sometimes frightening experience.

Print makers such as Currier and Ives exploited the nostalgia, publishing romantic images of a neat and tidy preindustrial world that had never, in fact, existed. Novelists as well wrote of a comforting if largely fictional lost world. Dickens, born in 1812, the most popular novelist of his time, never mentioned the railroads and telegraph that so characterized the new economic world he lived in.

Chapter Nine

CHAINING THE
LIGHTNING OF HEAVEN

T HE SPEED OF TRAVEL was not the only thing accelerating in the early nineteenth century. So was the speed of communications.

It is hard to imagine today, when satellites and undersea cables keep every part of the globe in instant communication with every other part, just how slowly news spread in the eighteenth century. The battles of Lexington and Concord, the opening events of the American Revolution, occurred on Wednesday, April 19, 1775. But news of the events reached New York only on Sunday, April 23, and Philadelphia on April 24. It was late at night on April 28 when an express rider finally brought the news to Williamsburg, Virginia. And it was on May 28, fully five and a half weeks later, that the British cabinet in London learned, to its horror, that the long-smoldering crisis in America had flashed into open war.

To be sure, there were ways to spread news more quickly than by horse. But they were impractical for general use. Queen Elizabeth I had ordered bonfires readied along the south coast of England to be used to signal that the Spanish armada had been sighted. At the end of the eighteenth century the Frenchman Claude Chappe erected a series of large semaphore stations across the French landscape with arms that were raised and lowered by pulleys. They signaled one another by using flags—much like Boy Scouts wigwagging—and could send messages in hours that would take days by horseback.

The French used this system (for which Chappe coined the word *tele-graph*—to write at a distance) extensively, but it was seldom used in the new United States. One was installed between Martha's Vineyard and Boston in 1800. But for the most part, the distances were too great and the available capital too small.

The need for faster communications was just as great as Europe's, however, and cheaper variations were used. In the 1830s a man, every business day, would climb to the top of the dome of the Merchant's Exchange on Wall Street, where the New York Stock and Exchange Board then held its auctions. There he would signal the opening prices to a man in Jersey City, across the Hudson. That man would signal them in turn to a man at the next steeple or hill, and the prices could reach Philadelphia in about thirty minutes. It was clumsy at best (and, of course, didn't work at all in bad weather).

As Wall Street's importance as a financial market grew and the thirst for the news of its prices became more intense, every expedient was tried. Even on Wall Street itself there was unremitting pressure to get the news instantly. That's why messengers on Wall Street are called runners. In the early days young boys were employed to constantly dash back and forth between the brokers' and customers' offices, and the exchange, with orders and the latest prices. (Today the few messengers left are still called runners, but they are mostly elderly, semiretired men who only amble along.)

The solution to the communications problem was already in the air. It had been known since the early eighteenth century that an electric current could be transmitted long distances down a wire, and if a means could be devised to vary that current in a regular way, information could be conveyed by it.

Numerous attempts were made to utilize this fact. In 1774 a system was devised in Geneva, Switzerland, using one wire for each letter of the alphabet. The electricity flowing through the wire would charge a pith ball attracting a bell and ringing it. This alphabetical carillon worked as a parlor demonstration but was hardly a practical system.

Only when much more efficient batteries and electromagnets were developed in the early nineteenth century and wire became much cheaper, thanks to new wire-making machinery, did the search for the electric telegraph get going in earnest. While William Fothergill Cooke and Charles Wheatstone in England developed a working system that was put into commercial use, it is the American Samuel F. B. Morse who came up with the system that was finally adopted around the world.

Morse, the son of Jedidiah Morse, a distinguished New England minister and author, trained as an artist. But while he had a minor genius for portraiture, he was more interested in painting large, "important" pictures for which he had only a pedestrian talent at best.

With much time on his hands, thanks to a lack of commissions, he began to think about electricity after a shipboard encounter with Charles Thomas Jackson, who had been doing research in the subject in Europe. On the spot, Morse conceived the telegraph. "If the presence of electricity can be made visible in any part of the circuit," he said, "I see no reason why intelligence may not be transmitted instantaneously by electricity."

Morse, ignorant of the technology involved and the literature regarding it, appears to have thought that the idea was original with him, not more than eighty years old already. In fact, the only part of the Morse telegraph system that was wholly original was his marvelously efficient code, which assigns dots and dashes according to the frequency with which the letters occur in English (E is simply ·, while X is - · · -).

With the help of one of America's most distinguished scientists, Joseph Henry, a professor at Princeton (and later the first director of the Smithsonian Institution), Morse built a working model in one of the rooms of New York University, consisting of batteries and seventeen hundred feet of wire coiled around a room, connected to electromagnets and telegraph keys at each end. When the key at one end was depressed, it closed the circuit, allowing electricity to flow down the wire, activating the magnet at the other end and causing that key to click down in turn.

Morse spent endless energy trying to devise a recording device to make the electricity "visible," only to discover that his code was so simple that it was easily interpreted by ear, and the letters could be simply written down by hand by a trained telegrapher.

Morse took on partners—Leonard Gale, a professor at NYU, and Alfred Vail, a gifted mechanic whose father owned a prosperous ironworks in New Jersey—to help refine his system. They applied to the government for funds to build a system large enough to demonstrate the technology's capability. But the government, typically, could not see the potential, and for six years they got nowhere. Only when they took on another partner, F. O. J. Smith (known to his friends as Fog), did they make progress. And that was because Smith was not only a member of Congress, but chairman of the House Committee on Commerce.

In 1843 Smith finally pried an appropriation of $30,000 out of Congress by inserting it into a bill in the frantic final minutes of a session. He then awarded himself the construction contract to build a telegraph line from Washington to Baltimore and wasted most of the money buying shoddy wire and trying to bury the line. The project started over, with the wire stretched on poles, and on May 24, 1844, Samuel Morse, in the Capitol Building, tapped out the code for "What hath God wrought!" and Alfred Vail in Baltimore repeated the signal accurately.

Once the practicality of the telegraph was demonstrated, it spread with astonishing swiftness. By the end of the 1840s nearly all major American cities were connected by telegraph, and a line reached San Francisco in 1861. The telegraph spanned the continent less than two decades after Morse sent his famous message. In 1866 Cyrus Field finally succeeded in laying a cable across the Atlantic Ocean, connecting Europe and America with instant communication. The long isolation of

America from the center of the Western world was over after more than 250 years.

By the time Morse died in 1872, rich in honors as well as money, it was possible to send a message from San Francisco to India in a few hours. In 1844 it would have taken perhaps half a year.

One of the reasons that the telegraph was able to spread so quickly was that it could use the pathways being made by the nearly equally fast-spreading railroads. And the telegraph in turn greatly contributed to the efficiency of the railroad. Most of the early lines were single tracked. That meant that if an oncoming train was expected, the train without the right of way had to wait on a siding until it passed. If enough time went by without the other train's appearance (breakdowns were commonplace in the early days of railroading, as were accidents), a conductor would walk several hundred yards in front of the train with a lantern to make sure there was no collision. That, of course, slowed the train to the speed of the conductor.

In 1851 an engineer on the Erie Railway, frustrated with waiting for an oncoming train, suddenly noticed the telegraph line that ran along the track and put two and two together. News of delayed trains and accidents could be telegraphed ahead, to minimize delays for other trains on the line. Within a few years an elaborate signaling system had been devised, allowing railroads to greatly speed up schedules and improve safety.

But no part of the developing American economy benefited more from the telegraph than did Wall Street. Because a market can never be larger than the area within which communication is effectively instant, the exchanges in Philadelphia and elsewhere continued to be important as securities markets. The telegraph quickly reduced them to insignificance.

Once the telegraph made instant communication possible, traders in Philadelphia and elsewhere could operate in the New York market just as easily as they could in their local one, and they immediately began to do so for the simple reason that the best prices, for both buyers and sellers, are always to be had in the largest market.

The reason was well understood at the time. "Money," James K. Medbery wrote in 1870, "always has a tendency to concentrate itself, and stocks, bonds, gold, rapidly accumulate at those points where the most considerable financial activity prevails. The greater the volume of floating wealth, the more conspicuous this peculiarity. It resulted from this law that New York City became to the United States what London is to the world. Eminent before, this chief metropolis of the seaboard now assumed an absolute financial supremacy. Its alternations of buoyancy and depression produced corresponding perturbations in every state, city, and village in the land."

THE TELEGRAPH ALSO PROFOUNDLY AFFECTED another means of communications that was in its infancy in the middle third of the nineteenth century: the modern newspaper.

There had been newspapers in the American colonies as early as 1690, when a Londoner named Benjamin Harris, who had fled England after being jailed for publishing sedition, published the first issue of a newspaper with the less-than-snappy title of *Publick Occurrences Both Foreign and Domestick* on September 25, 1690, in Boston. Harris promised that his newspaper would be "furnished once a moneth [*sic*] (or if any Glut of occurrences happen, oftener)." The first issue, however, was also the last, because the governor and council of Massachusetts promptly suppressed it. But newspapers soon appeared again both in Boston and elsewhere, and by the time of the Revolution were being published in all major colonial cities.

These early newspapers did not much resemble later ones. For one thing, the news was gathered at the same slow pace at which it traveled in the eighteenth century, with little sense of urgency about any but the most important stories. For another, they were very expensive. The flatbed press familiar to Benjamin Franklin (or Gutenberg, for that matter) could print only a very limited number of copies in a timely manner, and these went by subscription mostly to coffeehouses and libraries.

The biggest difference between the newspapers of the preindustrial world and those of today was politics. Most general-interest newspapers were the instruments of political factions, praising one party and excoriating all others. They were, in reality, little more than an editorial page wrapped in some highly tendentious news.

A Scots immigrant to New York, James Gordon Bennett, changed all that. Born in 1795 into one of Scotland's few Catholic families, Bennett was always a man apart, which can be an asset for a journalist. He was also remarkably ugly, with severely crossed eyes. When a young journalist interviewed him in the 1850s at his office across from New York's City Hall, he reported that Bennett "looked at me with one eye, [while] he looked out at the City Hall with the other."

Well educated in Aberdeen, he wrote his first piece of journalism about the Battle of Waterloo, when he was twenty, and four years later, sensing greater opportunity, immigrated to the United States. He worked at a series of newspapers from Boston to Charleston before settling in New York where, three times, he tried to found a newspaper that would expound Jacksonian principles. Each attempt was a failure.

Steam, however, was changing the newspaper business as it was changing everything else by the 1830s. The new rotary presses, powered by steam, could turn out thousands of copies of a newspaper a night and at a much lower price than had been possible before. Bennett decided to try something new. On May 6, 1835, with $500 in capital, an office in a dank cellar, and himself as the only employee, Bennett began publishing the *New York Herald*.

Bennett made the *Herald* nonpartisan in its news articles, sought always to be the first with the news, and sold it to a mass audience by having it hawked on the street at a penny a copy by the armies of news-boys that would quickly become a feature of the American urban scene for more than a hundred years. None of these ideas was original with Bennett. But it was he who put them all together for the first time. He also introduced a dazzling array of other journalistic innovations. He was the first to print a weather report and to cover sports regularly. He was the first to cover business news and stock prices in a general-interest newspaper. And while "respectable" papers weren't supposed to notice such things, when a beautiful prostitute was murdered in one of New York's more fashionable brothels, Bennett played the story for all it was worth.

The *Herald*'s circulation soared, and other papers were forced to fol-low suit as the city, and then the country, became transfixed with the story. Within a few years the *Herald* was among the city's most successful papers. Bennett traveled to Europe, where he signed up correspondents in London, Rome, and Paris to supply the *Herald* with exclusive copy, the world's first foreign correspondents. He fought Congress to establish the principle that out-of-town newspapers had as much right to the congres-sional press galleries as the local papers, the beginning of the Washing-ton press corps. He even coined the use of the word *leak,* to refer to the stories slipped to reporters by politicians for their own purposes.

As the telegraph began to spread across the country, Bennett exploited it to the hilt. When the Mexican War broke out, only two years after Morse's successful demonstration, Bennett organized a consortium of newspapers to fund a pony express from New Orleans to Charleston, which was connected to New York by telegraph. The reports the New York papers published were often days ahead of the official reports arriv-ing in Washington.

By the time of the Civil War the *Herald* was, by far, the largest and most influential newspaper in the country, and all other major papers had followed its model, profoundly transforming the newspaper busi-ness. Its daily circulation during the war reached as high as four hundred thousand, many times the total circulation of all American newspapers combined fifty years earlier.

Millions now counted on the newspapers to keep them informed of their ever-expanding world. "The daily newspaper," wrote the *North American Review* in 1866, only three decades after Bennett had founded the *Herald,* "is one of those things which are rooted in the necessities of modern civilization. The steam engine is not more essential to us. The newspaper is that which connects each individual with the general life of mankind."

By no means the least of the newspapers' influence was in advertising. In the preindustrial era, retail operations had, perforce, been very small, usually selling locally made merchandise. The railroads, telegraph, and newspapers made larger operations possible, and urban merchants quickly began to exploit the new opportunities. In 1846 A. T. Stewart, a Scots-Irish immigrant, opened the "Marble Palace" at 280 Broadway in New York City. Located just north of City Hall, it was the first commercial building to have a marble facade and featured a domed atrium and lavish appointments.

A. T. Stewart's store offered low markups, set prices, and "sales" that were advertised in the papers. It also featured what was called "free entrance," where customers were invited to browse rather than be attended by a clerk at all times. With its luxurious atmosphere and freedom to move about the store, Stewart's, for the first time, made shopping a pleasant pastime for those with the money and leisure to indulge in it rather than merely a necessity. The ready-made attributes of a middle-class lifestyle—furniture, draperies, carpeting, china, and prints—were offered at these new department stores, and the new middle class bought them in vast quantities to decorate their houses in the dense, cluttered, high Victorian style that reached the zenith of its popularity at mid-century.

By the 1860s, when Stewart opened his "Iron Palace"—one of the largest cast-iron structures in the world—a mile uptown at Broadway and Ninth Street, he was the largest single payer of customs duties in the country, thanks to a very considerable wholesale business to other merchants around the country.

Even in rural areas that were still beyond the reach of the railroad, new merchandising techniques opened up new markets. Peddlers increasingly took advantage of the improved roads to sell such newly available manufactured goods as pails and tubs, cloth, tools, and "Yankee notions" to housewives along their routes, lessening the loneliness that pervaded so much of rural America in the nineteenth century.

The new merchants also helped popularize Christmas as a secular holiday in this country. Most American Protestants (other than Anglicans) had not celebrated Christmas in colonial days. But as the new mobility brought these Protestant families into contact with those who did celebrate Christmas, many began to do so, often pushed by their children. Writers, such as the New Yorker Clement Clarke Moore ("A Visit from St. Nicholas"—by no coincidence New York City's patron saint—first published in 1823) and Charles Dickens, began to celebrate the nonreligious aspects of Christmas. The new merchants began decorating their stores for Christmas (the Christmas tree, brought to the English-speaking world by Prince Albert, became popular at this time), and, naturally, merchants emphasized the ancient custom of giving presents at this time of year.

By mid-century Christmas had become the major secular holiday it is today and would grow into the most important engine of the retail business.

———

IT WAS IN THE YEARS BEFORE the Civil War as well that the Industrial Revolution began to give to daily life a recognizably modern form. Besides far more rapid transportation, thanks to railroads and steamboats, and communication, thanks to the telegraph and the newspapers, the domestic comforts of home also increased markedly.

Before the Industrial Revolution, the last major improvement in domestic technology had been the chimney, which came into use in prosperous households in the high Middle Ages. As late as the 1820s houses were still heated by fireplaces and lit by candles at night. Water was hauled by bucket from a well, spring, or cistern. Cooking was done on the open hearth.

In the 1790s a Briton named William Murdock discovered that coal, when heated, gave off a gas that would burn with a bright yellow flame. Gaslight was demonstrated in Philadelphia as early as 1796. Baltimore passed an ordinance in 1816 encouraging the use of gaslight for street lighting, and the idea spread rapidly to other American cities. By the 1830s the main avenues and streets of American cities were lit, thanks to a network of pipes laid underground that fed the gas from a local gasworks. With the dangerous gloom that had enveloped the urban landscape until then abating, after-dark activity increased markedly in the cities.

But while people welcomed the new light source for the streets, they were much more wary of allowing it into their homes for fear of asphyxiation and explosion. Their fears were by no means unfounded, but the advantages of gaslight over candles overcame them, and by the 1850s its faint hiss and odd, dank smell filled the homes of the urban middle and upper classes. "Gas is now considered almost indispensable in the city," a New Yorker reported in 1851. "So much so, that scarcely a respectable dwelling house is now built without gas fixtures."

For the first time in history, interior illumination was cheap, so that it could be used in abundance, and people began to stay up later and read far more than previously. Books, magazines, and newspapers all increased sales markedly at this time, as did sheet music.

Nighttime activity was further encouraged by the spread of central heating. Falling prices for piping and ducting made it affordable, and hot air systems began to appear in houses in the 1830s. By the 1860s steam radiators were rapidly replacing the primitive hot air furnaces, and the American love affair with central heating was on in earnest. Foreign visitors were often appalled. "The method of heating in many of the best houses is a terrible grievance to persons not accustomed to it," wrote Thomas Golley Grattan, who had been British consul in Boston, "and a fatal misfortune to those who are. . . . An enormous furnace sends up, day and night, streams of hot air through apertures and pipes. . . . It meets you the moment the street door is opened to let you in, and it rushes after you when you emerge again, half-stewed and parboiled, into the welcome air."

The cast-iron cookstove as well proved a great improvement over the hearth, making the lives of women much easier, and became widespread.

Despite all these improvements, running a household was still a great deal of work, and large ones required large numbers of servants to function efficiently. There had been a "servant problem" early in the century as the number of households seeking to employ them increased far faster than the supply. But as young women began to move off the family farm and into the cities, and foreign immigration increased, especially in the 1840s, the price of servants' wages began to fall sharply. Even modest households could afford to have someone to help the wife.

By mid-century a typical upper-middle-class household employed a cook, a waiter (who did much of the heavy work, such as shoveling coal, as well as waiting on the table at mealtimes), and a maid to clean the house. The more affluent would also have an upstairs maid, a laundress, a houseman (who did the heavy work), a coachman, and a governess for the children. A skilled domestic, such as a good cook, could earn as much as $6 or $7 a week as well as room and board, very good wages for women at that time.

In many households, favorite servants were an integral part of the family, greatly loved and valued. Under these circumstances, domestic service, particularly for an unmarried woman, could be a pleasant life, especially compared with the alternative: a job in one of the new factories and a room, or part of a room, in the teeming, noisome slums that were spreading quickly in all northern cities at this time. And despite the vast growth of industry in the later nineteenth century, in 1900 the largest single category of employment tracked by the U.S. Census would still be domestic service.

As the cities grew in size by leaps and bounds in the first decades of the nineteenth century, the problem of supplying the inhabitants with water and disposing of sewage increased apace. In the early years of the century the affluent had rain barrels or cisterns, fed from their rooftops, but the rest had to haul water from the nearest well. This water was often grossly contaminated from the sewage from privies and the chamber pots that were emptied into the streets. Although not understood at the time, this was the source of the frequent epidemics of such diseases as yellow fever and cholera that ravaged American cities at this time.

Philadelphia was the first city to build a modern water supply that could be piped into houses and allow waste to be disposed of through sewers. In 1832 the first houses in America to be built with bathrooms were supplied by this system. New York, surrounded by salt water, had a far more difficult technological problem to deal with. Nonetheless, after building a forty-five-mile-long aqueduct to bring water in from the Croton River in Westchester County, New York opened the "Croton System" on July 4, 1842.

Philip Hone was agog. Months later he wrote in his diary that "nothing is talked of or thought of in New York but Croton water. . . . Fountains, aqueducts, hydrants, and hose attract our attention and impede our progress through the streets. . . . Water! Water! Is the universal note which is sounded through every part of the city, and infuses joy and exultation into the masses."

George Templeton Strong was positively giddy when his father installed Croton water in his house on Greenwich Street in 1843. Taking a bath no longer involved heating water on the stove and pouring it into a hip bath dragged into the kitchen for the purpose. "I've led a rather amphibious life for the last week," he wrote happily in his diary, "paddling in the bathing tub every night and constantly making new discoveries in the art and mystery of ablution. Taking a shower bath upside down is the last novelty."

Boston, ever vigilant against the possibility that people might be enjoying themselves on the Sabbath, banned bathing on Sundays.

———

BEFORE THE EARLY NINETEENTH CENTURY, even among the notoriously peripatetic population of the United States, people seldom ventured more than fifty miles from where they had been born, or, if they did, never saw their birthplace again. Now, in less than a lifetime, it had become possible to travel a hundred miles in a day, receive instant word from someone a thousand miles away, and read of events that were taking place, right then, halfway around the world. It was possible to have hot water run out of a tap, be warm on the coldest night, read a book at night without eyestrain.

These miracles of daily life that piled one upon the other in the first decades of the nineteenth century—railroads, telegraph, newspapers, heating, lighting, running water—induced a mood of optimism and a belief in progress that had not been known before. The sense that anything was possible pervaded what would come to be called the Victorian age, throughout the Western world. But in the United States, still only half formed and growing economically more quickly than any other major country in the world, that mood was palpable.

We have never completely lost it, even in the worst of times that were to come.

Chapter Ten

WHALES, WOOD, ICE, AND GOLD

ALTHOUGH GASLIGHT RAPIDLY ILLUMINATED the cities beginning in the 1820s, it wasn't available in the countryside, where most Americans still lived. Coal gas required a considerable (as well as messy, smelly, and sometimes dangerous) industrial process, and the gas was piped directly from the gasworks to the users. Because of the high cost of the infrastructure, only areas with a high population density could support a gasworks. But as reading and other evening activities increased, the demand for artificial light in rural areas increased as well. More and more, it was met with whale oil.

Men had hunted whales since Neolithic times, and the Basques, Norwegians, Netherlanders, and Scots all went whaling. In early days it was mostly offshore whaling, with the carcasses towed back for processing. But as offshore stocks declined, Europeans went farther and farther afield in search of whales.

Whales produced a cornucopia of products. Besides the meat, the blubber, when tried out, produced an oil that was an excellent fuel for lamps as well as a lubricant for machinery. And the baleen, the flexible, fringed structures up to twelve feet long that replaced teeth in most large whales, yielded great quantities of whalebone, a stiff but flexible material that was indispensable for corsets, buggy whips, and myriad other uses.

New England whaling began as early as 1645. The prime target of early whalers was the right whale, so called because it was abundant, relatively easy to hunt, and floated when it was dead. Thus it was the right whale to catch. Then in 1712 an offshore whaler was blown out to sea in a storm, and the crew managed to catch a sperm whale and bring it safely home.

Sperm whales are not like the other great whales. They have teeth, not baleen, and live mainly on giant squid caught at great depths. Their oil is superior to that of other whales and they have in their huge heads, as part of their echolocation system, a vast reservoir filled with spermaceti, a waxy substance that made the best candles.

Sperm whales were immensely profitable, and New Englanders began to specialize in them, building oceangoing whalers to pursue them on the main deep. By 1765 New England whalers were calling at the Azores and were operating off the coast of Brazil soon afterward. By the 1770s New England was exporting three to four hundred thousand pounds of spermaceti candles a year.

The British government had encouraged whaling, paying a bounty for whaling vessels more than two hundred tons. But the industry flourished after independence as well. By the nineteenth century American whalers were found throughout the world's oceans, and voyages often lasted two years, sometimes four. In 1800 some three hundred whalers operated out of such New England ports as Provincetown, Nantucket, New Bedford, and Marblehead. By the 1840s there were more than seven hundred American whaling ships searching the seas for whales and turning their home ports into boom towns as the ships unloaded as many as two thousand barrels of oil each at the end of their voyages. Many of the early nineteenth-century New England fortunes were based on whaling.

Unfortunately, the New Englanders were too good at catching

whales and caught them far faster than the great mammals could reproduce. As the world's stocks of the various species declined, the voyages became longer and longer (it was an American whaler who discovered the *Bounty* mutineers on Pitcairn Island in 1808). The price of whale oil rose steadily as demand continued to increase faster than supply.

But one of the underappreciated aspects of a free-market economy is how it deals so efficiently with shortages. As demand outstrips supply, prices rise. The price rise, in turn, causes both increased conservation of the scarce resource and an intensified search for additional supplies or substitutes in order to convert the high prices into profit. One editor of the day noted "the impetuous energy with which the American mind takes up any branch of industry that promises to pay well."

As the price of whale oil increased (it reached $2.50 a gallon in the 1850s, when $5 a week was a good wage for skilled labor) the search for other illuminants and lubricants increased as well. Camphene, derived from turpentine, is an excellent illuminant, but has a nasty habit of exploding. Coal tar—what is left over after the gas for gaslight is extracted from coal—was first distilled into kerosene in the 1850s, but the process is complex and expensive. Still, by the late 1850s a plant in New York City was producing five thousand gallons of kerosene a day from coal tar, so acute was the need and so high was the price of whale oil.

(Coal tar, while it was used only briefly as a source for kerosene, would prove to be a fecund source of chemicals with other economic uses, including insecticides, plastics, paints, and medicines. It would be the basis of a whole new industry—chemicals—that blossomed in the second half of the nineteenth century. Aniline dyes, discovered by the eighteen-year-old Englishman William Henry Perkin in 1856, were the first commercial chemical products to come from coal tar. They quickly destroyed the market for vegetable dyes derived from such plants as indigo and madder.)

The solution to the problem of finding a cheap, good illuminant came from a wholly unexpected source, rock oil. Petroleum (which means *rock oil* in Latin) has been known from ancient times but was largely a curiosity. Its principal use was as a cure-all medicine. (It would seem that nearly everything unappetizing—and crude oil is certainly that—and which is not deadly poisonous in small doses has been declared to be medicinal at one time or another.) In many areas of the world it bubbles up out of the ground on its own and could be gathered by skimming it off ponds or soaking it up with rags.

In 1853 a Dartmouth graduate named George Bissell happened to be visiting his old school when he saw in a professor's office a bottle of rock oil that had come from western Pennsylvania. He knew that the stuff was flammable and suddenly conceived of the idea that it could be turned into an illuminant. He organized a small group of investors and asked one of the country's leading chemists, professor Benjamin Silliman, Jr., of Yale, to look into the possibilities. Silliman reported that rock oil could easily be fractionated into various substances, including kerosene, by heating it. "Gentlemen," Silliman reported, "it appears to me that there is much ground for encouragement in the belief that your Company have in their possession a raw material from which by simple and not expensive processes, they may manufacture very valuable products."

But while Bissell and his investors (soon including Silliman himself, who bought two hundred shares) now knew that they could manufacture a highly saleable commodity from rock oil, the supply of rock oil was still very limited. An industry could not be based on what could be skimmed from ponds. Then Bissell had another epiphany. Shading himself under a druggist's awning in New York City on a hot summer's day in 1856, he noted an advertisement for a patent medicine made from rock oil that showed several derricks of the sort that were used to bore for salt. The rock oil used in the medicine, it happened, came from the oil obtained as a by-product of drilling for salt. Bissell wondered if it might be possible to use the technology of drilling to find oil.

The company sent a man named Edwin Drake to northwestern Pennsylvania, from where most rock oil in this country then came, and

Drake, with no small difficulty, finally found men who knew how to drill for salt and who were willing to drill for oil, widely regarded as a ridiculous idea. On August 27, 1859, outside Titusville, at sixty-nine feet, the first oil well in the world struck oil. Drake attached a pump to the well and began to pump up what seemed an unlimited quantity of rock oil. The biggest problem quickly became not finding oil but finding enough barrels to store it in.

Professor Silliman was far more right than he knew. Within a century, petroleum would become so central to the functioning of the world economy that armies would march to acquire or defend its sources.

———

ANOTHER DEPLETING RESOURCE that fueled the American economy in the years before the Civil War was the country's seemingly inexhaustible forests. Thomas Jefferson predicted that it would be a thousand years before the frontier reached the Pacific Ocean. But even in Jefferson's lifetime, all but three of the states east of the Mississippi had been admitted to the Union, and the eastern forest was being cut down at an astonishing rate. The rich farmlands of the Middle West could not be productive until the trees were cleared, and the demand for lumber rose at an ever-increasing rate as the country's population increased by a third every decade, from 5.3 million in 1800, to 31.4 million in 1860.

The statistics are staggering. In 1820 Michigan was nearly uninhabited by Europeans. By 1897 it had shipped 160 billion board feet of white pine lumber, leaving less than 6 billion still standing. State after state experienced similar deforestation.

But lumber was by no means the only use of the wood cut down by the forest industry in these years. Wood remained the major fuel in the forest-rich United States long after coal had supplanted it in forest-poor Europe. The growing number of steamboats, and after 1830, railroad locomotives, greatly increased the demand for firewood. The huge smokestacks unique to American locomotives of this period, shaped like inverted cones, were designed to deal with the greater flaming-cinder output of wood-burning fireboxes.

Firewood production had a far larger impact on the land than did trees felled for lumber. The latter amounted to about twenty-five thousand square miles of forest, but fuel cutters denuded fully two hundred thousand square miles between 1811 and 1867, enough to make almost five billion cords of firewood. (To get some idea of just how much firewood that is, consider that five billion cords, neatly stacked, would cover the state of Connecticut to a depth of four feet.)

The effect on the natural ecosystem of this epic deforestation, needless to say, was titanic. But agricultural production soared as the forests turned to fields. And most of the people of the time saw nothing wrong whatever with the transformation. "I never beheld a scene more delightful," a visitor to southern Ohio wrote in 1823, only twenty years after Ohio became a state, "than when I look down on that field. Three hundred acres of waving corn . . . And those fifteen or twenty men scattered over it—all at work. It was doubly interesting coming, as it did, out of nature's forest, only broken by the occasional cabins and the small patches of cleared land of the early settlers."

Indeed, most people at the time thought that this alteration of the landscape was biblically ordained, for Genesis commands human beings to fill up "the earth, and subdue it." Except for a few lonely voices such as George Perkins Marsh, whose *Man and Nature* was published in 1864, and the establishment of large urban parks in the rapidly expanding cities, what today is called the environmental movement did not yet exist.

Ohio's rising corn production fueled the meat-packing industry, which was centered in Cincinnati until Chicago surpassed it in 1860. In 1833 Cincinnati had processed eighty-five thousand hogs. Fifteen years later it handled five hundred thousand. The growth of American agricultural production in the pre–Civil War era is without parallel in world economic history. While statistics exist only for the years from 1839 on, corn production went from 378 million bushels that year to 839 million twenty years later. Wheat went from 85 million bushels to 173 million.

But as the country's productive farmland expanded rapidly, American stewardship of the land did not improve from the standards of the colonial era. Indeed it declined. In the early days, when the population was still very low, the best land was cleared for agriculture but the slopes and wetlands were largely left alone. Only in the plantation South, where one crop dominated an area, was erosion and worn-out soil a problem. As early as the 1780s Patrick Henry thought that "the greatest patriot is he who fills in the most gullies."

But because the supply of land seemed without end, the value placed on each unit was small. This is only common sense, at least in the short term (and, as Lord Keynes observed, in the long term we are all dead). One husbands the scarce, while the plentiful is used, or misused, freely. Caring for the land was an inescapable necessity in Europe, where there was no more to be had. But in America, new, unspoiled land nearly free for the taking was just over the next row of hills or river valley, and settlers could always move on. For three hundred years they did exactly that, with ever increasing speed.

But labor, scarce and expensive in pre–Civil War America as it had been in the colonial era, was a resource to be cherished. And American inventiveness devised many ways to make American agriculture more productive. The plows of colonial days were little different from those in use in medieval Europe and were made of wood. They worked well enough in the light soils of the eastern United States, but were useless in the rich, heavy soils of the developing Middle West.

Thomas Jefferson studied the plow and attempted to devise a better one. In 1797 Charles Newbold began manufacturing plows made of cast iron, and in 1814 Jethrow Wood designed a plow with interchangeable parts, making it much easier to repair. But even iron plows were useless in much of the Middle West because the soil would not turn over but rather fell back in place once the plow passed.

A blacksmith named John Deere was a Vermonter who had joined the New England diaspora and settled in the oddly named Grand Detour, Illinois. There, while engaged in fixing the broken plows of farmers, he began experimenting with new designs. In 1837 he made a plow using a piece of a steel circular saw blade, and it worked beautifully in the toughest Middle Western soils as the blade cut cleanly through and did not "scour." Deere promptly went into the manufacturing business, setting up a factory in Moline, Illinois, to turn out the new plows that soon spread throughout the growing farm belt. The company's motto regarding its founder, "He gave to the world the steel plow," would be in use as late as the mid-twentieth century, long after the horse-drawn single plow had vanished from the American farm.

But no one did more to transform American agriculture than Cyrus McCormick. Yankee ingenuity has never been a New England monopoly, and Cyrus McCormick was a Virginian from Rockbridge County in the Shenandoah Valley who supported the South and defended slavery until the end of the Civil War. Like Eli Whitney's father, McCormick's father was both a farmer and a mechanic, making and fixing equipment for other farmers. And like Whitney, McCormick was a born tinkerer. Working on his father's twelve-hundred-acre farm, McCormick began thinking about how wheat was harvested. Harvesting had long been the limiting factor in wheat production. There is a strictly limited period after the wheat has ripened in which it can be successfully harvested. And working with a scythe and cradle, a man could cut only about one acre of wheat per day. Obviously, there was no point in planting more wheat than the available labor could harvest.

McCormick broke the process of harvesting wheat into its separate components and devised a mechanical means for accomplishing each. Then he designed a machine that accomplished them all. By the time he was twenty-two, he had a working prototype of a mechanical harvester, powered by a wheel, called a driving wheel or bull wheel, that turned in contact with the ground as the machine moved behind the horse. Farmers were, to put it mildly, skeptical at first. McCormick didn't sell a single machine for ten years, and in 1842 he sold only seven. But when Great Britain, after the failure of the British harvest in 1845, repealed the "Corn Laws" that protected British farmers from international competition, the demand for American wheat grew quickly. McCormick seized the opportunity.

He built a factory in Chicago, then a city less than twenty years old, and began to mass-produce harvesters. In five years he sold five thousand. By 1860 Cyrus McCormick was a very rich man, reporting in that year's census that he had personal assets of $278,000, and real estate valued at $1.75 million.

With McCormick's reaper, one man could harvest eight acres a day, not one, and the American Middle West could become the bread basket of the world. In 1839 only eighty bushels of wheat were shipped out of the infant town of Chicago. Ten years later Chicago shipped two million.

The McCormick reaper not only greatly enlarged the potential size of American grain crops, it changed how Americans have earned their livelihoods. With the introduction of the reaper and the endless parade of mechanical agricultural equipment that followed, the percentage of American workers engaged in agriculture has steadily declined, even while agricultural output has continued to grow. The McCormick reaper thus helped crucially to supply the labor needed in the great expansion of American industry that followed the Civil War.

In 1851 McCormick exhibited his reaper at the Great Exposition in London, the first world's fair and one of the defining moments of the nineteenth century. At first skepticism abounded. The *Times* of London, with typical xenophobia, described McCormick's machine as "A cross between a flying machine, a wheel barrow, and a chariot. . . . An extravagant Yankee contrivance, huge, unwieldy, unsightly and incomprehensible."

After a demonstration in an English wheatfield, however, the *Times* changed its mind completely. "The reaping machine from the United States," it wrote on June 9, 1851, "is the most valuable contribution from abroad to the stock of our previous knowledge. . . . It is worth the whole cost of the Exposition."

Other American mechanical inventions dazzled the crowds at London's Crystal Palace that summer as well, and prizes were won by American plows, the Colt revolver, Goodyear rubber products, and the sewing machine of Elias Howe.

The last, when it was redesigned by Isaac Singer to make it much more reliable and versatile, quickly improved the lot of housewives and drastically lowered the cost of ready-made clothes. A shirt sewn by hand required more than fourteen hours to manufacture. With the new sewing machine, a seamstress could make one in only a little over an hour.

Many clothing workers (there were more than five thousand in New York City alone in 1853) feared that their livelihoods would be devastated by the sewing machine, but exactly the opposite happened. As the price of ready-made clothes plunged, thanks to the sewing machine, the increased demand for them more than made up for the fall in price. This is the reason industrialization has greatly enriched, not impoverished, the world's workers, at least in the long term. In the short term, of course, as new technology sometimes destroys the market for old skills, the economic pain can be terrible for whole classes of workers. The great American folk song "John Henry Was a Steel-Driving Man," which dates to the mid-nineteenth century, exemplifies exactly this problem, rendered into art.

———

IN THE EARLY YEARS of the nineteenth century this country developed another major industry that fueled a nearly worldwide trade. And it maintained a near monopoly on it, one that lasted for decades before vanishing as technology overtook it: the ice trade.

Ice was free for the taking during part of the year in New England and a precious commodity during another part and in other, warmer areas of the world. It was the genius of Frederic Tudor of Massachusetts that saw an opportunity to make money by bringing the ubiquitous supply of ice in wintertime New England to the bottomless demand for it in the summertime and the tropics.

There had been an ice trade of sorts in ancient Rome when snow was brought down in baskets from the high Apennines to the tables of the very rich. And New England farmers maintained ice houses dug into the earth in which they would store ice cut from ponds during the winter and insulated with straw for use in the summer to keep perishable foods, such as milk, cold. Tudor thought he could make money by bringing this ice to parts of the world where ice was never seen. He was about the only one who thought so.

On February 13, 1806, a ship chartered by Tudor named the *Favorite* sailed out of Boston harbor, and the *Boston Gazette* reported, "No joke. A vessel with a cargo of 80 tons of Ice has cleared out from this port for Martinique. We hope this will not prove to be a slippery speculation."

Unfortunately for Tudor, it proved to be exactly that. There was no ice house in Martinique to store what ice survived the voyage, and Tudor had not yet learned how to insulate the ice effectively on shipboard. Nor did the Martiniquais, who had done without ice for nearly two centuries, know exactly what to do with it, so they treated it as a curiosity.

Over the ensuing three decades, Tudor struggled up the learning curve of the ice trade, establishing ice houses in likely markets, perfecting insulation, and educating his potential customers. Ice harvesting, from New England ponds and rivers, was also by this time becoming a routine business, with special tools for cutting and hauling the ice to ever-larger ice houses on the shore.

In a perfect example of economy synergy, it turned out that the best insulating material for ice was what had always been a troublesome waste product of the lumber industry: sawdust. Previously thrown into the nearest stream to be carried away (and where it often caused dams and flooding), now it could be sold to the ice industry at $2.50 a cord.

By the 1830s ice had become a very profitable American export. In 1833 American ice was being shipped as far as Calcutta, when the *Tuscany,* which had sailed from Boston on May 12, reached the mouth of the Ganges on September 5. Calcutta, one of the hottest and most humid cities on earth, and then the capital of British India, was ninety miles up the Hooghly River, and the population awaited the ice with breathless anticipation. The *India Gazette* demanded that the ice be admitted duty free and that permission be granted to unload the ice in the cool of the evening. Authorities quickly granted the demands. Frederic Tudor managed to get about a hundred tons of ice to Calcutta, and the British there gratefully bought it all at a profit for the American investors of about $10,000.

By the 1850s American ice was being exported regularly to nearly all tropical ports, including Rio de Janeiro, Bombay, Madras, Hong Kong, and Batavia (now Jakarta). In 1847 about twenty-three thousand tons of ice was shipped out of Boston to foreign ports on ninety-five ships, while nearly fifty-two thousand tons was shipped to southern American ports.

Other cities with easy access to areas where ice could be harvested began their own ice trades. The Hudson River had as many as 135 ice houses on its shore, each capable of holding thousands of tons. Ice boxes became standard household equipment in middle- and upper-class houses in the 1840s and daily delivery of ice a routine. The ice man and the ice wagon entered into American folklore and legend, and the American love affair with both iced drinks and frozen desserts (still a notable American peculiarity to most Europeans, especially the British) was already in full swing.

By 1880 the extent of the ice trade was estimated at eight million tons annually, and mild winters invariably brought newspaper warnings of an impending "ice famine" the following summer and swiftly rising prices. Mechanical refrigeration ended the trade with the more distant foreign cities in the 1880s, when it could no longer compete with locally made ice. But the domestic trade continued to expand until well into the twentieth century, when the household refrigerator began replacing the ice man.

THE DEEP DEPRESSION that had started in 1837 began to lift in the mid-1840s. Federal government revenues, a good measure of economic activity, had been a miserable $8.3 million in 1843, the lowest in decades. But the following year they jumped up to $29 million. The Mexican War, which began in 1846, then stimulated the economy strongly, as wars always do, and prosperity returned. The Mexican War also, of course, added vast new territories to the United States when Mexico relinquished all claims to Texas north of the Rio Grande. Further, it ceded almost a million square miles of what is now Arizona, Utah, Nevada, California, and parts of Colorado, Wyoming, and New Mexico in early 1848 in exchange for $15 million and the assumption by the United States of several million dollars' worth of debts owed by Mexico to American citizens.

That same year the division of the Oregon Territory along the forty-ninth parallel was peacefully agreed to with Great Britain. The United States now had a coastline on the Pacific Ocean nearly as long as its Atlantic one. But much of the territory west of Texas and the states bordering the Mississippi was unsettled and largely unknown.

Even before the treaty that ended the Mexican War was fully negotiated, an event that would deeply affect the United States had taken place in what was still Mexican territory: the discovery of gold at Sutter's Mill in California.

On January 24, 1848, a man named James Marshall was inspecting the millrace of a sawmill that he had just constructed for his employer,

John Sutter, on the American River, not far from what is today the city of Sacramento. He had turned the water into the millrace the night before to clear it of debris, and now something "about half the size and the shape of a pea" caught his eye. "It made my heart thump," he remembered later, "for I was certain it was gold." He turned to his workmen and said, "Boys, by God, I believe I have found a gold mine."

The California gold strike was not the first one in the United States. A young boy named Conrad Reed had found a large gold nugget (it weighed seventeen pounds) in a stream on his father's farm in Cabarrus County, North Carolina, in 1799. As word spread, others began looking for gold in streams and digging for it. North Carolina produced enough gold in the ensuing decades that the first American gold coins were minted there in the 1830s, at a private mint established by Christopher Bechtler, a German immigrant, in Rutherford County. (The Constitution forbids states to coin money, but does not forbid individuals to do so, although the coins, of course, are not legal tender. The Bechtler coins, minted in denominations of $1, $2.50, and $5, were well and honestly made and proved very popular in the hard-money-starved economy of the antebellum South.)

But the California gold strike was of an entirely different order of magnitude from that in North Carolina. In 1847 the United States had produced 43,000 ounces of gold, most of it as a by-product of base-metal mining. In 1848 it produced more than ten times as much, thanks to California, and in 1849, as word spread around the world, it produced 1,935,000 ounces. In 1853 California spewed out more than 3 million ounces, worth some $65 million, $4 million more than total federal revenues that year.

Gold is strange stuff. For one thing, it is extremely heavy. By definition, a cubic centimeter of water weighs one gram. Thus a cubic foot of water would weigh 62.43 pounds. A cubic foot of granite weighs about 170 pounds. Gold weighs more than 1,200 pounds per cubic foot. For another, gold is inert, not combining with other elements (that is why it does not tarnish). Thus gold is usually found in its pure state, sometimes in very small flakes or dust, often embedded in stone such as quartz, sometimes in large nuggets of pure gold.

When gold nuggets, flakes, and dust, erode out of hillsides, they are washed downstream, but because gold is so very heavy, the metal tends to settle out whenever the current slows down, such as in an eddy or on the inside of a bend in the stream, and there the gold concentrates.

Thus gold, unlike nearly all other metals, can often be successfully mined with very little capital investment. In the early days of a gold strike, the miners often need little more than a few tools, a pan, and a willingness to do hard work. In 1848 there was no shortage of men more than willing to abandon their slow-but-steady employment as farmers, teachers, bank clerks, and a thousand other jobs, to seek instant wealth in the California gold fields.

The result was one of the most remarkable migrations of the nineteenth century or, indeed, any century. As the news spread (the attempt by James Marshall and John Sutter to keep the news a secret was, of course, an utter failure), whole towns were depopulated as men rushed to the gold fields. San Francisco, then a town of fewer than a thousand, was virtually emptied and its harbor became choked with shipping as arriving sailors deserted to moil for gold. As news spread farther, Hawaii, Oregon, South America, Australia, and China were equally affected, as men by the thousands headed to California.

News took a long time to reach the East Coast, however, and it was late summer before rumors of gold in California began to circulate. And it was December 8 before official word was released. That day President James K. Polk sent Congress a message informing it of the gold strike and sending along a guaranteed-to-get-everyone's-attention piece of proof: a nugget that weighed fully twenty pounds. At $20.66 an ounce (the official ratio of the United States dollar to gold since 1837), the nugget was worth about $4,800. In 1848 that was enough for a large family to live in very considerable upper-middle-class comfort for a year or more.

Something very close to mass hysteria was the result. In 1849 about ninety thousand Americans set off for California, and as many followed in 1850. That is not far short of 1 percent of the population. And most of them, not surprisingly, were men. When California became a state, less than three years after the gold strike, its population was 92 percent male.

It was not an easy journey. In 1849 there were only three ways to get to California from the eastern United States. One was overland, a difficult journey of six months through unsettled and sometimes hostile territory. Another was to go around Cape Horn by sailing ship, a journey that also took about six months. It was usually more comfortable but more expensive as well. The third means was via Panama. Steamships began running regularly to Panama, and passengers would then have to make their way across the rain-soaked, fever-ridden isthmus to the Pacific and hope to find a passage north to California, such as on the steamer *California*, which had been dispatched around the Horn for the purpose.

In theory, the trip via Panama could take as little as seventy days, but often the wait in Panama City was weeks long. Ogden Mills, who would make a fortune in banking in the gold rush, found three thousand people waiting for passage to the gold fields and no northbound ships at all. He finally took a passage south looking for a ship to charter and had to go as far as Callao, in Peru, to find one. His journey from New York to California ended up taking six months as well.

Politically, the California gold rush pushed the country's center of gravity sharply westward. In 1850 the population center of the United States—east of Baltimore in 1790—was located near Parkersburg, Virginia (now West Virginia). But as early as 1851, as gold fever was still epidemic, John L. B. Soule wrote in the *Terre Haute Express,* "Go west, young man, go west!" a phrase that was quickly picked up by Horace Greeley and usually attributed to him thereafter.

The country's connections to its new, distant state, more than a thousand miles west of the westernmost point of Texas, became a major concern. Clipper ships, whose great virtue was speed (and whose great drawback was lack of cargo space) began to be built in Boston and New York shipyards in increasing numbers to carry people to California and reduce the time needed to go around the Horn. The appropriately named *Flying Cloud* made the trip from New York to San Francisco in an astonishing eighty-nine days, half the time of normal sailing ships. In 1853 some 120 clippers were launched. But when a railroad was built across Panama and more reliable steamship service was established on the Pacific leg of the journey, the clippers could not compete and the number built rapidly declined, the last of them sliding down the ways in 1859.

An all-American route was also pressed. In 1853 the United States bought an additional twenty-nine thousand square miles from Mexico in what is now southern Arizona because it was thought at the time that that would be the best route for a southern railroad. In 1860 the Pony Express reduced communication time with California to about ten days, and in 1861 a telegraph link reduced it to minutes.

———

THE EFFECT OF THE CALIFORNIA GOLD STRIKE on the American economy was quite as great as on its politics.

In the mid-nineteenth century gold was money. After the end of the Napoleonic Wars, in 1821, Great Britain went on the gold standard. That meant that the Bank of England stood ready to buy or sell unlimited quantities of pounds sterling for gold at the rate of 3 pounds, 17 shillings, 10½ pence per ounce. (That ratio had been determined more than a century earlier by Sir Isaac Newton, of all people, who enjoyed the well-remunerated but largely no-show job of Master of the King's Mint.)

Because the United Kingdom dominated the world economy and world trade in the nineteenth century, and the Bank of England was, in effect, the world's de facto central bank, all major trading nations soon pegged their currencies to gold. World trade was conducted on a gold (or pound sterling) basis. But while the United States was, externally, on the gold standard, internally monetary chaos continued and, indeed, increased. In the 1850s there were more than seven thousand kinds of more or less valid banknotes in circulation and more than five thousand that were fraudulent or counterfeit.

By 1860 more than two thousand banks were in operation in the United States. Some were large, conservative, and sound, headquartered in the big eastern cities. Most, however, were located in small towns and dependent on local economies. Still others were known as wildcat banks because their headquarters (the only place their notes could be redeemed for gold and silver) were located "out among the wildcats" where they were, quite deliberately, hard to find.

As California gold flowed into the American economy, the money supply increased markedly. The minting of gold coins by the federal government increased, as did the issuance of banknotes based on gold reserves. Because the country had no central bank, there was no mechanism to regulate the money supply or to use monetary policy to control what Alan Greenspan would famously call "irrational exuberance." The result was a huge, but unsustainable, boom.

There was just over 9,000 miles of railroad trackage in the United States in 1850, but a decade later there was 30,626. Pig iron production soared from 63,000 tons in 1850 to 883,000 tons a mere six years later. Increasingly, the iron ore came from the Marquette Iron Range in the Upper Peninsula of Michigan, the first of the iron ore deposits found around Lake Superior that would prove in the next half century to be the largest and richest in the world. The production of coal, only beginning to replace wood as the main source of fuel in American transportation and industry, doubled in these years.

As always when the American economy boomed, capital poured in from Europe to finance development. American securities held overseas in 1847 amounted to $193.7 million. A decade later they were $383.3 million.

Much of this new capital was brokered through Wall Street, which cemented its position as the financial center of the country in the 1850s. "Every beat of this great financial organ," wrote the *Louisville Courier* in 1857, "is felt from Maine to Florida, and from the Atlantic to the Pacific." Out-of-state banks increasingly kept large deposits in New York City banks to facilitate their customers' business needs in New York, as the city came to dominate the southern cotton trade and Middle Western wheat trade with Europe. And twenty-seven new banks opened in New York itself in just the period from 1851 to 1853.

As the American economy rapidly expanded in the early 1850s, the froth, as it always is, was most obvious in New York's financial market. New exchanges opened up to handle penny stocks that the New York Stock and Exchange Board would not touch, and the number of corporations formed in the 1850s equaled the number created in the entire first half of the century. By 1856 there were 360 railroad stocks, 985 bank stocks, and 75 insurance stocks being regularly traded on Wall Street, as the average volume on the Street increased by a factor of ten.

By 1857, however, the boom was running out of steam. California gold production was leveling off; the Crimean War and poor harvests in Europe, which had stimulated demand for American exports, were over. New York wharves were now crowded with ships that could find no cargoes. Six thousand cotton looms in New England sat idle by that summer.

James Gordon Bennett wrote in the *Herald* on June 27 that year, "What can be the end of all this but another general collapse like that of 1837, only upon a much grander scale? . . . Government spoilation, public defaulters, paper bubbles of all descriptions, . . . millions of dollars, made or borrowed, expended in fine houses and gaudy furniture; hun-

186 — AN EMPIRE OF WEALTH

dreds of thousands of silly rivalries of fashionable parvenus, in silks, laces, diamonds and every variety of costly frippery are only a few among the crying evils of the day."

As the summer began to wane, prices on Wall Street began to reflect the sagging national economy. The weaker banks and brokers began to collapse. When the steamer *Central America* foundered off North Carolina in a hurricane on September 12, it took four hundred passengers with it. But, of more concern to the market, $1.6 million in California gold went down with the ship as well, and panic hit Wall Street. It soon spread to Europe in the first truly international market crash, a sure sign of the growing importance of the American economy to the world. By mid-October most banks in the country (and all the major New York City ones) had suspended payment in specie. It was only temporary, as they built their reserves and called in loans, and most had resumed payment in gold and silver by December, but the economy would be mired in a new depression for the next three years. Federal revenues fell by a third from 1857 to 1858.

THE AMERICAN ECONOMY in the first six decades after the adoption of the Constitution had proved one of the wonders of the world. The country's territory had more than tripled, and its population had increased eightfold. But the size of the economy had expanded by a factor of eighteen or more. A string of small states, with largely agricultural economies, had expanded across half a continent. American manufactures had grown from trivial to one of the world's leading concentrations of industry. A transportation and communications net had been created that was the largest on earth.

But it was still an unfinished economy. While the United States had more railroad trackage than any country in the world by 1860, for instance, it still imported much of its rails and rolling stock from England, as it did most of its steel, a metal that was becoming increasingly important.

More, sectional political forces were increasingly pulling the country apart across a North-South divide, despite an ever more integrated national economy. As it turned out, these forces could not be successfully contained by political means, despite decades of effort. What Sir Winston Churchill called the most inevitable conflict ever fought by the English-speaking peoples would prove the most consequential event in American history.

For while the American Union had been created by the Revolution, the American *nation* would be forged only upon the awful anvil of the Civil War.

PART III

THE EMERGING COLOSSUS

Sixty years ago there were no great fortunes in America, few large fortunes, no poverty. Now there is some poverty (though only in a few places can it be called pauperism), many large fortunes, and a greater number of gigantic fortunes than any other country in the world.

— James Bryce, Viscount Bryce
The American Commonwealth, 1894

The rich only select from the heap what is most precious and agreeable. They consume little more than the poor, and in spite of their natural selfishness and rapacity . . . they divide with the poor the produce of all their improvements. They are led by an invisible hand to make nearly the same distribution of the necessaries of life which would have been made, had the earth been divided into equal portions among all its inhabitants.

— Adam Smith
The Theory of Moral Sentiments, 1759

Transition

THE CIVIL WAR

T HE AMERICAN CIVIL WAR was the largest war fought in the Western world in the century between the Battle of Waterloo, fought on June 18, 1815, and the outbreak of World War I on August 1, 1914. Spread across half a continent, the troops moved by railroads and commanded by telegraph, the people informed by large-circulation newspapers, it was also the first great conflict of the industrial era.

The carnage was without precedent. In the single day of September 17, 1862, at the Battle of Antietam, the Union Army had casualties of 2,108 killed and 9,549 wounded. That was more casualties than the United States Army had sustained in the entire Mexican War, which had lasted two years. The total military deaths in the war on both sides—officially 498,333—exceeded 3 percent of the American male population in 1860, four and a half times our percentage losses in World War II.

Because the Civil War was far more like the great conflicts of the twentieth century than such earlier struggles as the Napoleonic Wars, both sides faced demands on their government finances and the economies that supported them that no nation had faced before. The fact that the North, with a far larger economy and a government fiscal system already in place, was better able to meet these demands played no small part in the war's eventual outcome.

Because of the depression that had started in 1857, the federal government had been operating in deficit since that time. In 1860 the national debt stood at $64,844,000 and the Treasury was nearly depleted. In December of that year, as the Deep South states began to secede one by one, there was at one point not even enough money on hand to meet the payroll.

Abraham Lincoln appointed Salmon P. Chase as secretary of the treasury. Chase, a man of great intellect if no sense of humor, and a former Ohio senator and governor, knew that he faced unprecedented problems. At the outbreak of the war, on April 15, 1861, the federal government had been spending around $172,000 a day in all departments. Three months later, at the time of the first Battle of Bull Run, the War Department alone was spending a million dollars a day. By the end of the year, spending would be up to $1.5 million a day.

How could such expenses be met? Governments have only three ways to raise money to pay their bills. They can tax, they can borrow, and they can print. The federal and Confederate governments both resorted to all three. The particular mix of the three means that was used by North and South, however, turned out to be crucial.

Since the demise of the Second Bank of the United States, the federal government had financed deficits mostly by borrowing, often short-term, from banks. The banks would then hold the bonds in their reserves or sell them to their largest customers. Because it lacked a central bank, the government had no means of its own in place to borrow money or transfer sums from one area of the country to another.

By July 1, 1861, three weeks before the first Battle of Bull Run, the debt had risen to $91 million. Immediately afterward, when the battle had shown that the war was likely to be a protracted one, Chase raised $50 million from Wall Street bankers in anticipation of federal bonds that came to be known as seven-thirties because they paid an interest rate of 7.30 percent. (The rate was apparently chosen because it meant that the bonds would pay 2 cents a day for every hundred dollars invested).

Fifty million was a huge underwriting for Wall Street at that time, and Chase estimated that a year later the national debt would stand at $517 million, a figure quite without precedent in American history. The secretary realized that the fiscal demands of modern warfare could not be met in the old way.

Fortunately for Chase (and the country), he was acquainted with a young Philadelphia banker named Jay Cooke, whose father had been an Ohio congressman. Cooke was made the agent of the federal government to sell a new issue of bonds called five-twenties (so-called because they could be redeemed in no less than five years nor more than twenty and, meanwhile, would pay 6 percent interest, in gold).

Cooke bypassed the banks and went directly to the people. He advertised heavily in newspapers and passed out handbills. He arranged for the Treasury to issue the bonds in denominations as small as $50, and allowed customers to pay on the installment plan. Thus he deliberately tried to involve the average citizen in buying government securities. In the words of Senator John Sherman of Ohio (the older brother of General William Tecumseh Sherman), Cooke made the virtues of these investments stare "in the face of the people in every household from Maine to California." Cooke thereby invented the bond drive, an important part of every major war since.

This had a profound effect on how Americans handled their assets. In the 1860s only a small percentage of the population had bank accounts and less than 1 percent owned securities of any sort. Most families kept whatever surplus cash they had under the mattress. By the end of the war, however, Cooke had sold bonds to about 5 percent of the population of the loyal states, turning them into mini-capitalists. Equally important, the dead capital hidden in their mattresses had been liberated for productive purposes. Cooke's bond drive was so successful that by May 1864 the government was actually raising money as fast as the

Navy and War departments could spend it, about $2 million a day at that point.

Largely thanks to Cooke, the North was able to throw much of the cost of the war onto the future, raising fully two-thirds of its revenues in the war years by selling bonds. The Confederacy, with a far smaller middle class, few large banks, and little financial expertise, was able to raise only about 40 percent of its revenues through borrowing. The situation for the South was made worse by the fact that the South had an economy notoriously lacking in liquidity. Thus the South's wealth could not be easily translated into money and spent on war matériel. While the South had 30 percent of the country's total assets at the outbreak of the war, it had only about 12 percent of the circulating currency and 21 percent of the banking assets. The word *land-poor* was not invented until Reconstruction days, but it also described the Southern economy in 1861.

The Northern bond sales caused a breathtaking rise in the national debt. It had stood at a minuscule 93 cents per person in 1857, before depression hit. Eight years later it stood at $75 per person. It would not be that high again until World War I, when the economy would be far larger. This caused a great increase in the amount of money that flowed annually through the federal government. Before the Civil War, the United States government had never spent more in a single year than $74.2 million (in 1858). Since the Civil War, it has never spent less than $236.9 million (1878). In 1865 alone it spent $1.297 billion, the first time in history that any nation had a billion-dollar budget.

A tax system based almost wholly on the tariff was obviously as inadequate to meet the emergency as had been the old way of borrowing money. In August 1861 the first American income tax was shepherded through Congress by Salmon Chase (who, ten years later while sitting as chief justice, would find it unconstitutional). It taxed all income "whether derived from any kind of property, rents, interests, dividends, salaries, or from any trade, employment or vocation carried on in the United States or elsewhere, or from any source whatever."

It originally called for a tax of 3 percent on incomes of more than $800 (then a middle-class income), rising to 5 percent on incomes of more than $10,000, a very comfortable income indeed at that time. In 1864 the income on taxes in excess of $10,000 was doubled to 10 percent.

Virtually everything else was taxed as well. Stamp taxes were imposed on legal documents and licenses, excise taxes on most commodities. The tax on liquor reached $2.50 a gallon, when the price, untaxed, would have been about 20 cents. The gross receipts of railroads, ferries, steamboats, and toll bridges were taxed. Advertising was taxed. The tariff was sharply raised. Altogether the federal government raised fully 21 percent of its revenues by taxation. And yet, while the level of taxation was far higher than Americans had previously experienced in their history, there was little evasion as, indeed, there would be little during later great conflicts. People, it seems, are willing to pay without complaint very high taxes in time of war or other serious national emergency. The South, with its less developed and cash-poor economy, was able to raise only about 6 percent of its revenues by taxation.

Thus the Confederacy had to rely for more than half its revenues on the third way for a government to met expenses: printing money. As early as May 1861 the Confederate government was issuing Treasury notes that would not be redeemable in gold and silver until two years after the signing of a peace treaty establishing independence. By the end of the war the South had issued more than a billion and a half dollars in paper money. Nor was the government in Richmond the only one in the South resorting to the printing press to pay its bills. City and state governments also issued notes. And because the South had no good paper mills or state-of-the-art printing facilities, counterfeiters had a field day.

The consequences of issuing large quantities of what economists call fiat money—money that is money only because the government says it is money—are inevitable and were as well known then as now. The first thing to happen is that Gresham's law kicks in. Good money—gold and silver in this case—disappears into mattresses as people hoard it, while they spend the money perceived to be less valuable or trustworthy.

The second thing to happen is that inflation takes off. As printing-press money flooded the Southern economy, inflation increased rapidly, more than 700 percent in the first two years of the war alone. As the war continued, the inflationary spiral deepened and the Southern economy began to spin out of control while the currency became essentially worthless.

Hoarding, shortages, and black markets spread inexorably as support for the war effort eroded and living standards fell sharply. The movie of *Gone With the Wind* is not good history in many, many ways, but the scene of the black butler, hatchet in hand, pursuing a scrawny rooster through the pouring rain, saying it is going to be Christmas dinner for the white folks epitomizes the facts of life for hundreds of thousands of Southern households in the last, desperate years of the war.

The North as well resorted to the printing press. In December 1861 the nation's banks were forced to stop paying out specie, and the federal government followed suit shortly afterward. The country had gone off the gold standard, and Wall Street panicked. "The bottom is out of the tub," Lincoln lamented. "What shall I do?"

What he did was issue paper money. With congressional authorization, the Treasury began issuing greenbacks, so-called because they were printed in green ink on the reverse. (Salmon P. Chase, his eye ever on achieving the White House, put his own face on the $1 bill in hopes of becoming better known.) By 1865 the country had issued $450 million in greenbacks. This was a vast sum by the standards of the time, but it amounted to only about 11 percent of federal spending in those years. And while there was, inevitably, a surge of inflation, it was a manageable 75 percent or so.

While the federal government was resorting to the printing press to finance a part of the war, Congress took advantage of the situation to reform both the American banking system and the chaotic supply of paper money. In 1863 (considerably amended in 1864) Congress established a system of nationally chartered banks. These banks had to have at least $50,000 in capital, a relatively large sum in those days, of which $30,000 had to be invested in U.S. Treasury securities. These banks were allowed to issue banknotes, but only those designed and engraved under the direction of the federal government, and these notes had to be backed 100 percent by pledged Treasury bonds.

It was thought that the state-chartered banks would take national charters, but few in fact did. So in March 1865 Congress narrowly passed a bill laying a 10 percent tax on the face value of banknotes issued by state-chartered banks. This had the effect of both driving the state banks to take national charters (there were only two hundred state banks remaining in 1866) and finally ending wildcat banking and a money supply consisting of thousands of different issues. By the end of the Civil War there were only two forms of paper money in the country, national banknotes, backed by bank reserves, and greenbacks.

———

ALTHOUGH THE FEDERAL GOVERNMENT had no hesitation in paying its bills with greenbacks, and requiring people to accept them by making them legal tender, the federal government itself did not accept them in payment of taxes. Taxes had to be paid in gold, and international trade continued to operate strictly on a gold basis.

This meant, of course, that there had to be some method of converting greenbacks into gold. While the federal government required that greenbacks trade at par with gold, that simply didn't comport with economic reality, and the law was ignored. The New York Stock and Exchange Board quickly began trading gold. But because, not surprisingly, the price of gold, as measured in greenbacks, tended to fluctuate inversely with the fortunes of the Union Army, the exchange banned the trading the next year as unpatriotic.

The curb brokers, who traded stocks out on Broad Street, then organized a place called Gilpin's News Room (although it is not quite clear who Gilpin was) in which to trade gold. Anyone willing to pay an annual $25 fee was welcome to trade there. Respectable merchants who needed gold for business purposes or to hedge against fluctuations in the price of greenbacks used Gilpin's, as did the hundreds of pure speculators hoping to make a fortune out of the vicissitudes of a war being fought for their country's very existence. These speculators were not very popular, often being referred to as "General Lee's left wing in Wall Street." Abraham Lincoln publicly wished that "every one of them had his devilish head shot off."

It made no difference to them what they were called; there was simply too much money to be made by the lucky or the prescient. And they went to a good deal of trouble to assure that prescience, employing agents stationed with both armies to keep them informed. Indeed, they were often better informed than was Washington, and Wall Street learned the outcome of the Battle of Gettysburg before President Lincoln did.

On June 17, 1864, Congress tried to stem the speculation by making it illegal to trade gold anywhere but in a broker's office. The effect of this law, besides closing Gilpin's and forcing gold trading out onto Broad Street, where it couldn't be regulated at all, was to increase the spread between gold and greenbacks. (The spread had reached its height just before Gettysburg, when it took 287 greenback dollars to buy 100 gold dollars.) The law was repealed just two weeks later and Gilpin's reopened.

That fall several members of the Wall Street establishment, including the very young J. P. Morgan and Levi P. Morton (who would later be governor of New York and vice president under Benjamin Harrison), established the New York Gold Exchange. The trading floor featured a large clocklike dial with a single hand that indicated the current price of gold. While it had higher standards and better enforced rules than Gilpin's (which quickly shut down), the New York Gold Exchange was still no place for the fainthearted.

Wall Street in general prospered almost beyond calculation in the Civil War years. Although the outbreak of the war caused panic, as the sudden onset of a great war always does, it was soon clear that the business of the Street—the trading of securities—would greatly increase. As the national debt soared by a factor of forty, bond trading soared equally. More, it was clear that much of the money the government was spending would go to firms such as iron mills, gun foundries, railroads, telegraph companies, and textile and shoe manufacturers. The profits of these firms would be invested in Wall Street and their capital needs met by it.

Before long the greatest expansion of business in the history of the Street was under way and Wall Street quickly blossomed into the second largest securities market on earth, second only to London. Fortunes were made in the next few years. In 1864 J. P. Morgan, only twenty-seven years old, had a taxable income of $53,287, fifty times what a skilled worker might expect to earn in a year. So busy were the brokers that the lunch counter was invented to afford them a quicker meal than could be had by going home. Fast food is, perhaps, not the least of the country's legacies from the Civil War.

The New York Stock and Exchange Board, which changed its name to the New York Stock Exchange in 1863, continued to hold its twice-daily sit-down auctions, but they were nowhere near adequate to handle the new business that flooded into the Street, and new exchanges opened up to handle the overflow, while business on the curb expanded dramatically.

The Mining Exchange, which had collapsed after the panic of 1857, was soon back in business, trading such stocks as Woolah Woolah Gulch Gold Mining and Stamping Company. In 1865 the Petroleum Board opened to handle stocks of the new companies exploiting the Pennsylvania oilfield. The most important of the new exchanges began, rather unpromisingly, in a basement known as the Coal Hole. It was soon doing more volume than the New York Stock Exchange and in 1864 reorganized as the Open Board of Brokers. It didn't use the old sit-down auction format, but rather adopted the continuous auction, each stock traded at a particular spot on the floor marked by a post (so-named after the lampposts on Broad Street, where the curb brokers traded individual stocks).

By 1865 the annual volume on the Street was above $6 billion. "Many brokers earned from eight hundred to ten thousand dollars a day in commissions," James K. Medbery wrote in 1870 in *Men and Mysteries of Wall Street*. "The entire population of the country entered the field. Offices were besieged by crowds of customers. . . . New York never exhibited such wide-spread evidences of prosperity. Broadway was lined with carriages. The fashionable milliners, dress-makers, and jewelers reaped golden harvests. The pageant of Fifth Avenue on Sunday, and of Central Park during the week-days was *bizarre,* gorgeous, wonderful! Never were there such dinners, such receptions, such balls. Anonyma startled the city with the splendor of her robes and the luxury of her equipages. Vanity Fair was no longer a dream."

———

DURING THE CIVIL WAR, the industrialization of the American economy, already well under way, expanded exponentially. The unprecedented demands of what became the largest army in the world and a navy second only to Britain's naturally greatly fueled the increased production. So did the tariff, which was raised to previously unseen levels to help pay for the war. As a result, American industry, still often less efficient than Britain's, was able to take away markets, and imports dropped sharply. In 1860 American imports had been valued at $354 million. Two years later they were only $189 million, despite the quickly expanding economy.

The difference was more than made up by American manufactures. In 1859 there had been 140,433 manufacturing firms in the United States. A decade later there would be 252,148. The domestic production of iron railroad rails, a good measure of industrial might as the nineteenth century continued, went from 205,000 tons in 1860 to 356,000 five years later (by 1870 production would be 620,000 tons). But industries less central to the economy were also boosted by the war. The process patented by Gail Borden for canning condensed milk in 1856 saw a great increase in demand, and this helped to spark a boom in the entire food processing industry.

The labor shortages caused by so many men being in the army fostered still more mechanical ingenuity than usual and were partially offset by continued immigration from Europe—more than eight hundred thousand people in the war years.

The South, too, saw industry grow because of the war, but from a much smaller base and with far greater constraints. The only full-scale iron mill in the South was the Tredegar Iron Works outside Richmond, and it greatly increased production. It had seven hundred workers in 1861 and twenty-five hundred by January 1863. But it could never get more than about a third of the supply of pig iron it needed to operate at full capacity, although Alabama was a major producer of iron ore.

Still, a cannon foundry was established at Macon, Georgia, and several bronze foundries there and elsewhere. The Confederacy built the largest gunpowder factory in North America, located in Augusta, Georgia. The South managed to keep its forces supplied with ammunition and was self-sufficient in small arms. Josiah Gorgas, who headed the Confederate Ordinance Bureau, wrote proudly in his diary in 1863, "Where three years ago we were not making a gun, pistol nor a sabre, no shot nor shell—[nor] a pound of powder—we now make all these in quantities to meet the demands of our large armies." When Lee surrendered at Appomattox, his troops were out of food, indeed they were half starved, but had an average of seventy-five rounds of ammunition per man and sufficient artillery shells.

The Confederate Navy laid plans for constructing 150 ships. It didn't meet that target, of course, but its construction of about 50 ships, including 21 ironclads, was no small accomplishment under the circumstances.

And it had no small effect on the war and the future of the American economy.

Confederate raiders, some constructed in Britain, such as the legendary CSS *Alabama,* roamed the seas snapping up Union vessels. The result was that American shipowners hastened to reflag their vessels with British colors to make them immune to capture, and many of these vessels never came back.

The United States had been a major maritime power since the earliest colonial days, for most of that period second only to Britain itself. But the Civil War precipitated a long decline in terms of freight carried in American bottoms. The country's merchant marine has never really recovered. In 1860, of the 8,275,000 tons of shipping clearing American ports that year, 5,921,000 tons, more than 71 percent, was American-owned. In 1865 less than 48 percent was American-owned. By 1890 only 22 percent was American. A major casualty of the Civil War was what had been the first great American industry.

Despite the prodigies of industrial improvisation the South accomplished during the war, most of its industries were eventually destroyed by the war along with much of the rest of its economy. The bonds and paper money issued by the Confederate and state governments became worthless, wiping out the region's liquid capital. Its agriculture—the very heart and soul of the Southern economy—had been devastated as much of the enslaved workforce abandoned the fields as soon as they could. The cotton crop, unable to be exported because of the Northern blockade (although much was clandestinely smuggled out to keep New England mills operating) rotted in warehouses.

With the end of the war and slavery, a new system of agriculture had to be developed in the South. The former slaves now controlled the use of their labor, but had not land, equipment, or expertise in dealing with a free economy. The former slaveowners for the most part kept their ownership of the capital assets needed to produce crops, such as the land and cotton gins, but often lacked the cash to pay farmhands' wages.

Numerous arrangements were tried, but soon a system of sharecropping, in which laborers were paid with a share of the crop, arose in the Deep South (it was largely unknown elsewhere in the country) and it would dominate southern agriculture until after World War II. But it never excluded other types of arrangements, nor was sharecropping limited to black workers. Many poor white families were also sharecroppers, and about 25 percent of black farming families owned the land they worked by 1880, a remarkable accomplishment only fifteen years after the end of slavery.

But the South retained the essential attributes of what today is called a Third World country: ownership of the means of production by a small, privileged elite; desperate poverty and grinding toil for most of the population; and an economy based on agriculture and extractive industries rather than manufacturing and services.

Worse, while the abolition of slavery was the greatest accomplishment of the Civil War, the virulent racism engendered by it persisted unabated. With the end of Reconstruction, southern whites reasserted political control, and blacks were largely disenfranchised for nearly a century. The intimate but uneasy relationship between the races in the South discouraged nonsouthern entrepreneurs from moving to the region to utilize such comparative advantages as low cost of living and cheap labor. Meanwhile a steady stream of talented and ambitious southerners migrated north in search of the much greater opportunities available there. For eighty years after its catastrophically failed attempt to achieve independence, the South would remain a Third World country, inside one that would develop the largest and most dynamic First World economy on earth.

Somewhat paradoxically, it was the Armageddon of the Civil War, with all its cost in blood and treasure, that brought this new dynamism into being. The very size of the struggle induced a solemn and profound pride in what had been saved thereby: the American Union. The Civil War transformed the United States (a phrase that, before the war, had

been grammatically construed in the plural) from a collection of associated states into a nation, one whose name was construed as singular. The
ancient motto of the country, *E pluribus unum,* had been achieved, at the
cost of half a million dead. Among the Great Powers today, only ethnically homogeneous Japan has fewer centrifugal forces at work in its body
politic.

Salmon P. Chase felt the new attitude as early as 1863. "We began
without capital," he wrote that year, "and if we should lose the *greater* part
of it before this [war] is over, labor will bring it back again and with a
power hitherto unfelt among us."

The fact that the war had been financed internally, and with such vast
sums, brought home just how rich and powerful the nation had become.
America is "to-day the most powerful nation on the face of the globe,"
Congressman Godlove S. Orth told an audience in Lafayette, Indiana, in
1864. "This war has been the means of developing resources and capabilities such as you never before dreamed that you possessed."

The people knew they possessed them now, and with the end of the
Civil War, as the country's military forces shrank quickly to insignificance, they would, in the next three decades, use those new resources
and capabilities to astonish the world.

Chapter Eleven

CAPITALISM RED
IN TOOTH AND CLAW

I N THE HALF CENTURY between the end of the Civil War and the beginning of World War I in Europe, the American economy changed more profoundly, grew more quickly, and became more diversified than at any earlier fifty-year period in the nation's history.

In 1865 the country, although already a major industrial power, was still basically an agricultural one. Not a single industrial concern was listed on the New York Stock Exchange. By the turn of the twentieth century, a mere generation later, the United States had the largest and most modern industrial economy on earth, one characterized by giant corporations undreamed of in 1865. The country, an importer of capital since its earliest days, had become a world financial power as well, the equal of Great Britain.

While displaced from the center of the American economy, agricultural production also greatly increased, as farmers and ranchers poured into the Great Plains via the railroads that were laid across them in these years. By 1890 the Census Bureau declared that the frontier—a primary feature of American political geography since 1607 and a profound influence on the American character—had ceased to exist. There was still much land that was unoccupied, but it was fragmented, and there was no longer an identifiable line across the continent that marked the end of civilization and the beginning of the once boundless wilderness. The United States was now a continental nation in geopolitical reality as well as nominal geographic fact.

The need to devise the new rules and institutions that would allow this new economy to flourish, and to assure that its fruits were equitably enjoyed by all segments of society, would dominate domestic American politics for the next century, just as preserving the Union and slavery had dominated the politics of the antebellum period. Many of the devices adopted to govern the new economy in this time would come through governmental and legislative action, especially in the latter decades. But, in fact, just as many would emerge out of the private sector, as the lawyers, bankers, brokers, railroaders, labor leaders, and industrialists sought to advance their own long-term self-interests, which were by no means always identical.

Increasingly involved in the debate over economic policy and the rules of the game as the nineteenth century drew to a close were intellectuals who had previously had no role beyond abstract economic theory. Often they sought to speak for society as a whole, rather than the self-interests of a particular group. Inevitably, of course, they spoke for their own self-interests, although it usually seems they were unaware that they had any. Some such as Karl Marx and Henry George remained theorists, but ones who attracted large popular followings. (Henry George, however, running as a reformer, was very nearly elected mayor of New York in 1886, and finished well ahead of the Republican candidate, Theodore Roosevelt.) Others, such as Charles Francis Adams and his brother Henry, were principally writers and journalists. Many of these intellectuals, however, had a very limited acquaintance with the real economic world they sought to influence.

It was, in short, a typically messy democratic process, but, as democratic processes usually do, it worked in the long term. No society in history had ever needed to govern a highly dynamic industrially based economy in a nation that was constitutionally a federal republic of limited powers. The United States learned how to do so, using, largely unconsciously, the great insights of the Founding Fathers: that men are not angels, that they are driven by self-interest, and that that self-interest can be exploited for the general good by an interlocking system of divided powers.

Although sometimes racked by severe depressions, the American economy would flourish so abundantly, in the long term, in the next 140 years precisely because the American nation devised a highly effective system of checks and balances for governing that economy in the decades after the Civil War.

IMMEDIATELY AFTER THE WAR, however, nothing characterized American politics and thus the American economy so much as corruption. There were, in effect, no cops on the beat, and the result, for a few years, was capitalism red in tooth and claw. It was often entertaining, at least for those who were not directly involved, but it was no way to run an economy. Capitalism without regulation and regulators is inherently unstable, as people will usually put their short-term self-interests ahead of the interests of the system as a whole, and either chaos or plutocracy will result. As Herbert Hoover explained, "The trouble with capitalism is capitalists. They're too damn greedy."

Nowhere was this corruption more pervasive than in New York, and especially on Wall Street. Before the Civil War, the financial market had been small enough that formal regulation was largely unnecessary; the various players could keep an eye on one another. While chicanery and outright fraud were hardly unknown, the players in the game were mostly professionals who knew what they were getting into. With the vast inflow of securities due to the war and an equally large increase in the number of people trading there, however, the situation changed.

But there was no mechanism to provide oversight. The federal government was not thought to have any role in regulating markets at this time, and the state and city governments were cesspools of corruption. As early as 1857 George Templeton Strong wrote despairingly in his diary, "Heaven be praised for all its mercies, the Legislature of the state of New York has adjourned." A few years later Horace Greeley in the *New York Tribune* wrote that it was not possible "that another body so reckless, not merely of right but of decency—not merely corrupt but shameless—will be assembled in our halls of legislation within the next ten years."

Greeley was wrong. In 1868 the New York State Legislature actually passed a law the effect of which was to legalize bribery. "No *conviction* [for bribery]," the act read, "shall be had under this act on the testimony of the other party to the offense, unless such evidence be *corroborated* in its material parts by other evidence." In that preelectronic age, that meant that as long as the official took his bribe in private and in cash, there was no possibility of his being convicted.

The state courts were no better than the legislature. Judges had been elected in New York State since the 1840s, when the state adopted a thoroughly Jacksonian constitution. This made them dependent on the political machines, as, indeed, they still are. The results were predictable. "The Supreme Court is our *Cloaca Maxima,*" George Templeton Strong, a very successful lawyer himself, wrote in his diary, "with lawyers for its rats." In 1868 the popular English *Fraser's Magazine* wrote that "in New York there is a custom among litigants as peculiar to that city, it is to be hoped, as it is supreme within it, of retaining a judge as well as a lawyer." Nowhere was this more true than in what became known on Wall Street as the Erie Wars, the struggle for control of the Erie Railway.

The Erie had had a very unusual history among early American railroads. It had been chartered as the result of a political deal and intended from the beginning to be a great trunk line. To secure for the Erie Canal project the political support of the "Southern Tier" of counties along the border with Pennsylvania, they were promised an "avenue" of their own. The railroad, chartered in 1832, was to be that avenue.

Political forces kept the Erie out of Buffalo, its logical western terminus, but where it might have competed with the canal, and from passing through New Jersey so as to terminate across the Hudson from New York City, its logical eastern terminus. Instead, what was, at 451 miles, briefly the longest railroad in the world ran from the tiny town of Dunkirk, New York, on the shore of Lake Erie, to the equally tiny town of Piermont, New York, on the Hudson River north of the New Jersey state line.

The line had gone bankrupt during its seventeen-year-long construction, and by the time it was finally finished it had an extraordinarily complex capital structure involving common stock, preferred stock, and convertible bonds. These securities were a speculative favorite on Wall Street, especially with a speculator named Daniel Drew, who was also, from time to time, the Erie Railway's treasurer and sat on its board.

Drew was one of the great characters in Wall Street history. Uneducated but profoundly—and genuinely—devout, he had begun his career as a drover, supplying the New York cattle market. He soon moved into Wall Street and steamboats. By the 1860s he was very rich, worth at one point, at least by his own reckoning, $16 million. He founded what is now Drew University and paid for the construction of several churches. But when not praising the Lord, he practiced with boundless enthusiasm every trick in Wall Street (some of which he invented himself) to separate the unwary from their money. Known as the "speculative director," Drew liked nothing better than to manipulate Erie securities.

The Erie, despite its chronic financial problems (it went bankrupt a second time in 1859), was a very large enterprise, with forty-four hundred employees that year, thousands of cars, and millions of dollars in revenue. Its economic potential was considerable. From his seat on the board, Drew had access to inside information, which he used without compunction, as did many of the other directors. The result, according to a ditty on the Street, was:

> Daniel says up—Erie goes up.
> Daniel says down—Erie goes down.
> Daniel says wiggle-waggle—Erie bobs both ways.

Railroading has always been a difficult business to make money in because it has very high fixed costs, which continue whether there is good business or bad. For that reason, market share is crucial to profitability on the lines where there is competition, as every additional passenger or ton of freight adds to income without adding much to expenses. Because of the need for market share, price wars were common between competing railroads in the nineteenth century (and are common between airlines today for precisely the same reasons).

But there are natural limits to these price wars, as below a certain point, rate cuts become economically suicidal. The board of the Erie, however, was largely unconcerned with such mundane, long-term matters as profitability or even viability. They were far more interested in short-term trading profits on the Street. This made the Erie the wild card of New York railroading (and caused Charles Francis Adams to dub the line the "Scarlet Woman of Wall Street"). An increasingly powerful figure in that market, Cornelius Vanderbilt, wanted to do something about it.

Vanderbilt had left the employ of Thomas Gibbons in 1829 and struck out on his own in the steamboat business. He was soon the greatest shipowner in the country, and in 1837 the *Journal of Commerce* first used the honorary title by which he has been known to history ever since: Commodore. The Commodore's business model was simplicity itself: (1) run the most efficient, lowest-cost organization possible; (2) compete fiercely by means of price until the opposition is either broke and can be bought out or pays you to stop competing; and (3) live up to your agreements. Matthew Hale Smith, both a lawyer and a Congregational minister, wrote of him in 1870 that "the Commodore's word is as good as his bond when it is freely given. He is equally exact in fulfilling his threats."

Vanderbilt was willing to accept payment to stop competing on a particular run because steamboats, unlike railroads, can operate anywhere there is water enough to float them. If paid to leave the Hudson, say, that made it just that much easier to compete on Long Island Sound or the New York to Philadelphia route. Some did not understand. The *New York Times* editorial page, in the 1850s, first used the image, if not quite the words, of the medieval "robber barons" in criticizing Vanderbilt's tactics. The robber barons, supposedly, lived along the Rhine and charged traffic a fee to pass their castles unmolested. (Whether they existed in fact or were a nineteenth-century construct is another matter.)

Charging passersby for safe passage was, of course, a pure extortion racket, for no wealth was created by it, merely transferred. But Vanderbilt was doing no such thing. In 1859 *Harper's Weekly,* far less economically obtuse than the *Times,* explained. "It has been much the fashion," it wrote, "to regard these contests as attempts on his part to levy blackmail on successful enterprises. It is hardly fair for any man to undertake to decide what are the particular motives of his neighbor in undertaking a specific work, if the work be in itself legitimate and fair. He must be judged by the results; and the results in every case of the establishment of opposition lines by Vanderbilt has been the *permanent reduction of fares*. Wherever he 'laid on' an opposition line, the fares were instantly reduced, and however the contest terminated, whether he bought out his opponents, as he often did, or they bought him out, the fares were never again raised to the old standard. This great boon—cheap travel—this community owes mainly to Cornelius Vanderbilt." Even the *Times* would soon come around to this view of the Commodore.

The term *robber baron,* of course, came to stand for the men, of whom Vanderbilt was one of the first, who built great industrial and transportation empires in the late nineteenth-century American economy. While many of these men were capable of ruthlessness, gross dishonesty, and self-aggrandizement (and others were honest men who scrupulously stayed within what were often inadequate laws), none of them merely transferred wealth to themselves from others by their activities. They all built vast wealth-creating enterprises. Regardless, the term is obviously here to stay.

By the early 1860s Vanderbilt began to move into railroads, a technology he had always disliked (he had been very nearly killed in one of the earliest major railroad accidents in this country). He acquired controlling interest in the New York and Harlem Line and the Hudson River Railroad, the only two railroads that had direct access to Manhattan Island. When his interests in these two roads were attacked by Wall Street speculators in 1863, including Daniel Drew, Vanderbilt quickly proved himself the utter master of the Wall Street game, cornering the speculators in Harlem twice and the Hudson once in a matter of weeks, earning millions, and establishing a reputation such as no man has had on Wall Street since. A British journalist of the day described him as "a Gaetulian lion among the hyenas and jackals of the desert."

In 1867 Vanderbilt was invited to become president of the New York Central and soon planned to merge it with his Hudson River Railroad, creating a line that would extend all the way from New York City to Buffalo and that would compete directly with the Erie Railway. Vanderbilt had no doubt that he could compete successfully. But he wanted the Erie run in a businesslike manner so that both lines could make money by agreeing on a division of traffic (which was perfectly legal at the time but today would be a combination in restraint of trade).

He determined to throw Drew off the board at the election of October 8, 1867, and get people who agreed with him elected instead. Drew, an old friend as well as a competitor since steamboating days, went to see the Commodore and promised to behave himself and even to work in Vanderbilt's interests. Vanderbilt relented, and Drew not only stayed on the board but was named treasurer of the Erie once again, a position he had not had since the mid-1850s. Also elected to the board were two newcomers to Wall Street, Jay Gould and Jim Fisk.

Drew quickly broke his promises to Vanderbilt by organizing a pool to bull the price of Erie stock upward. And when Vanderbilt asked the board to agree to divide the New York City traffic among the Erie, the New York Central, and the Pennsylvania, the Erie board voted it down with only Frank Work—the Commodore's man on the board—in favor.

Vanderbilt, thwarted, decided to strike back. If he couldn't influence the board to behave like businessmen, he would use what Charles Francis Adams called "the brute force of his millions" to buy control of the Erie. There were officially 251,050 shares of Erie common. But Drew also controlled 28,000 unissued shares that were in Drew's possession as collateral for a loan, and some of the many issues of Erie bonds could be converted into common stock. And Drew, as treasurer, was in a perfect position to issue still more convertible bonds and to convert them as necessary to prevent Vanderbilt from winning control by buying a majority of the common stock.

Vanderbilt went to Justice George G. Barnard, described by a contemporary as "a Tammany healot numbered among the Vanderbilt properties." Barnard quickly issued an injunction forbidding Daniel Drew as treasurer from converting bonds into stock and personally from selling any stock in his possession. The Commodore thought he had the situation under control and ordered his brokers to go into the market and buy all the Erie stock that was offered.

But Drew got one of *his* judges, named Gilbert and sitting in Brooklyn, to order that the Erie continue converting bonds into stock on request. As a broker, E. C. Stedman, explained, "Since they were forbidden by Barnard to convert bonds into stock, and forbidden by Gilbert to refuse to do so, who but the most captious could blame them for doing as they pleased?"

Within a few days, the Erie had issued bonds that were converted into one hundred thousand shares of new stock and thrown on the market. "If this printing press don't break down," Jim Fisk offered, "I'll be damned if I don't give the old hog all he wants of Erie."

When Vanderbilt found out that the Erie board was printing new shares of stock as fast as his brokers could buy them, he went back to Justice Barnard and had him issue arrest warrants. The sheriff was dispatched to bring in the Erie board. The board, alerted to their impending arrests, fled.

As William Worthington Fowler, a broker who wrote a best-selling memoir of Wall Street in 1870, described it, a policeman on the beat "observed a squad of respectably dressed, but terrified looking men, loaded down with packages of greenbacks, account books, bundles of papers tied up with red tape, emerge in haste and disorder from the Erie building. Thinking perhaps that something illicit had been taking place, and these individuals might be plunderers playing a bold game in open daylight, he approached them, but soon found out his mistake. They were only the executive committee of the Erie Company, flying the wrath of the Commodore, and laden with the spoils of their recent campaign." Indeed they were: they had $7 million of the Commodore's money stuffed in a carpetbag. They quickly fled to New Jersey, where they were safely beyond the immediate reach of the New York law.

The action then moved to Albany, where each side tried to bribe the legislators, who "flocked to Albany like beeves to a cattle mart. All were for sale, and each brought a price proportioned to his *weight*." Jay Gould, according to the *New York Herald,* arrived in Albany with a suitcase filled with thousand-dollar bills, and when arrested produced $500,000 in bail money on the spot.

The Commodore could have won a bribing contest, but he realized that it might be a pyrrhic victory in the long run, for public opinion was turning against the idea of his owning the Erie as well as the Hudson River, Harlem, and New York Central lines. This would have given him near monopoly power in the New York transportation market. Vanderbilt decided to deal and sent a note to Drew. He insisted on two things. The worthless stock (it had been declared to be not "good delivery" by all the major exchanges) had to be taken off his hands at a price near what he had paid for it, and Daniel Drew had to sever all ties to the Erie. To get their agreement, Jay Gould was made president of the Erie and Jim Fisk treasurer.

The Erie Wars had galvanized the public, and the story was given even more space in the newspapers than the concurrent impeachment trial of President Andrew Johnson. But while the public was vastly entertained, most members of the New York commercial establishment were horrified.

Unlike speculators, who look no further than the next big killing, brokers earn their money one commission at a time and need a market with as much predictability as possible so as to enjoy the greatest number of customers possible. If the amount of a stock that has been issued could be doubled or halved without even a moment's notice by the management, how could anyone know what the value of a share was? The *Commercial and Financial Chronicle,* an influential weekly, put its finger on the heart of the problem. The capitalists were not wholly at fault because "the letter of the law is very deficient in its regulation of the management of corporate interests." The *Chronicle* published a suggested law to remedy the situation, requiring that directors get consent of the stockholders for new issues, that no issues of stock be made without notice, that a register of all issues be available for inspection at a financial institution, and that infractions be punished criminally.

In the late 1860s there was no way to get such a law enacted because of the corrupt state legislature. The stock exchanges, however, owned by the brokers, could act on their own and, within a month, did so. On November 30, 1868, the Open Board of Brokers and the New York Stock Exchange issued identical sets of regulations requiring listed companies to register all listed securities within two months in a registry open to public inspection, and to give thirty days' public notice of intent to issue new securities. The Erie Railway, now the plaything of Jay Gould and Jim Fisk, refused and was delisted.

Shortly afterward, these two stock exchanges merged under the name of the New York Stock Exchange. Finally there was an institution on Wall Street large enough and powerful enough to act as an effective regulator. Jay Gould quickly found out that he had no choice but to comply if there was to be a decent market for Erie securities, and he registered the company's stock and bonds on September 13, 1869.

It was just a start, of course. James K. Medbery, in his *Men and Mysteries of Wall Street* in 1870, knew what was at stake. "It remains for the brokers of the Stock Exchange," he wrote, "to decide whether they will seek the petty profits of a speculation marred by grave faults, or will cast their influence still farther and with more strenuous emphasis against the encroachment of the cliques. The former means isolation. The latter will be prelusive of an expansion in international relations which will make New York imperial, and Wall Street what its pivotal position demands and allows, the paramount financial center of the globe."

Wall Street chose to go down the second path, and the New York Stock Exchange began to exert its new power to effectively regulate stock trading. The Street was still no place for fools—it never will be—but compared with just a few years earlier, Wall Street had become a modern, reasonably well-regulated financial market. The stocks traded on Wall Street at this time had a total market capitalization of about $3 billion, while London's market was more than $10 billion. But Wall Street was catching up fast. Successful self-regulation would guide Wall Street for the next sixty years as it eclipsed London and became the world's leading financial center.

———

THE CORRUPTION OF THE POSTWAR PERIOD was by no means limited to New York capitalists, government, or New York railroads. Indeed, the greatest railroad project in the nation's history—the transcontinental railroad built between 1864 and 1869—also set off the greatest financial and political scandal of the nineteenth century.

The transcontinental railroad had been envisioned ever since California had joined the Union in 1850, but the deepening political crisis between North and South had prevented any action. In 1862, with only loyal states now represented in Congress, the Pacific Railroad Act was passed creating the Union Pacific Railroad, the first corporation chartered by the federal government since the Second Bank of the United States in 1816. The project was intended as much as a symbol that the Union would endure (hence the name of the railroad company) as it was a commercial project.

The company, from the beginning, was to receive a great deal of government help because it would have been impossible to raise capital in the open market for a railroad across more than a thousand miles of unsettled land. Railroads in the East (and in Europe) had largely connected areas of economic activity, greatly increasing the amount of that activity. In the West they often created these areas of activity, as people and commerce followed the paths of the railroads. The original capital of the Union Pacific was to be one hundred thousand shares at a par value of $1,000 each, a huge capitalization of $100 million. The Union Pacific and the Central Pacific, which was to build eastward from Sacramento, would be granted a right-of-way of two hundred feet on each side on public lands. In addition the railroads would receive, for every mile of track completed, title to sixty-four hundred acres of land to be sold to settlers, alternating with sections of land retained by the federal government.

There was more. For each mile completed, the Union Pacific would receive from $16,000 to $48,000 in government bonds, depending on the difficulty of the terrain, and government loans, in the form of first mortgage bonds, repayable over thirty years, for construction costs. It was soon clear, however, that more help would be needed. In 1864 a new bill allowed the two railroads to sell first mortgage bonds (subordinating the government mortgage bonds) and doubled the land grants to the railroads to 12,800 acres per mile of track laid.

Despite the enormous government subsidies, it was still a very risky venture, and the management quickly moved to ensure their own profits, if not those of the regular stockholders. They set up a construction company, owned by themselves, with a fancy French name, Crédit Mobilier, and hired it to build the railroad. Needless to say, they charged the railroad top dollar, and often then some, for their services. Although the chief engineer, Peter Dey, had estimated that the initial section, west of Omaha, could be built for an average of $30,000 a mile, when Crédit Mobilier asked for $60,000, the president of the line, Thomas C. Durant, ordered Dey to resubmit his proposal, making the sum needed $60,000 a mile. Rather than go along, Dey resigned what he called "the best position in my profession this country has ever offered to any man." Others, needless to say, were not so scrupulous.

As many as ten thousand men—immigrant Irish, freed blacks, mustered-out soldiers, Chinese immigrants—worked on the two roads as they wended their ways across the plains, mountains, and deserts of the West toward a rendezvous at Promontory Point, Utah. The crews sometimes laid rails at the astonishing rate of four a minute. The casualty rate was appalling. Many were killed in accidents, but many too were killed in the drunken brawls that regularly erupted in the camps that moved along with the railhead.

Even at the time, the transcontinental railroad was perceived as one of the great epics of that age of engineering miracles in which they lived. William Tecumseh Sherman called it the "work of giants." The Western poet Joaquin Miller thought that "there is more poetry in the rush of a single railroad across the continent than in all the gory story of the burning of Troy."

Although physically one of the marvels of the age, financially the Union Pacific was wrecked by its construction, thanks to Crédit

Mobilier, which was wildly profitable. In 1867 Crédit Mobilier paid its first dividend to its stockholders, amounting to 76 percent of their investment. Future dividends ranged up to 350 percent. In the second dividend of 1868 alone, a person holding $10,000 par value in Crédit Mobilier stock received $9,000 in cash, $7,500 in Union Pacific bonds then selling at par, and forty shares of Union Pacific stock worth $1,600, a return on capital of 181 percent.

Oakes Ames, a stockholder of Crédit Mobilier as well as a congressman from Massachusetts, thought that other members of Congress should be cut in on the action to make sure there would be no trouble from that quarter, and many were. They had to pay for the stock, but as they were allowed to do so out of the munificent dividends, that was not a burden. Among those receiving stock were Schuyler Colfax, who would be President Grant's first vice president, and Henry Wilson, who was his second; others were James A. Garfield, who would be elected president in 1880, and James G. Blaine, who would be the Republican nominee for president in 1884.

This did not go unnoticed. In January 1869 Charles Francis Adams, who would himself be president of the Union Pacific Railroad in the 1880s, wrote in the *North American Review* that Crédit Mobilier was nothing but a Pacific Railroad ring. "The members of it are in Congress; they are trustees for the bondholders; they are directors; they are stockholders; they are contractors; in Washington they vote the subsidies, in New York they receive them, upon the plains they expend them, and in the 'Crédit Mobilier' they divide them."

Adams's charges were only allegations, as he had no proof, and were easily brushed off. But, as so often happens, thieves fell out, and a disgruntled Crédit Mobilier stockholder, Henry S. McComb, thinking himself cheated, sued. The court papers were leaked to Charles A. Dana, editor of the *New York Sun,* and on September 4, 1872, the story ran under a six-column lead:

THE KING OF FRAUDS
HOW THE CREDIT MOBILIER BOUGHT
ITS WAY THROUGH CONGRESS
Colossal Bribery

A congressional committee convened to investigate the scandal recommended that Oakes Ames and one other congressman—one of the few Democrats involved—be expelled, but they were only censured. The Republicans, however, suffered heavily in the election of 1874 and lost control of the House. Ulysses S. Grant, himself personally honest to a fault, would go down in history as presiding over the most corrupt administration in the nation's history.

IT WAS A NEWSPAPER that brought the Crédit Mobilier scandal to public attention, a sure sign of the growing power of the new medium to influence the political events it covered. Lord Macaulay had dubbed the press the "fourth estate" as early as 1828, but it was only after the newspapers had become a mass medium, read by millions, that they truly became a powerful force in the body politic in both Britain and the United States. The exposure of gross incompetence in the Crimean War by William Howard Russell (later Sir William) of the *Times* of London brought down the government of Lord Aberdeen and led to thorough reform in the British army.

The year before the Crédit Mobilier scandal broke, the so-called Tweed Ring, which had built a modest courthouse north of New York's City Hall for $12 million—fully 20 percent more than the vastly larger and more ornate Houses of Parliament had cost to build in London in the 1840s—had been exposed by the *New York Times*. A wave of reform had swept over New York State and City as a result. A new law regarding bribery was placed as an amendment in the state constitution, where it was safely out of the reach of the legislature.

The judiciary and the profession of law itself were also reformed, thanks mainly to the lawyers. "The abused machinery of law is a terror to property owners," George Templeton Strong had written in his diary. "No banker or merchant is sure that some person calling himself a 'receiver,' appointed *ex parte* as the first step in some frivolous suit he never heard of, may not march into his counting room at any moment, demand possession of all his assets and the ruinous suspension of his whole business . . . No city can long continue rich and prosperous that tolerates abuses like these. Capital will flee to safer quarters."

The basic problem was that technology had outrun the law. The New York lawyer David Dudley Field, in one of the great intellectual achievements of the nineteenth century, had written the Field Code of Civil Procedure, adopted in New York State in the 1840s and widely copied elsewhere. (Used by Britain as the basis of reforming its laws in the 1870s, it is today the basis of civil law throughout the common-law world.)

Before the telegraph and railroads, judges needed wide discretion to act in order to protect life and property. But by the 1860s this discretion was being widely abused. In a speech that was printed in many newspapers, William Maxwell Evarts, one of the country's leading lawyers, who had served as attorney general in the last days of the Andrew Johnson administration, said regarding his early career that "for a lawyer to come out from the chambers of a judge with an *ex parte* writ that he could not defend before the public, would have occasioned the same sentiment towards him as if he came out with a stolen pocket book."

Evarts and other leading lawyers organized the New York State Bar Association in 1870 to police the profession and to work to make needed reforms in the law, such as restricting judges from interfering in cases not before them. Lawyers in other states quickly followed suit and formed their own bar associations. The following year, as the Tweed Ring collapsed, the Bar Association contributed $30,000 to help pay the expenses of prosecuting corrupt judges, such as Justice Barnard, who was successfully impeached, and helped push extensive reform of the law through the suddenly virtuous legislature.

Scandals such as the Tweed Court House, Crédit Mobilier, and all the ones that have followed right up to Watergate and the Enron scandal of our own day have proved to be the engines of reform. It is impossible to anticipate every form of corruption that might develop in a constantly evolving free-market economy and in a government of limited powers. And the law will always lag well behind the ideas, both good and bad, that people, driven by self-interest, will develop for quickly exploiting new opportunities as they appear. The reason that the nineteenth century seems so peculiarly scandal-ridden, perhaps, is that there was so much economic, technological, and social change in that century.

With the reforms brought about by the scandals of the immediate post–Civil War era, the American economy and American politics settled into a more law-abiding period while the economy continued to evolve at breakneck speed. But many of the men who would play a central role in shaping the new economy, men such as J. P. Morgan, John D. Rockefeller, and Andrew Carnegie, came of age in the era of unprecedented government corruption and would never be able to conceive of government as a suitable instrument for reforming and regulating the economy.

To them, government was part of the problem, not the solution. They thought it was up to men like themselves to put the new American economy on a sound and honorable basis and keep it there. A rising political movement that would come to be known by many different names—the left, liberalism, populism, progressivism, and so on—would have a very different view.

Chapter Twelve

DOING BUSINESS WITH
GLASS POCKETS

W HILE THE NATIONAL ECONOMY was exploding in size in the post–Civil War era, monetary and banking regulation did not keep pace. Although the American economy became the largest in the world in these years and would come to rival the economy of all Europe in size, the United States remained without a central bank and thus without a mechanism to regulate the nation's money supply. The ghost of Thomas Jefferson's hatred of banks still haunted the American economy, although that economy now bore no resemblance whatever to the nation of yeoman farmers that Jefferson had envisioned.

The state-chartered banks, at first nearly extinguished when they were deprived of the ability to issue banknotes, made a strong comeback at this time. They regained the ability to create money by simply crediting the borrower's checking account (a British innovation in banking). Reduced to fewer than 200 banks at the end of the Civil War, by 1900 there would be 4,405 state-chartered banks in operation, most of them small and financially weak.

The new national banking system functioned well in the Northeast, where the economy was most developed and liquid capital most available. But because national banks at that time were forbidden to branch or to operate across state lines, their number as well rapidly increased, to 3,731 by the turn of the century. While usually larger and stronger than the state banks, the national banks were often equally dependent on a single local economy. One of the greatest strengths of the American economy, its immense size and diversity, was denied to the banking sector with which all other sectors necessarily had to deal. It would, in time, prove a near fatal weakness.

And in the South and West, many areas lacked the resources to meet the requirements of a national charter. The states of Mississippi and Florida did not have a single national bank between them. Worse, national banks were forbidden to lend money on the collateral of real estate, the one asset these areas had in abundance. The very basis of the money supply became a major political issue in the late nineteenth century as the gold standard divided the country.

With the coming of the Civil War, both the government and the banks had gone off the gold standard, the government issuing millions of nonredeemable greenbacks to help it pay its bills. The National Banking Act, and the tax on banknotes issued by state-chartered banks, gave the country a uniform currency for the first time by the end of the war. But as long as the greenbacks circulated, the United States was not, internally, on a gold standard, as their price could, and did, vary with respect to gold. By the late 1860s it took around $135 in greenbacks to buy $100 in gold.

International trade, however, *was* on the gold standard. This meant that merchants buying or selling abroad had to buy gold to pay tariffs and had to hedge in the gold market to ensure that fluctuations in the price of greenbacks did not impact their profits. Jay Gould, one of the shrewdest men ever to operate in Wall Street, thought he saw opportunity in this situation. In 1869 he decided to corner gold.

A corner is nothing more than control of the entire supply of a commodity—whether pork bellies, shares of a railroad company, or gold—for a period of time. Anyone needing to buy that commodity during that time must pay the price demanded by the holder of the corner or do without. Traders who have sold short—in other words, sold what they do not own in hopes of a fall in price—have no choice but to buy when they are required to deliver, for, as Daniel Drew is supposed to have famously explained,

> *He who sells what isn't his'n*
> *Buys it back or goes to prison.*

Attempted corners were commonplace in the Wall Street of the 1860s, and successful ones in various stocks happened every year. But attempting a corner in gold—the heart and soul of the nineteenth-century world monetary system—was an act of financial audacity unequaled before or since. For one thing, the federal government held millions in gold and could break any attempted corner at will. But Gould thought he could handle the honest but hopelessly naive President Grant.

He arranged to have Major General Daniel Butterfield, a Civil War hero (and, incidentally, the composer of "Taps") appointed subtreasurer at New York and thus the man who would have to order any sale of gold from the government vaults. Asked later if the wires had been tapped to learn of government intentions, Gould's partner Jim Fisk said, "Tap the wires? Nonsense! It was only necessary to tap Butterfield to find out all we wanted." Meanwhile, throughout the summer of 1869, Gould lobbied President Grant hard not to authorize any sale of gold, citing the need of American farmers to export their crops at good prices. Meanwhile, he and his allies began accumulating the metal on Wall Street.

There was surprisingly little gold in the floating supply—the amount immediately available to the market at a given time—no more than about $20 million. The Gold Room on Wall Street at that time was doing about $70 million a day in trading, but much of that was what was called "phantom gold," gold bought on paper-thin margins. As one Wall Streeter testified with some exaggeration, "if a man has a thousand dollars, he can go and buy five millions of gold if he feels so inclined." Gould, president of one of the largest railroads in the country, had more than enough resources to buy up the floating supply many times over as he lured more and more shorts into his net.

The corner climaxed on September 24, 1869, known ever since as Black Friday. This was the first day in Wall Street history, but not the last, to earn the appellation "Black." It was perhaps the most exciting single day in the whole history of the Street. The Gold Room itself was pandemonium as traders fought frantically to protect their interests. Around the entire country, business almost halted as men gathered in brokers' offices and banks to watch the price of gold in New York ratchet upward on the newly invented stock tickers.

Outside, on Broad Street, matters were little better. An eyewitness reported that "Broad Street was thronged by some thousands of men . . . and within an hour staid businessmen, coatless, collarless, and some hatless, raged in the street, as if the inmates of a dozen lunatic asylums had been turned loose. Up the price of gold went steadily amid shouts, screams, and the wringing of hands."

But President Grant had finally become aware of what was going on, and the Treasury ordered the sale of $4 million in gold at 11:42 A.M., an order that arrived in Butterfield's office minutes later. The corner, however, was already broken. The price of gold (in greenbacks) had stood at 160 at 11:40 that morning. By noon it was at 140 and falling. The next day the *New York Herald* reported that "for the remainder of the day the Gold Room and all the approaches thereto were like the vicinity of a great fire or calamity after the climax has passed. A sudden quiet and calm came over the scene."

We will never know if Gould made or lost money that day, for the mess was more swept under the rug than straightened out. Contracts that specified payment in gold—as Gold Room contracts necessarily did—were unenforceable at law, so it was possible to default without legal consequence, and many traders did just that. But because the gold corner was a buyers' panic, as traders tried desperately to cover their short sales, there were no long-term consequences for the economy as a whole. It is sellers' panics that have so often marked the start of depressions, as people rush to unload stocks and bonds at any price and to take their money out of banks that they see as unsafe.

Sellers' panics produce, by their nature, a sudden surge in demand for money as investors and depositors seek liquidity, and money, of course, is the ultimately liquid asset. Because there was no central bank empowered to regulate the money supply and to provide the liquidity needed to protect the banking system in times of stress, however, these sellers' panics greatly exacerbated the downward swings of the business cycle. Basically sound financial institutions collapsed by the hundreds when they were unable to meet the sudden demand for money. Often they took the life savings of families and the liquid assets of businesses with them.

The years immediately following the Civil War were a time of enormous economic expansion, a classic American boom. Railroad mileage doubled in a mere eight years, while wheat production did likewise. But in 1873 Jay Cooke, the Philadelphia banker who had invented the bond drive to help finance the Civil War and became the most famous banker in the country as a result, unexpectedly announced he was insolvent in September.

Panic swept Wall Street, where banks and brokerage houses, unable to convert their assets into cash quickly enough, failed by the dozens and the New York Stock Exchange was forced to close down for ten days because it could not maintain an orderly market. A deep depression stalked the land for the next six years.

The depressions of this time began to reach ever deeper into the American economy because a much larger percentage of the working population had become dependent on regular wages and national markets. Subsistence farmers who sold their surpluses locally could weather financial depressions fairly easily. Industrial workers and farmers who borrowed from banks on crops in the ground and sold to large grain companies could not.

THE NEW INDUSTRIAL and trading corporations were increasingly corporate in form, and the corporation became crucial to the American economy by the last third of the nineteenth century because enterprises dramatically increased in size at that time. In the eighteenth and early nineteenth centuries the economy had been characterized by individually and family-owned and operated enterprises. Organizations with more than a hundred employees were a rarity. By the time of the Civil War, however, several railroads were employing thousands, and industrial companies were growing rapidly as well. The Bath Iron Works of Maine, the largest industrial employer in 1860, had forty-five hundred workers.

Because railroads were very capital-intensive enterprises, they were mostly organized as corporations from the beginning. And as the railroads grew and spread across the land, their suppliers and, increasingly, their freight customers became larger and also became corporations.

In the earliest days of independence, obtaining a corporate charter had required a specific act of a state legislature, with all the politics that involved. But beginning in the early nineteenth century, states began passing general incorporation statutes, allowing companies to obtain charters under certain circumstances automatically. Legislatures began surrendering the power to form corporations not for altruistic reasons, of course, but simply because it was no longer possible for them to handle the demand for corporate charters.

In the entire colonial period, there had been only 7 companies incorporated in the British North American colonies. But in just the last four years of the eighteenth century, 335 businesses incorporated in the new

United States. Between 1800 and 1860, the state of Pennsylvania alone incorporated more than 2,000.

In 1811 New York State became the first to pass a general incorporation statute, although it was originally restricted to companies seeking to manufacture particular items, such as anchors and linen goods. The types of businesses eligible to incorporate soon included all forms of transportation and nearly all forms of manufacturing and financial services as well, however.

The corporate form had numerous advantages over partnerships. Partnerships automatically terminated at the death of one of the partners, but corporations could live forever (although early ones were often limited to a term of years). And in a partnership, any partner can sign a contract, binding on all the partners, whereas a corporation could hire management to handle the business of the firm. Most important, a corporation can sue (and be sued) and buy, own, and sell property as an entity. That is why Chief Justice John Marshall described a corporation as "an artificial being," one that was "invisible, intangible, and existing only in contemplation of the law."

Corporations can also merge. Many of the original railroads were local affairs, aimed at solving particular transportation bottlenecks. They were often financed by people living in the neighborhood, who bought their stock, chose their management, and kept an eye on things. But these small railroads quickly merged into larger affairs as they sought efficiency and economies of scale. The New York Central, which originally ran from Buffalo to Albany, parallel to the Erie Canal, was formed in 1853 from the merger of nine local roads.

As the railroads became bigger, they also became more remote from their stockholders, who in turn became much more numerous. Many stockholders of the new, larger railroads were more interested in speculative profits on Wall Street than in the company itself. This often allowed management to run the company for its own benefit, as the managers of the Union Pacific had, rather than for the benefit of the stockholders. Devising ways to force corporate management to act as the fiduciaries they had become would fall largely to the private sector at this time.

———

AS THE NINETEENTH CENTURY began to draw to a close, an ancient problem in corporate affairs became more and more acute: accounting. As enterprises became larger and more complex, accountants began devising more and more tools to keep track of the money and to enable managers to see exactly where and how it was being spent (or misspent). The great corporate enterprises of the Gilded Age were only possible because of these new accounting tools. And this evolution of accountancy continues unabated today. Cash flow, for instance, is now regarded as one of the most important indicators of corporate prospects. But the very phrase *cash flow* was coined only in 1954.

As the distance between managers and owners increased, their interests in accounting diverged as well. The stockholders wanted timely information so they could evaluate the worth of their holdings and compare the results of their company with the competition to determine how good a job the management was doing. The managers, naturally, wanted to make the books, and thus themselves, look as good as possible. It was common in the nineteenth century for managers to part company with the truth and lurch over into fraud. It is hardly unknown today.

Many publicly traded companies did not release figures at all. When the New York Stock Exchange asked the Delaware, Lackawanna, and Western Railroad for information about its finances, it was told to mind its own business. The "Railroad makes no report," it curtly informed the exchange, and "publishes no statements."

Even when a railroad issued a report, it was likely to be, in the words of a contemporary, "a very blind document, and the average shareholder . . . generally gives up before he begins." The Erie Railway, because some of its many bond issues were backed by New York State, was required to file an annual report with Albany. But it could frame that report pretty much as it chose. The managers of the Erie—one of the most mismanaged railroads in history—had no hesitation in using very creative accounting indeed to hide their own shenanigans. Horace Greeley in 1870 harrumphed in the *New York Tribune* that if the new annual report of the Erie was accurate, then "Alaska has a tropical climate and strawberries in their season."

The weekly *Commercial and Financial Chronicle* put its finger on the problem and foretold its solution. "The one condition of success in such intrigues," it noted, "is secrecy. Secure to the public at large the opportunity of knowing all that a director can know of the value and prospects of his own stock, and the occupation of the 'speculative director' is gone. . . . The full balance sheet . . . showing the sources and amounts of its revenues, the disposition made of every dollar, the earnings of its property, the expenses of working, of supplies, of new construction, and of repairs, the amount and form of its debt, and the disposition made of all its funds, ought to be made up and published every quarter."

Wall Street took to this new idea immediately. After all, brokers and bankers needed to know the true worth of a company as much as the stockholders did. Henry Clews, a very influential broker in the post–Civil War era, led the push for both regular reports and for independent accountants to certify them as accurate. The big Wall Street banks, which were becoming ever more powerful, and the New York Stock Exchange increasingly required companies that needed capital or wanted to be listed on the exchange to conform to what would come to be called "generally accepted accounting principles" and to have their books certified by independent accountants.

Most accountants had been mere corporate employees before the last third of the nineteenth century. As late as 1884 only 81 self-employed accountants were listed in the city directories (the nineteenth-century equivalent of the phone book) of New York, Philadelphia, and Chicago. Just five years later there were 322. And these accountants were beginning to organize. In 1882 the Institute of Accountants and Bookkeepers was formed in New York and began issuing certificates to those who could pass a strict examination. In 1887 the American Association of Public Accountants came into being, the ancestor of today's main governing body of the profession.

In 1896 New York State passed legislation establishing the legal basis of this new profession, and, incidentally, using the phrase *certified public accountant* for the first time to designate those who met the criteria of the law. The legislation, and the phrase, were quickly copied by the other states. By the beginning of the First World War, the system was universal throughout the American capitalist economy.

The one major fiscal area where generally accepted accounting principles and independent accountants have remained rare is in government. Indeed, a century later, most state governments, while often requiring their creatures to adhere to GAAP, do not do so themselves. And the federal government—the largest fiscal entity on earth—still keeps its books in much the same way as it did in the nineteenth century. With no countervailing forces, such as the Wall Street banks and the Stock Exchange, to exert the needed pressure, the "managers" of government—the legislators, governors, and presidents—have been able to put their self-interests ahead of those of the "stockholders."

———

AS THE INDUSTRIAL CORPORATIONS grew and proliferated, their need for capital increased as well. Increasingly, it was supplied not by the British, who had been by far the most important suppliers of capital in the pre–Civil War era, but by Wall Street, through its rapidly growing investment banks. No one epitomized the new Wall Street power center more than its most important banker of this time, J. P. Morgan.

Morgan, unlike so many of the financial titans of the Gilded Age, was born rich, in Hartford, Connecticut, in 1836. His grandfather had been a founder of the Aetna Fire Insurance Company and had invested in railroads, steamboats, and real estate as Hartford became one of the wealthiest cities, per capita, in the country. His father, Junius Spencer Morgan, became a partner of the American banking firm of George Peabody and Company in London in the 1850s. He took over the firm on Peabody's retirement, renaming it J. S. Morgan and Company. It quickly evolved into one of the more important international banking houses.

Morgan traveled frequently in Europe as a child, visiting the great museums that had no equivalent in the United States at the time and acquiring a deep interest in art. Very well educated—Morgan attended the University of Göttingen in Germany when few Americans of his generation who were planning a career in business went to college at all—he spoke both French and German fluently.

Morgan thus grew up in a much wider world than most of his generation and was thoroughly familiar with international banking at its highest levels from childhood. Deeply influenced by his father and Peabody, he believed that personal integrity was central to success in banking. At the end of his life, he was asked at a congressional hearing, "Is not commercial credit based primarily upon money or property?"

"No, sir," replied Morgan. "The first thing is character."

"Before money or property?"

"Before money or property or anything else," Morgan insisted. "Money cannot buy it. . . . Because a man I do not trust could not get money from me on all the bonds in Christendom."

By the 1870s Morgan was well established on Wall Street and a partner in the respected firm of Drexel, Morgan and Company, located at the southeast corner of Wall and Broad Streets, where the Morgan Bank has been ever since. Morgan soon set himself apart from the average Wall Street banker.

In 1878 William H. Vanderbilt, who had inherited the bulk of his father's vast fortune, including no less than 87 percent of the New York

Central Railroad, wanted to diversity his holdings. But selling a large chunk of a company as prominent as the New York Central without depressing the price of the stock was no easy task.

Morgan took it on, however, and succeeded in selling 150,000 shares in London at the very good price of $120 a share. And he managed to do it so quietly that no notice was taken. It was regarded as a masterful job of underwriting and, more, it made Morgan a major power in American railroading as he held the proxies of the new English shareholders and represented them on the company board.

Morgan intended to use his new power to bring order to the railroad business, which was greatly in need of it. There were still many small railroads that continued to operate independently, although because of their size they were unable to benefit from economies of scale or to use the latest technology. And because most of the major trunk lines had been assembled out of many smaller lines, they often had bizarrely complex capital structures.

Further, railroads had often engaged in cutthroat competition that resulted in wasteful, sometimes ruinous, overbuilding. Even the mighty Pennsylvania Railroad and the New York Central had been at each other's throats. The Pennsylvania was building a line on the west side of the Hudson to compete with the Central, whose main tracks were on the east side of the river. The Central, in turn, was building a new line that paralleled the Pennsylvania's through that state's Allegheny Mountains.

Morgan won the agreement of both lines to negotiate and invited the company presidents to talks on board his majestic yacht *Corsair*. As the *Corsair* steamed up and down the Hudson River between New York and West Point, Morgan got the two lines to abandon their duplicating lines. (The tunnels that had been dug through the Pennsylvania mountains for the New York Central were abandoned, then, seventy years later, were incorporated into the Pennsylvania Turnpike.) Morgan's prestige soared after the *Corsair* agreement, and much profitable new business came to his firm as a result.

Although he was a banker, not a railroad man, he quickly became the most powerful force in the industry, as he reorganized the Baltimore and Ohio, the Chesapeake and Ohio, the Erie, and other major railroads in these years. With more rational corporate structures and routes, they became much more economically viable. Morgan put partners of J. P. Morgan and Company on their boards to see to it that the roads did not slip back into the bad old ways of the past. Altogether, fully one-third of the railroad trackage in the United States passed through reorganization and, usually, the control of Wall Street banks, in the last two decades of the century. What emerged was a mature industry.

———

THE EXPANSION OF AMERICAN RAILROADS after the Civil War was nothing less than extraordinary. With 30,626 miles of track in 1860, the United States already had a larger railroad system than any country in the world. But by 1870 it had 52,922 miles; in 1880, 93,262 miles; in 1890, 166,703 miles. In 1900 it had 193,346 miles, a more than sixfold increase in forty years. While the annual value of manufactured goods in the United States increased by seventeen times between 1865 and 1916, the annual freight ton-mileage of the railroads increased by thirty-five times.

By the turn of the century the railroads tightly knitted together an economy that was now fully national in scope, and nearly every town of any size was served by a railroad. The major cities were usually served by several. This new railroad net presented a new problem, however, one that had been wholly unanticipated. The reckoning of time had always been a local matter, set by local noon, which, at the latitude of New York, is about one minute later for each eleven miles one travels westward. Thus when it was noon in New York, it was 11:55 AM in Philadelphia, 11:47 in Washington, and 11:35 in Pittsburgh. The state of Illinois used twenty-seven different time zones; Wisconsin, thirty-eight.

When travel had been at the speed of a horse, this did not matter, but with railroads it did, as scheduling was a nightmare. Congress, fearing local reaction, had dithered for years about dealing with the problem, so the railroads acted on their own. In 1883 they established standard time, dividing the country into four time zones, Eastern, Central, Mountain, and Pacific, with noon an hour later in each. On November 18, 1883, the nation's railroads started operating on this standard time. And while there were complaints that the country should operate on "God's time— not Vanderbilt's," within a few years standard time had, indeed, become standard, and it was hard to imagine a world without it.

The ability of the railroads to impose something so fundamental as standard time indicates just how powerful they had become by the late nineteenth century. Railroading was by far the country's largest industry. It employed more than a million workers in 1900, out of a total population of nearly seventy-six million, earning an annual average of $567 each. Being run by human beings, the railroads, naturally, did not hesitate to exercise their market power to their own advantage.

The railroads had developed a trunk and branch system, with a main line connecting major cities and numerous spurs to smaller communities. Often there was stiff competition on these trunk lines. Between Chicago and New York, for instance, one could choose the New York Central, the Pennsylvania, or the Erie. On the trunk lines, the railroad alternated between competing fiercely for business, cutting rates to the bone, and forming cartels to divide the business and set prices. These cartels usually broke down quickly for lack of an enforcement mechanism.

But most of the branch lines were monopolies, and the railroads could—and most certainly did—increase their profits by charging customers on these branch lines more, sometimes much more.

The customers, not surprisingly, resented it. The Granger movement, the National Grange of the Patrons of Husbandry, which began in 1867 as a social and educational organization, soon evolved into a lobbying organization to represent farmers against the railroads. By 1875 there were twenty thousand local granges throughout the Midwest and eight hundred thousand members, most with an intense dislike of the local railroad.

A movement this big, of course, quickly got the attention of politicians in the state capitals, who started regulating the railroads, often setting up commissions to oversee the business. But these commissions proved ineffective, partly because the very well-heeled railroad lobbyists saw to it that they had very limited powers.

Often they were little more than a political cover for allowing the railroads to do as they pleased. In 1895 one critic of the California Railroad Commission wrote, "The curious fact remains that a body created sixteen years ago for the sole purpose of curbing a single railroad corporation [the Central Pacific] with a strong hand, was found to be uniformly, without a break, during all that period, its apologist and defender."

It was soon clear that individual states could do little to effectively regulate railroads that now sprawled over many states. And when the Supreme Court ruled in 1886 in *Wabash Railway v. Illinois* that states had no power over railroads that were carrying goods across a state line, the fight to regulate the railroads moved, of necessity, to Washington.

Congress was now confronted with the same political problem that had faced the state legislatures. They had tens of thousands of votes at stake on the side of the farmers, small merchants, and manufacturers, and hundreds of thousands of dollars at stake on the side of the railroads.

What resulted, after a year of intense political sausage making, was, in the words of one historian, "a bargain in which no one interest predominated except perhaps the legislators' interest in finally getting the conflict . . . off their backs and shifting it to a commission and the courts." A federal railroad commission, similar to the ones that had largely failed at the state level, was the only proposal under serious consideration.

Assuming their interests were properly considered, the railroads had no objections. Indeed, a vice president of the Pennsylvania Railroad said as early as 1884 that "a large majority of the railroads in the United

States would be delighted if a railroad commission . . . could make rates upon their traffic which would insure them six per cent dividends, and I have no doubt, with such a guarantee, they would be very glad to come under the direct supervision and operation of the national government."

What the railroads wanted, in other words, was a government-sponsored and enforced cartel in place of the many private ones that kept failing.

The bill that finally emerged, establishing the Interstate Commerce Commission, required that "all charges . . . shall be reasonable and just," but didn't define what that might mean. Further, it forbade different rates on trunk lines and branches and for short hauls and long hauls. But the commission had to use the courts to enforce any orders. The railroads had no trouble finding their way around the new rules when they wanted to. Even though the ICC was strengthened in the next decade, it quickly evolved into exactly what the railroads had wanted in the first place, a government-sponsored and enforced cartel. This is not surprising as most of its technical expertise had, necessarily, to be hired from the railroad industry itself, a problem that has plagued all government regulatory bodies since.

But whatever its inadequacies, the Interstate Commerce Commission marked a major turning point in the history of the American economy. For the first time, the federal government was attempting to regulate a portion of that economy.

More was quickly to follow, although, as with the Interstate Commerce Commission, the intent and the results were often not in accord, at least in the short term. In 1890, still pressured to curb the power of the railroads, Congress passed the Sherman Antitrust Act. Both houses of Congress that year were controlled by pro-business Republicans and a Republican, Benjamin Harrison, sat in the White House, but the Sherman Act passed with only a single dissenting vote and was signed into law.

The reason was that it was notably short on specifics. "Every contract, combination in the form of trust or otherwise," it stated, "or conspiracy, in restraint of trade or commerce among the several states, or with foreign nations, is hereby declared to be illegal." Further, it made it a crime to "monopolize or attempt to monopolize, or combine or conspire . . . to monopolize any part of the trade or commerce among the several states."

The pro-business Harrison, Cleveland, and McKinley administrations, and the deeply conservative Supreme Court of the 1890s, assured that the Sherman Act would be nearly a nullity in its early years. But, in the words of historian Charles Warren, the Sherman Antitrust Act helped awaken "Congress to the realization of the vast power wrapped up in the Commerce Clause," and an activist president in the next decade would use it to make the federal government a major factor in regulating the American economy.

So much so, in fact, that before his death in 1913, J. P. Morgan, who had been a leader in the movement to require corporations to use standardized bookkeeping and independent accountants to certify their financial reports, would complain that Theodore Roosevelt expected men to conduct business with "glass pockets." Like most people—in government quite as much as in business—he valued transparency in the affairs of others more than in his own.

Chapter Thirteen

WAS THERE EVER
SUCH A BUSINESS!

NOTHING SO EPITOMIZED THE ECONOMY of the late nineteenth-century Western world as steel. Its production became the measure of a country's industrial power, and its uses were almost without limit. Its influence in other sectors of the economy, such as railroads and real estate, was immense. But steel was hardly an invention of the time. Indeed, it has been around for at least three thousand years. What was new was the cost of producing it.

Pig iron, the first step in iron and steel production, is converted into bar iron by remelting it and mixing it with ground limestone to remove still more impurities. Cast iron is then created by pouring this into molds, producing such items as frying pans, cookstoves, and construction members. Cast iron was widely used in urban construction in the antebellum period, but it had serious drawbacks. Extremely strong in compression, cast iron makes excellent columns. But, because it is very brittle, it is weak in tension, making it unsuitable for beams. For them, wrought iron was needed.

Wrought iron is made by melting pig iron and stirring it repeatedly until it achieves a pasty consistency and most of the impurities have been volatilized. The laborers who worked these furnaces were known as puddlers and were both highly skilled and highly paid. After the metal is removed from the puddling furnace, it is subjected to pressure and rolled and folded over and over—in effect, it is kneaded like bread dough—until it develops the fibrous quality that makes wrought iron much less brittle than cast iron and thus moderately strong in tension. Wrought iron is quite soft compared to cast iron but it is also ductile, able to be drawn out and hammered into various shapes, just as copper can be.

Wrought iron, of course, was much more expensive to produce than cast iron but could be used for making beams, bridges, ships, and, most important to the nineteenth-century economy after 1830, railroad rails. The Industrial Revolution simply could not have moved into high gear without large quantities of wrought iron.

Steel, which is iron alloyed with just the right amount of carbon under suitable conditions, has the good qualities of both cast and wrought iron. It is extremely strong and hard, like cast iron, while it is also malleable and withstands shock like wrought iron. And it is far stronger in tension than either, and thus makes a superb building material.

But until the mid-nineteenth century, the only way to make steel was in small batches from wrought iron, mixing the iron with carbon and heating it for a period of days. Thus its use was limited to very high-value items such as sword blades, razors, and tools, where its ability to withstand shock and take and hold a sharp edge justified its high cost. At mid-century, roughly 250,000 tons of steel were being made by the old methods in Europe, and only about 10,000 tons in the United States.

Then, in 1856, an Englishman named Henry Bessemer (later Sir Henry) invented the Bessemer converter, which allowed steel to be made directly and quickly from pig iron. As so often happens in the history of technological development, the initial insight was the result of an accidental observation. Bessemer had developed a new type of artillery shell, but the cast-iron cannons of the day were not strong enough to handle it. He began experimenting in hopes of developing a stronger metal, and one day a gust of wind happened to hit some molten iron. The oxygen in the air, combining with the iron and carbon in the molten metal, raised the temperature of the metal and volatilized the impurities. Most of the carbon was driven off. What was left was steel.

Bessemer, realizing what had happened, immediately set about designing an industrial process that would duplicate what he had observed accidentally. His converter was a large vessel, about ten feet wide by twenty feet high, with trunnions so that its contents could be poured. It was made of steel and lined with firebrick. At the bottom, air could be blasted through holes in the firebrick into the "charge," as the mass of molten metal in the crucible was called, converting it to steel in a stupendous blast of flame and heat. With the Bessemer converter, ten to thirty tons of pig iron could be turned into steel every twelve to fifteen minutes in what is one of the most spectacular of all industrial processes.

The labor activist John A. Fitch wrote in 1910 that "there is a glamor about the making of steel. The very size of things—the immensity of the tools, the scale of production—grips the mind with an overwhelming sense of power. Blast furnaces, eighty, ninety, one hundred feet tall, gaunt and insatiable, are continually gaping to admit ton after ton of ore, fuel, and stone. Bessemer converters dazzle the eye with their leaping flames. Steel ingots at white heat, weighing thousands of pounds, are carried from place to place and tossed about like toys. . . . [C]ranes pick up steel rails or fifty-foot girders as jauntily as if their tons were ounces. These are the things that cast a spell over the visitor in these workshops of Vulcan."

One of the visitors to Henry Bessemer's steelworks in Sheffield, England, in 1872, was a young Scottish immigrant to America, Andrew Carnegie. He was mightily impressed—so impressed, in fact, that in the next thirty years he would ride the growing demand for steel to one of the greatest American fortunes.

Carnegie had been born in Dunfermline, a few miles northwest and across the Firth of Forth from Edinburgh, in 1835. His father was a hand weaver who owned his own loom, on which he made intricately patterned damask cloth. Dunfermline was a center of the damask trade, and skilled weavers such as William Carnegie could make a good living at it.

But the Industrial Revolution destroyed William Carnegie's livelihood. By the 1840s power looms could produce cloth such as damask much more cheaply than handlooms. While there had been 84,560 handloom weavers in Scotland in 1840, there would be only 25,000 ten years later. William Carnegie would not be one of them.

The elder Carnegie sank into despair, and his far tougher-minded wife took charge of the crisis. She had gotten a letter from her sister, who had immigrated to America, settling in Pittsburgh. "This country's far better for the working man," her sister wrote, "than the old one, & there is room enough & to spare, notwithstanding the thousands that flock to her borders." In 1847, when Andrew was twelve, the Carnegie family moved to Pittsburgh.

The Carnegies were in the first wave of one of the great movements of people in human history, known as the Atlantic migration. At first most of the immigrants came from the British Isles, especially Ireland after the onset of the Great Famine of the 1840s. Later Germany, Italy, and Eastern Europe provided immigrants in huge numbers, more than two million in 1900 alone.

In its size and significance the Atlantic migration was the equal of the barbarian movements in late classic times that helped bring the Roman Empire to an end. But while many of the barbarian tribes had been pushed by those behind them, the more than thirty million people who crossed the Atlantic to settle in America between 1820 and 1914 were largely pulled by the lure of economic opportunity.

Many, such as the land-starved Scandinavians who settled in the Upper Middle West, moved to rural areas and established farms. But most, at least at first, settled in the country's burgeoning cities, in the fast-spreading districts that came to be called slums (a word that came into use, in both Britain and America, about 1825). For the first time in American history, a substantial portion of the population was poor. But most of the new urban poor were not poor for long.

These slums, by modern standards, were terrible almost beyond imagination, with crime- and vermin-ridden, sunless apartments that often housed several people, sometimes several families, to a room and had only communal privies behind the buildings. In the 1900 census, when conditions in the slums had much improved from mid-century, one district in New York's Lower East Side had a population of more than fifty thousand but only about five hundred bathtubs.

Such housing, however, was no worse—and often better—than what the impoverished immigrants left behind in Europe, and as Mrs. Carnegie's sister—and millions like her—reported back home, the economic opportunities were far greater. The labor shortage so characteristic of the American economy since its earliest days had not abated. So the average stay for an immigrant family in the worst of the slums was less than fifteen years, before they were able to move to better housing in better neighborhoods and begin the climb into the American middle class.

The migration of people to the United States in search of economic opportunity has never ceased, although legal limits were placed on it beginning in the early 1920s. And this vast migration did far more than help provide the labor needed to power the American economy. It has given the United States the most ethnically diverse population of any country in the world. And because of that, it has provided the country with close personal connections with nearly every other country on the globe, an immense economic and political advantage.

The Carnegies moved into two rooms above a workshop that faced a muddy alleyway behind Mrs. Carnegie's sister's house in Allegheny City, a neighborhood of Pittsburgh. Mrs. Carnegie found work making shoes, and Mr. Carnegie worked in a cotton mill. Andrew got a job there as well, as a bobbin boy earning $1.20 a week for twelve-hour days, six days a week.

Needless to say, it didn't take the bright and ambitious Andrew Carnegie fifteen years to start up the ladder. By 1849 he had a job as a telegraph messenger boy, earning $2.50 a week. This gave him many opportunities to become familiar with Pittsburgh and its business establishment, and Carnegie made the most of them. Soon he was an operator, working the telegraph himself and able to interpret it by ear, writing down the messages directly. His salary was up to $25 a month.

In 1853, in a classic example of Louis Pasteur's dictum that chance favors the prepared mind, Thomas A. Scott, general superintendent of the Pennsylvania Railroad, a frequent visitor to the telegraph office where Carnegie worked, needed a telegraph operator of his own to help with the system being installed by the railroad. He chose Carnegie, not yet eighteen years old. By the time Carnegie was thirty-three, in 1868, he had an annual income of $50,000, thanks to the tutelage of Thomas Scott and numerous shrewd investments in railway sleeping cars, oil, telegraph lines, and iron manufacturing. But after his visit to Bessemer's works in Sheffield, he decided to concentrate on steel.

———

IT HAD BEEN PURE CHANCE that had brought the Carnegie family to Pittsburgh, but its comparative advantages would make it the center of the American steel industry.

Set where the Allegheny and Monongahela rivers join to form the Ohio and provide easy transportation over a wide area, Pittsburgh had been founded, as so many cities west of the mountains were, as a trading post. Shortly after the Revolution, Pittsburgh began to exploit the abundant nearby sources of both iron ore and coal and specialize in manufacturing. While the rest of the country still relied on wood, coal became the dominant fuel in Pittsburgh, powering factories that were turning out glass, iron, and other energy-intensive products. As early as 1817, when the population was still only six thousand, there were 250 factories in

operation, and the nascent city, with already typical American booster-
ism, was calling itself the "Birmingham of America." Because of the
cheap coal, Pittsburgh exploited the steam engine long before it began to
displace water power elsewhere, and most of its factories were steam-
powered by 1830.

There was, however, a price to be paid for the cheap coal, which produces far more smoke than does wood. About 1820, when Pittsburgh was
still a relatively small town, a visitor wrote that the smoke formed "a cloud
which almost amounts to night and overspreads Pittsburgh with the
appearance of gloom and melancholy." By the 1860s even Anthony Trollope, London-born and no stranger to coal smoke, was impressed with the
pall. Looked down on from the surrounding hills, Trollope reported, some
of the tops of the churches could be seen, "But the city itself is buried in a
dense cloud. I was never more in love with smoke and dirt than when I
stood here and watched the darkness of night close in upon the floating
soot which hovered over the house-tops of the city." As the Industrial Revolution gathered strength, other American cities became polluted with
coal smoke and soot, but none so badly as Pittsburgh.

The most important coal beds in the Pittsburgh area were those surrounding the town of Connellsville, about thirty miles southeast of the
city. What made Connellsville coal special was that it was nearly perfect
for converting into coke. Indeed it is the best coking coal in the world.

Coke is to coal exactly what charcoal is to wood: heated in the
absence of air to drive off the impurities, it becomes pure carbon and
burns at an even and easily adjusted temperature. And either charcoal or
coke is indispensable to iron and steel production. As the iron industry in
Pittsburgh grew, it turned more and more to coke, the production of
which was far more easily industrialized than was charcoal.

By the time Andrew Carnegie was moving into steel, Henry Clay
Frick, who had been born in West Overton, Pennsylvania, not far from
Connellsville, in 1849, was moving into coke. Like Carnegie, Frick was
a very hardheaded businessman and willing to take big risks for big

rewards. And like Carnegie, he was a millionaire by the time he was thirty. Unlike Carnegie, however, he had little concern with public opinion or the great social issues of the day. Carnegie always wanted to be loved and admired by society at large. Frick was perfectly willing to settle for its respect. Unlike Carnegie, he rarely granted newspaper interviews and never wrote articles for publication.

By the 1880s the Carnegie Steel Company and the H. C. Frick Company dominated their respective industries, and Carnegie was by far Frick's biggest customer. In late 1881, while Frick was on his honeymoon in New York, Carnegie, who loved surprises, suddenly proposed a merger of their companies at a family lunch one day. Frick, who had no inkling the proposal was coming, was stunned. So was Carnegie's ever-vigilant mother, now in her seventies. The silence that ensued was finally broken by what is perhaps the most famous instance of maternal concern in American business history.

"Ah, Andra," said Mrs. Carnegie in her broad Scots accent, "that's a very fine thing for Mr. Freek. But what do we get oout of it?"

Needless to say, Carnegie had calculated closely what he would get out of it. First, the Carnegie Steel Company would get guaranteed supplies of coke at the best possible price; second, he would get the surpassing executive skills of Henry Clay Frick; and third, he would further the vertical integration of the steel industry in general and his company in particular.

Vertical integration simply means bringing under one corporation's control part or all of the stream of production from raw materials to distribution. It had been going on since the dawn of the Industrial Revolution (Francis Cabot Lowell had been the first to integrate spinning and weaving in a single building) but greatly accelerated in the last quarter of the nineteenth century as industrialists sought economies of scale as well as of speed to cut costs.

Carnegie and Frick shared a simple management philosophy: (1) Innovate constantly and invest heavily in the latest equipment and techniques to drive down operating costs. (2) Always be the low-cost producer so as to remain profitable in bad economic times. (3) Retain most of the profits in good times to take advantage of opportunities in bad times as less efficient competitors fail.

One such opportunity arose in 1889, by which time Frick was chairman of the Carnegie steel companies (Carnegie himself never held an executive position in the companies he controlled, but as the holder of a comfortable majority of the stock, he was always the man in charge). That year Frick snapped up the troubled Duquesne Steel Works, paying for it with $1 million in Carnegie company bonds due to mature in five years. By the time the bonds were paid off, the plant had paid for itself five times over.

Much of the technological advances that Carnegie was so quick to use came from Europe's older and more established steel industries, just as, nearly a century earlier, the American cloth industry had piggybacked on Britain's technological lead. As one of Carnegie's principal lieutenants, Captain W. M. Jones, explained to the British Iron and Steel Institute as early as 1881, "While your metallurgists as well as those of France and Germany, have been devoting their time and talents to the discovery of new processes, we have swallowed the information so generously tendered through the printed reports of the Institute, and we have selfishly devoted ourselves to beating you in output."

And beat them they did. In 1867 only 1,643 tons of Bessemer steel was produced in the United States. Thirty years later, in 1897, the tonnage produced was 7,156,957, more than Britain and Germany combined. By the turn of the century the Carnegie Steel Company alone would outproduce Britain. It would also be immensely profitable. In 1899 the Carnegie Steel Company, the low-cost producer in the prosperous and heavily protected American market, made $21 million in profit. The following year profits doubled. No wonder Andrew Carnegie exclaimed at one point, "Was there ever such a business!"

And steel was also transforming the American urban landscape. When stone was the principal construction material of large buildings, they could not rise much above six stories, even after the elevator was perfected in the 1850s, because of the necessary thickness of the walls. It was church steeples that rose above their neighbors and punctuated the urban skyline. But as the price of steel declined steadily as the industry's efficiency rose—by the 1880s the far longer-lasting steel railroad rails cost less than the old wrought-iron rails—more and more buildings were built with steel skeletons and could soar to the sky. Between the 1880s and 1913 the record height for buildings was broken as often as every year as "skyscrapers" came to dominate American urban skylines in an awesome display of the power of steel.

CAPITAL AND LABOR are equally necessary to industrial production on a grand scale, such as that of steel, for neither can create wealth without the other. The problem has always been one of deciding how to allocate the wealth created between them. Before the Industrial Revolution, capital and labor lived on intimate terms, often within the same family. Much of the labor was supplied by apprentices, who earned little beyond room and board, but acquired skills they could exploit later as adults.

What labor was hired in the open market was often highly skilled and could command good wages. That, of course, did not mean that there were no disagreements. The earliest recorded strike in what is now the United States took place in 1768 when journeymen tailors walked out in New York City. In 1798 the first union, the Federal Society of Journeyman Cordwainers, was formed in Philadelphia. (*Journeymen,* as the word implies, are those who are paid by the day; *cordwainer* is a now obsolete term for a shoemaker.) By the 1820s unions of these types were joining together in tradesmen associations, representing all, or many, skilled workers in a given area. By the 1850s, with the formation of the International Typographical Union that represented printers in both the United States and Canada, labor organizations of national and even international scope began to emerge.

But as the factory system spread and the division of labor became more finely broken down, more and more unskilled workers sought jobs in factories, first in such industries as spinning and cloth weaving and later in steel and other heavy industries as they exploded in size after the Civil War. Unlike the skilled workers, such as the puddlers, these men (and women in the cloth industry) had little individual bargaining power.

This has long been the heart of the problem in the eternal tug-of-war between capital and labor in the American economy. Capital speaks with one voice, either because the company is dominated by one person, such as Andrew Carnegie, or the stockholders hire management to speak for them. But labor, at first, was fragmented, and individual laborers often had no choice but to take what was offered.

Even after the craft labor unions formed, they did little to help the unskilled workers. The skilled workers looked down on the unskilled, many of whom were recent immigrants, seeing them not as allies against management but as a burden, likely to bring down their own wages. This produced a deep split in the ranks of labor that would not be fully healed until the 1950s.

There were several attempts to form organizations that would represent all workers, such as the National Labor Union, formed in 1866, and, most famously, the Knights of Labor, organized three years later, which had more than seven hundred thousand members by the mid-1880s. It accepted workers of all skill levels and even admitted black workers, although they were in separate locals. The Industrial Workers of the World, the so-called Wobblies, formed in 1905, never numbered its membership in more than tens of thousands, but had great influence, thanks to its innovative tactics and bold proposals.

Unfortunately, many of these wide-based labor organizations had other agendas besides improving the wages and working conditions of their members. They increasingly adhered to socialist ideas imported from Europe that, perhaps not surprisingly, found little support among the population of a nation that had been founded and built by generations of individualists bent on their own economic advancement.

Socialism, in all its many forms, is based on class and the idea that the various social classes are fixed, and therefore the members of each class have economic interests that are in common and opposed to the other classes. But the so-called classes in democratic countries are, in fact, nothing more than lines drawn by intellectuals across what are, in the real world, economic continua. For generations now, more than 90 percent of Americans have defined themselves as "middle class."

And no country in history has developed a social structure more rewarding of individual economic success than the United States. Ward McAllister, the self-appointed arbiter of New York society in the Gilded Age—he coined the phrase "the four hundred"—described that group's membership. It consisted, he wrote, of those, "who are now *prominently* to the front, who have the means to maintain their position, either by gold, brains or beauty, gold being always the most potent 'open sesame,' beauty the next in importance, while brains and ancestors count for very little." No wonder so many intellectuals have been chronically disaffected with American society.

Andrew Carnegie himself, of course, was the ultimate example of what millions of other immigrants hoped that they, or their children, might become. Another immigrant of Carnegie's generation followed a different path to immortality. Born in 1850, Samuel Gompers, like Carnegie, was born British and poor. His parents were Jewish, and his father worked as a cigar maker and was active in the union and socialist movements there. In 1863 the family immigrated to New York, where Gompers soon began working as a cigar maker himself and quickly became involved in union matters. By the 1880s he was heading the Cigar Makers International Union.

Gompers believed in organizing unions on craft lines, giving primacy to national organizations, rather than locals, and to reaching labor's goals by economic action—strikes, boycotts, picketing, and so on—not political action. He was convinced that "pure and simple" unionism would lead to success and to "industrial emancipation." Gompers was a socialist, in theory, but realized that the only way to achieve that distant goal was to see that labor became strong enough to bargain with management as an equal first.

In 1886 he took the cigar workers out of the Knights of Labor and formed the American Federation of Labor, an organization of craft unions. He would be president of the AFL for the rest of his life (except for the year 1895) and would be the most famous labor leader in the country. By 1900 about 10 percent of American workers were members of unions, a larger percentage than is found in the private sector today.

Because there were almost no rules dealing with the inevitable conflict between labor and management, there were bound to be clashes that turned violent. And because management was in a far better position to influence government at this time, government almost always came down on the side of the companies in a crisis. In 1877, at the bottom of the depression of the 1870s, management of most of the eastern railroads coordinated to cut pay by 10 percent for all workers, suddenly and without warning. Workers on the Baltimore and Ohio struck, seized possession of the rail yards, and refused to allow freight trains to leave.

The strike quickly spread to other railroads, including the other three major trunk lines that linked the East Coast with the Middle West. When the governor of Pennsylvania called out the state militia, it dispersed the strikers on the Pennsylvania Railroad in Pittsburgh, killing twenty-six of them. An enraged mob then forced the militia to seek refuge in a roundhouse at the Pennsylvania yard and set it afire. The militia managed to fight their way out but then left the city of Pittsburgh in the hands of the strikers and looters, who destroyed more than $5 million in railroad property. President Hayes felt he had no choice but to dispatch regular army troops to restore civil order.

Because of such violence, and, of course, because of their own self-interest, many of the more established citizens feared unions, regarding labor leaders, many of whom were immigrants such as Gompers, as foreigners with dangerous, un-American ideas. Andrew Carnegie, however, often championed the rights of workers in the frequent articles he wrote and published. But these articles dealt with abstractions of social and economic policy. Where Carnegie's personal interests were involved, Carnegie had no hesitation to oppose and break unions at his own plants, although he typically left the dirty work to others.

In 1889 a strike in the face of great demand for steel in a booming economy and a panicky Carnegie executive in charge of negotiations had resulted in a contract highly favorable to the union that represented workers at the Homestead Works outside Pittsburgh. Carnegie was determined that when the contract came up for renewal in 1892, the union be broken. Always concerned with his reputation, he gave Henry Clay Frick carte blanche to do what was necessary and then left for Scotland.

Frick built an eleven-foot-high fence, three miles long, around the entire plant and equipped it with watch towers, searchlights, and barbed wire. It was immediately dubbed "Fort Frick." He also arranged with the Pinkerton Detective Agency to supply three hundred men to defend the plant in the event of a lockout.

When the union rejected Frick's offer (as Frick had hoped and intended it would), he announced that the plant would deal only with individual workers, not the union, and began shutting down the plant. The workers struck. He tried to sneak the Pinkertons in on barges towed up the Monongahela River into the plant, but they were spotted by workers, who immediately broke through the fence and seized the plant. (So much for Fort Frick.) A battle raged all day as the Pinkertons tried to land, with casualties on both sides. Finally a truce was negotiated, allowing the Pinkertons to withdraw. Three of them were killed during the withdrawal, however, and the governor of Pennsylvania sent in six thousand militia to restore order. Under their protection, Frick was able to hire nonunion workers.

A blaze of publicity more or less destroyed Carnegie's carefully cultivated friend-of-the-working-man reputation. But when Frick was attacked a few days later in his office by an assassin named Alexander Berkman, he earned both the sympathy and even admiration of the country. Despite two bullet wounds to the neck and three stab wounds, Frick fought back ferociously and managed to subdue his attacker with the help of his office assistants. He then refused anesthetic as the doctor probed for the bullets and insisted on finishing his day's work.

The assassin was entirely unconnected with the labor dispute, but was, inevitably, associated with it, and public sympathy for the union drained away. "It would seem," said Hugh O'Donnell, the strike's leader, "that the bullet from Berkman's pistol, failing in its foul intent, went straight through the heart of the Homestead strike." By November the strike was over and the company had won an unequivocal victory in economic terms.

Violence in labor disputes abated toward the end of the nineteenth century, but it would be another generation, until the 1930s, before the interests of labor would be fully protected by government action and labor could bargain with management on something like equal terms.

WHILE THE LATE-NINETEENTH-CENTURY American economy was increasingly built by and with steel, it was increasingly fueled by oil. In 1859, the year Edwin Drake drilled the first well, American production amounted to only 2,000 barrels. Ten years later it was 4.25 million and by 1900, American production would be nearly 60 million barrels. But while production rose steadily, the price of oil was chaotic, sinking as low as 10 cents a barrel—far below the cost of the barrel itself—and soaring as high as $13.75 during the 1860s. One reason for this was the vast number of refineries then in existence. Cleveland alone had more than thirty, many of them nickel-and-dime, ramshackle operations.

Many people, while happy to exploit the new oil business, were unwilling to make large financial commitments to it for fear that the oil would suddenly dry up. The field in northwestern Pennsylvania was very nearly the only one in the world until the 1870s, when the Baku field in what was then southern Russia opened up. There would be no major new field in the United States until the fabulous Spindletop field in Texas was first tapped in 1902.

But a firm named Rockefeller, Flagler, and Andrews, formed to exploit the burgeoning market for petroleum products, especially kerosene, took the gamble of building top-quality refineries. Like Carnegie, it intended to exploit being the low-cost producer, with all the advantages of that position. The firm also began buying up other refineries as the opportunity presented itself.

The firm realized that there was no controlling the price of crude oil but that it could control, at least partly, another important input into the price of petroleum products: transportation. It began negotiating aggressively with the railroads to give the firm rebates in return for guaranteeing high levels of traffic. It was this arrangement that often allowed the firm to undersell its competitors and still make handsome profits, further strengthening the firm's already formidable competitive position.

In 1870 one of the partners, Henry Flagler, convinced the others to change the firm from a partnership to a corporation, which would make it easier for the partners to continue to raise capital to finance their relentless expansion while retaining control. The new corporation, named Standard Oil, was capitalized at $1 million and owned at that time about 10 percent of the country's oil refining capacity. By 1880 it would control 80 percent of a much larger industry.

The expansion of Standard Oil became one of the iconic stories of late-nineteenth-century America, as its stockholders became rich beyond imagination and its influence in the American economy spread ever wider. Indeed, the media reaction to Standard Oil and John D. Rockefeller in the Gilded Age is strikingly similar to the reaction to the triumph of Microsoft and Bill Gates a hundred years later. It is perhaps a coincidence that Rockefeller and Gates were just about the same age, their early forties, when they became household names and the living symbols of a new and, to some, threatening economic structure.

The image of Standard Oil that remains even today in the American folk memory was the product of a number of writers and editorial cartoonists who often had a political agenda to advance first and foremost. The most brilliant of these was Ida Tarbell, whose *History of the Standard Oil Company,* first published in *McClure's* magazine in 1902, vividly depicted a company ruthlessly expanding over the corporate bodies of its competitors, whose assets it gobbled up as it went.

That is by no means a wholly false picture, but it is a somewhat misleading one. For one thing, as the grip of Standard Oil relentlessly tightened on the oil industry, prices for petroleum products *declined* steadily, dropping by two-thirds over the course of the last three decades of the nineteenth century. It is simply a myth that monopolies will raise prices once they have the power to do so. Monopolies, like everyone else, want to maximize their profits, not their prices. Lower prices, which increase demand, and increased efficiency, which cuts costs, is usually the best way to achieve the highest possible profits. What makes monopolies (and most of them today are government agencies, from motor vehicle bureaus to public schools) so economically evil is the fact that, without competitive pressure, they become highly risk-aversive—and therefore shy away from innovation—and notably indifferent to their customers' convenience.

Further, Standard Oil used its position as the country's largest refiner not only to extract the largest rebates from the railroads but also to induce them to deny rebates to refiners that Standard Oil wanted to acquire. It even sometimes forced railroads to give it secret rebates not only on its own oil, but on that shipped by its competitors as well, essentially a tax on competing with Standard Oil. (This is about as close as the "robber barons" ever came to behaving like, well, robber barons.) It thus effectively presented these refiners with Hobson's choice: they could agree to be acquired, at a price set by Standard Oil, or they could be driven into bankruptcy by high transportation costs.

The acquisition price set, however, was a fair one, arrived at by a formula developed by Henry Flagler, and consistently applied. Sometimes, especially if the owners of the refinery being acquired had executive talents that Standard wished to make use of, the price was a generous one. Further, the seller had the choice of receiving cash or Standard Oil stock. Those who chose the latter—and there were hundreds—became millionaires as they rode the stock of the Standard Oil Company to capitalist glory. Those who took the cash often ended up whining to Ida Tarbell.

None of this, of course, was illegal, and that was the real problem. In the late nineteenth century people such as Rockefeller, Flagler, Carnegie, and J. P. Morgan were creating at a breathtaking pace the modern corporate economy, and thus a wholly new economic universe. They were moving far faster than society could fashion, through the usually slow-moving political process, the rules needed to govern that new universe wisely and fairly. But that must always be the case in democratic capitalism, as individuals can always act far faster than can society as a whole. Until the rules were written—largely in the first decades of the twentieth century—it was a matter of (in the words of Sir Walter Scott)

> *The good old rule, the simple plan*
> *That they should take who have the power*
> *And they should keep who can.*

Part of the problem is that there is a large, inherent inertia in any political system, and democracy is no exception. Politicians, after all, are in the reelection business, and it is often easier to do nothing than to offend one group or another. So while the American economy had changed profoundly since the mid-nineteenth century, the state incorporation laws, for instance, had not. As an Ohio corporation, Standard Oil was not allowed to own property in other states or to hold the stock of other corporations. As it quickly expanded throughout the Northeast, the country, and then across the globe, however, Standard Oil necessarily acquired property in other states and purchased other corporations.

The incorporation laws, largely written in an era before the railroads and telegraph had made a national economy possible, were no longer adequate to meet the needs of the new economy. To get around the outdated law, Henry Flager, as secretary of Standard Oil, had himself appointed as trustee to hold the property or stock that Standard Oil itself could not legally own. By the end of the 1870s, however, Standard owned dozens of properties and companies in other states, each, in theory, held by a trustee who was in some cases Flagler and in other cases other people. It was a hopelessly unwieldy corporate structure.

In all probability, it was Flagler—a superb executive—who found the solution. Instead of each subsidiary company having a single trustee, with these trustees scattered throughout the Standard Oil empire, the same three men, all at the Cleveland headquarters, were appointed trustees for all the subsidiary companies. In theory, they controlled all of Standard Oil's assets outside Ohio. In fact, of course, they did exactly what they were told.

Thus was born the business trust, a form that was quickly imitated by other companies that were becoming national in scope. The "trusts" would be one of the great bogeymen of American politics for the next hundred years, but, ironically, the actual trust form of organization devised by Henry Flagler lasted only until 1889. That year New Jersey—seeking a source of new tax revenue—became the first state to modernize its incorporation laws and bring them into conformity with the new economic realities. New Jersey now permitted holding companies and interstate activities, and companies flocked to incorporate there, as, later, they would flock to Delaware, to enjoy the benefits of a corporation-friendly legal climate. Standard Oil of New Jersey quickly became the center of the Rockefeller interests, and the Standard Oil Trust, in the legal sense, disappeared.

———

WITH THE GROWTH of American industry, the nature of American foreign trade changed drastically. The United States remained, as it remains today, a formidable exporter of agricultural and mineral products. Two new ones were even added in the post–Civil War era: petroleum and copper. But it also became a major exporter of manufactured goods that it had previously imported. In 1865 they had constituted only 22.78 percent of American exports. By the turn of the twentieth century they were 31.65 percent of a vastly larger trade. The percentage of world trade, meanwhile, that was American in origin doubled in these years to about 12 percent of total trade.

Nowhere was this more noticeable than in iron and steel products, the cutting edge of late-nineteenth-century technology. Before the Civil War the United States exported only $6 million worth of iron and steel manufactures a year. In 1900 it exported $121,914,000 worth of locomotives, engines, rails, electrical machinery, wire, pipes, metalworking machinery, boilers, and other goods. Even sewing machines and typewriters were being exported in quantity.

Europe had long imported raw materials from the United States and elsewhere and exported finished goods to America and the rest of the world. To alarmist economic commentators—all too often a redundancy then as now—it seemed that an American colossus had suddenly appeared to snatch this profitable trade away, threatening to reduce once-mighty Europe to an economic backwater. Books with such ominous titles as *The American Invaders, The Americanization of the World,* and *The "American Commercial Invasion" of Europe* began to fill the bookstores of the Old World in the 1890s.

The new American economy also created enormous new personal fortunes, of an order of magnitude quite undreamed of before. Indeed, nothing has so consistently characterized the American economy throughout its history as the tendency of new fortunes to supplant old ones. When John Jacob Astor died the richest man in America in 1848, he left $25 million. Commodore Vanderbilt left $105 million less than thirty years later. Andrew Carnegie sold out in 1901 for $480 million. Fifteen years later, John D. Rockefeller was worth $2 billion.

Mark Twain noted the trend as early as 1867 when he reported that New York's old "Knickerbocker aristocracy," "find themselves supplanted by upstart princes of Shoddy, vulgar with unknown grandfathers. Their incomes, which were something for the common herd to gape and gossip about once, are mere livelihoods now—would not pay Shoddy's house rent." That has not changed. Except Rockefeller and Hearst, not a single name that was legendary for wealth in the Gilded Age—Twain's "princes of Shoddy"—is to be found today on the Forbes Four Hundred list. And the Rockefeller fortune, vast as it remains, is barely a tenth of the fortune that Bill Gates has created in just the last twenty years.

This country has never developed an aristocracy, because the concept of primogeniture, with the eldest son inheriting the bulk of the fortune, never took hold. Thus great fortunes have always been quickly dispersed among heirs in only a few generations. The American super rich are therefore always nouveau riche and often act accordingly, giving new meaning in each generation to the phrase *conspicuous consumption.* In the Gilded Age, they married European titles, built vast summer cottages and winter retreats that cost millions but were occupied only a few weeks a year.

Their main residences were equally lavish. And while every American town and city had its millionaires' row, where the banker and the factory owners lived, nothing could compare with what New York, the country's richest, most money-minded city, produced. There, by the turn of the twentieth century, a parade of mansions, each larger and grander than the next, ran for nearly three miles up Fifth Avenue, from the Forties to the Nineties. It was one of the wonders of the age that created it and attracted visitors from around the world to gawk at this symbol of America's unlimited wealth. Today, like the fortunes that created them, all but a few of the houses have vanished. Those that remain are consulates, schools, and museums.

What have not vanished are the public monuments the rich also built both to glorify their names and to justify their fortunes. The giving of vast sums to eleemosynary institutions by the very rich is a uniquely American practice; the European upper classes have no such tradition. It began early in the nineteenth century with people like George Peabody (the Peabody Museums at Harvard and Yale, among much else), Peter Cooper (the Cooper Union, still the only major college in the United States to charge no tuition), and John Jacob Astor, whose Astor Library is today the core of the New York Public Library, the second largest library in the country and the largest privately financed library in the world.

As the nineteenth century began to wane, the people who were building great fortunes began to found or endow museums, concert halls, orchestras, colleges, hospitals, and libraries in astonishing numbers in every major city. Carnegie had written that "a man who dies rich, dies disgraced," and gave away nearly his entire fortune, building more than five thousand town libraries among numerous other beneficences. Henry Clay Frick gave his incomparable art collection to the city of New York, along with his Fifth Avenue mansion to house it and $15 million to maintain it. John D. Rockefeller, who as a committed Baptist had always tithed even before he became the richest man in the world, donated millions almost beyond counting to worthy causes across the country.

J. P. Morgan's art collection, the largest ever in private hands, is now mostly in the Metropolitan Museum, the Wadsworth Atheneum in Hartford, and the Morgan Library, which also houses one of the world's great collections of manuscripts and rare books.

The United States in its early days had been a cultural backwater, and artists and writers routinely went to Europe to study. By the turn of the twentieth century, the United States was as great a cultural and intellectual power as it was an economic one, largely thanks to the often poorly educated men who are remembered today as the robber barons.

———

THE INDUSTRIAL EMPIRES that were created by the robber barons appeared more and more threatening in their economic power as they merged into ever-larger companies. In the latter half of the 1890s, this trend toward consolidation accelerated. In 1897 there were 69 corporate mergers; in 1898 there were 303; the next year 1,208. Of the seventy-three "trusts" with capitalization of more than $10 million in 1900, two-thirds had been created in the previous three years.

In 1901 J. P. Morgan created the largest company of all, U.S. Steel, merging Andrew Carnegie's empire with several other steel companies to form a new company capitalized at $1.4 billion. The revenues of the federal government that year were a mere $586 million. The sheer size of the enterprise stunned the world. Even the *Wall Street Journal* confessed to "uneasiness over the magnitude of the affair," and wondered if the new corporation would mark "the high tide of industrial capitalism." A joke made the rounds where a teacher asks a little boy about who made the world. "God made the world in 4004 B.C.," he replied, "and it was reorganized in 1901 by J. P. Morgan."

But when Theodore Roosevelt entered the White House in September 1901, the laissez-faire attitude of the federal government began to change. In 1904 the government announced that it would sue under the Sherman Antitrust Act—long thought a dead letter—to break up a new

Morgan consolidation, the Northern Securities Corporation. Morgan hurried to Washington to get the matter straightened out.

"If we have done anything wrong," Morgan told the president, fully encapsulating his idea of how the commercial world should work, "send your man to my man and they can fix it up."

"That can't be done," Roosevelt replied.

"We don't want to fix it up," his attorney general, Philander Knox, explained. "We want to stop it."

From that point on, the federal government would be an active referee in the marketplace, trying—not always successfully, to be sure—to balance the needs of efficiency and economies of scale against the threat of overweening power in organizations that owed allegiance only to their stockholders, not to society as a whole.

In 1907 the federal government took on the biggest "trust" of all, Standard Oil. The case reached the Supreme Court in 1910 and was decided the following year, when the Court ruled unanimously that Standard Oil was a combination in restraint of trade. It ordered Standard Oil broken up into more than thirty separate companies.

The liberal wing of American politics hailed the decision, needless to say, but in one of the great ironies of American economic history, the effect of the ruling on the greatest fortune in the world was only to increase it. In the two years after the breakup of Standard Oil, the stock in the successor companies doubled in value, making John D. Rockefeller twice as rich as he had been before.

Chapter Fourteen

A CROSS OF GOLD

B ECAUSE THE UNITED STATES was a highly industrialized nation by the 1890s, the depression that began in 1893 brought unparalleled economic suffering to the American population. In 1860 there had been four farm workers for every factory worker, but by 1890 the ratio had dropped to two to one. That meant that one American family in three was now dependent on a regular paycheck for food, shelter, and clothing.

In the spring of that year the Philadelphia and Reading Railroad and the National Cordage Company—the so-called Rope Trust—both unexpectedly declared themselves insolvent, and panic swept Wall Street. The economic carnage quickly spread through the rest of the economy. By the end of that year some fifteen thousand companies had failed, along with 491 banks. The gross national product fell by 12 percent, and unemployment rose rapidly from a mere 3 percent in 1892 to 18.4 percent two years later.

Until the 1870s the very word *unemployed* had applied to anyone without an occupation, from five-year-old children to housewives to people who lived off the income of their investments. But in 1878, as the depression of that decade was ending, a Massachusetts survey redefined the unemployed to mean men over eighteen who were "out of work and seeking it." By the mid-1890s they numbered in the millions, and hunger stalked the streets of the now vast slum districts of American cities, while only private charities were available to alleviate the misery.

The immediate cause of the new depression—as in most previous ones in this country—had been overexpansion due to the lack of a central bank to tap the brakes when needed. But there was also an underlying cause: a monetary policy that tried to do two incompatible things at the same time.

———

THE YEAR FOLLOWING THE GOLD PANIC of 1869, the House of Representatives had held hearings on the matter. Chaired by Representative James A. Garfield, the hearings first brought the Ohio congressman to national prominence. Garfield, the soundest of sound-money men, wrote in the report the committee issued that "So long as we have two standards of value recognized by law, which may be made to vary in respect to each other by artificial means, so long will speculation in the price of gold offer temptations too great to be resisted."

Garfield, in other words, wanted to return to the gold standard and retire the greenbacks. The merchants dealing in international trade, many of whom had seen ruin staring them in the face on Black Friday, also wanted to return to the gold standard. So did Wall Street's increasingly powerful banks and those who were involved in the country's heavy industry. These were, of course, the people who by then dominated the Republican Party. But there were many more who opposed a return to the gold standard.

The gold standard has one big advantage as a monetary system: it makes inflation nearly impossible. If a country increases its paper money supply beyond what the market will bear, holders of banknotes will begin turning them in for gold, and gold will move abroad as foreign central banks begin not to trust the currency and cash it in for what they do trust: gold.

But inflation is always popular with debtors because it allows them to repay their debts in cheaper money. And in areas like the devastated South, whose banking assets and other liquid wealth had been largely destroyed in the war, a monetary system based on the gold standard meant continued depression, whereas "easy money" would have helped revival. In fact, the late nineteenth century was marked by a very slow but steady deflation.

As a result of its effects, the gold standard was popular in the Northeast, where finance, foreign trade, and industry were centered, but unpopular among small farmers on the frontier and in the South. There, much of the population regarded the gold standard as nothing more than a plot by "Wall Street" to drive them into bankruptcy. In 1876 the Greenback Labor Party nominated a candidate for president (the aged Peter Cooper of New York, paradoxically one of the richest men in the country; he was, perhaps, the very first "limousine liberal"). In 1878 the party attracted 1,060,000 votes in congressional elections, enough to elect fourteen congressmen.

Although the government had stopped printing greenbacks (other than to replace those worn out by use) at the end of the Civil War, it minted silver dollars out of the rapidly increasing amounts of silver being mined in the West, giving the country a "bimetallic standard." Then, in 1873, it stopped minting these dollars as Congress voted to return to the gold standard by 1879. This was immediately labeled, by those who opposed the gold standard, "the crime of '73." Both sides of the issue put unrelenting pressure on Congress, which, as democratic legislatures usually do, especially when dealing with complex and abstruse economic issues, tried to have it both ways.

The country returned to the gold standard on schedule, on January 1, 1879, and the Treasury was required to maintain a gold reserve of $100 million to meet any demand for the precious metal. The previous year, Congress had voted to keep the $346,681,000 worth of greenbacks that were still in circulation, but made them redeemable in gold, as were the silver coins. Further, it had also passed the Bland-Allison Act. This required the Treasury to purchase between $2 million and $4 million a month of silver on the open market and turn it into coin at a ratio of sixteen to one with gold. In other words, Congress declared by fiat that sixteen ounces of silver were worth the same as one ounce of gold. This new silver coinage, of course, had the effect of considerably increasing the country's money supply, the classic recipe for inflation.

At first, sixteen to one was approximately the actual price ratio between gold and silver. But as the great silver strikes in the West, such in Coeur d'Alene, Idaho, and the fabulous Comstock Lode in Nevada, first discovered in 1859, came into production, the price of silver began to drop in the marketplace. By 1890 it had reached about twenty to one. That year, Congress passed the Sherman Silver Act, mandating that the Treasury buy 4.5 million ounces of silver a month, just about all the silver the country was producing, and coin it.

With the gold standard keeping the value of the dollar steady, while the silver policy greatly increased the money supply, the government managed, in effect, both to guarantee and to forbid inflation. Inevitably, Gresham's law kicked in. With silver worth one-twentieth the price of gold in the market but one-sixteenth the price when coined as money, people naturally spent the silver and kept the gold, which began to trickle out of the Treasury.

In the 1880s the government was running very large budget surpluses, which masked the schizophrenic monetary policy. But when the crash of 1893 marked the onset of a new depression, the trickle of gold out of the Treasury rose to a flood. As government revenues plunged—they declined from $386 million to $306 million between 1893 and 1894—Congress hastily repealed the Sherman Silver Act. But people and, more importantly, foreign governments had lost faith in the dollar, and the demand for gold at the Treasury increased markedly. The government issued bonds to buy more gold to replenish the gold reserve, but the metal continued to flow out.

Before long the situation was critical. The Treasury gold reserve dipped below the $100 million required by law in 1894 and was replenished by the proceeds of a $50 million bond issue in January of that year. But it fell to just $68 million the following January. A week later it was down to $45 million. Congress refused to allow President Cleveland, a staunch supporter of the gold standard, to sell another bond issue to replenish the vanishing gold reserve.

The government was paralyzed. Soon it was possible to almost literally watch gold flee the country as bullion worth millions of dollars was loaded on ships in New York and headed for European central banks. Bets were being made on Wall Street as to exactly when the Treasury would run out of gold and the country would be forced off the gold standard.

Badly alarmed, J. P. Morgan, by then the country's undisputed leading banker, took the train to Washington to see that that did not happen. President Cleveland, while personally a backer of sound money and the gold standard, was all too aware that he headed a party a large element of which both wanted the country to be rid of the gold standard altogether and hated "Wall Street" and all its works. He refused to see Morgan. But with the situation deteriorating by the hour, Cleveland decided the next morning that he had no choice but to hear what Morgan had to say.

The president still had hopes that he could persuade Congress to authorize a new bond issue, but that, of course, would take time. A clerk informed Cleveland that the subtreasury in New York had only $9 million in gold remaining in its vaults. Morgan said that he knew of drafts for $12 million that might be presented to the Treasury at any moment. If that happened, he warned, "It will be all over by three o'clock."

"Have you anything to suggest?" Cleveland, out of options, asked. Morgan did. Issuing more bonds in the domestic market, he argued, would do no good in the long term anyway as the gold would simply recycle back out of the Treasury. But he and August Belmont, Jr., the

American representative of the Rothschilds, who was also present at the White House that day, would raise $100 million in gold in Europe that would stem the run on the Treasury. More, Morgan's lawyers had uncovered an obscure Civil War–era law, still in force, that allowed the government to issue bonds with which to buy coin without further congressional action.

Astonishingly, Morgan was willing to guarantee that the gold would not flow back to Europe, at least in the short term. It was an extraordinary act of financial courage. But thanks to Morgan's already awesome reputation, and the use of sophisticated foreign exchange techniques, he was able to keep his word. In June 1895 the Treasury's gold reserve stood at $107.5 million. More important, economic recovery had begun. Morgan had saved the gold standard.

Morgan and Belmont, needless to say, were vilified by the enemies of the gold standard, and the Democratic Convention of 1896 was dominated by them. William Jennings Bryan, a former congressman from Nebraska, now editor in chief of the fiercely pro-silver *Omaha World-Herald,* gave the delegates the red meat they craved on the issue in one of the most famous speeches in American history.

Bryan assured the delegates at the outset that "The humblest citizen in all the land, when clad in the armor of a righteous cause, is stronger than all the hosts of error." His cause was the abolition of the gold standard, and he laid out in sonorous prose how it harmed the interests of the farmers and laborers and served only the interests of, in Thomas Carlisle's phrase, "the idle holders of idle capital."

That was the great question, he told the delegates. "Upon which side will the Democratic Party fight—upon the side of 'the idle holders of idle capital' or upon the side of 'the struggling masses'?"

As his magnificent voice reached effortlessly to every corner of the convention hall in Chicago, he held the audience in the palm of his hand as he reached the end. "Having behind us the producing masses of this nation and the world, supported by the commercial interests, the laboring interests, and the toilers everywhere, we will answer their demand for a gold standard by saying to them, You shall not press down upon the brow of labor this crown of thorns; you shall not crucify mankind upon a cross of gold."

The delegates went wild upon the conclusion of what the novelist Willa Cather—who was present—called "that never-to-be-forgotten speech." For half an hour pandemonium reigned, and in the end Bryan was nominated for president at the age of only thirty-six. He remains to this day by far the youngest man ever to be nominated by a major party.

William McKinley, the Republican nominee, campaigned from his porch in Canton, Ohio, giving speeches to crowds brought in by train. Bryan, on the other hand traveled indefatigably on the first whistle-stop campaign in American history. The line between the two parties could not have been sharper. The Populists, moreover, more radical than the Democrats, endorsed Bryan rather than fielding their own candidate.

"We have petitioned," Bryan had declared in his great speech, "and our petitions have been scorned; we have entreated, and our entreaties have been disregarded; we have begged and they have mocked when our calamity came. We beg no longer; we entreat no more; we petition no more. We defy them!"

Meanwhile a Republican newspaper editorialized that "The Jacobins are in full control in Chicago [where Bryan was nominated]. No large political movement in America has ever before spawned such hideous and repulsive vipers."

The candidates, as always, left the name calling to their supporters. But Bryan's economic ideas seriously alarmed even many Americans of modest means and many more with personal ambitions. Many eastern and midwestern Democrats, alarmed at what they regarded as Bryan's rabble-rousing, supported McKinley.

Early in the campaign, however, it looked like Bryan's platform might be a winning formula. The Dow-Jones Industrial Average, developed just that spring by Charles Dow, editor of the young *Wall Street*

Journal, to measure the stock market as a whole, slumped by a third over the course of the summer.

Then the economic recovery from the depression picked up steam as summer advanced, greatly helping the party that ran on the slogan of "Sound Money, Protection, and Prosperity." The Dow-Jones, a rough barometer of the political as well as financial weather, recovered as fall continued.

In November McKinley won with 52 percent of the vote, sweeping the wealthiest and most economically developed parts of the country: the entire Northeast and Middle West, along with the Upper Plains states, plus California and Oregon. Bryan took the South and the rest of the western states.

Many of the Democrats who had abandoned Bryan for the more conservative McKinley never came back. The Republicans would be the majority party for the next generation, losing the White House only when the GOP split in 1912.

But Bryan, while losing (he would lose again in 1900 and 1908, dragging a cross of gold through the political wilderness) identified clearly the future of American national politics. "The sympathies of the Democratic Party," he had told the delegates in his speech, ". . . are on the side of the struggling masses who have been the foundation of the Democratic Party. There are two ideas of government. There are those who believe that, if you will only legislate to make the well-to-do prosperous, their prosperity will leak through on those below. The Democratic idea, however, has been that if you legislate to make the masses prosperous, their prosperity will find its way up through every class which rests upon them."

The choice was clear, but the country, in fact, chose both approaches. American politics is the politics of the center, not the extremes, and this country is given to splitting differences or, when possible, having it both ways at once. Over the next hundred years, as political dominance was enjoyed by each party in turn, the country would employ both trickle-down and trickle-up economic policies. The result has been nearly wholly salutary, if, as politics in a democracy always is, philosophically untidy.

———

OTHER THAN THE GOLD STANDARD, no issue so involved the body politic in this peaceful era as that of taxation. The federal government had relied on the tariff as its principal source of revenue from the days of Alexander Hamilton until the Civil War forced it to tax everything that it could, including incomes.

With the wartime revenues no longer needed after victory, many of the new federal taxes were cut or eliminated. The tariff, however, was not. The greatly enlarged industrial base had grown nestled in the protective bosom of the tariff, and it fought fiercely to maintain it. Meanwhile, the longtime locus of opposition to a high tariff, the South, had lost all political influence until the end of Reconstruction in the late 1870s. By then the idea of high tariffs was Republican Party doctrine. No matter how mature and efficient American industry became—and by the turn of the century it was the most efficient in the world—the managers, stockholders, and workers of industrial companies all fought fiercely and successfully to maintain the tariff at a level far above the government's revenue needs.

As a result, the government ran a string of twenty-eight straight surpluses beginning in 1866, unparalleled in American history. In the prosperous year of 1882, government revenues ran ahead of outlays by a staggering 36 percent. By the turn of the century the huge Civil War debt had been cut by nearly two-thirds in absolute dollars, and as a percentage of GDP had dropped more steeply still, from about 50 percent to well under 10 percent.

The Civil War income tax had been cut in 1867 to a uniform 5 percent on incomes more than $1,000. Three years later the rate was reduced again, and in 1872 the tax was eliminated altogether. There had been little if any advocacy of an income tax before the Civil War, but all government programs, once established, develop a political constituency, and the income tax was no exception.

Its advocates had logically if not politically compelling arguments on their side. Indirect taxes, such as excise taxes and the tariff, are taxes on consumption and are thus inescapably regressive. That is to say they fall most heavily on those least able to afford them. The poor necessarily spend a far higher percentage of their incomes on necessities than do the rich and therefore pay a far higher percentage of their incomes in these consumption-based taxes.

Senator John Sherman, Republican of Ohio, and no radical by a long shot, explained it in the debate over eliminating the income tax in 1872. "Here we have in New York Mr. Astor," he said, "with an income of millions derived from real estate . . . and we have along side of him a poor man receiving a thousand dollars a year. What is the discrimination of the law in that case? It is altogether against the poor man—Everything that he consumes we tax, and yet we are afraid to tax the income of Mr. Astor. Is there any justice in it? Why, sir, the income tax is the only one that tends to equalize these burdens between the rich and the poor."

Sherman was right. But, as always with taxes, it was political power, not equity, that prevailed. The congressmen of seven northeastern states, which collectively paid about 70 percent of the income tax, voted sixty-one to fourteen to eliminate the tax, while fourteen southern and western states, which paid only 11 percent of the tax, voted five to sixty-one to retain it. In other words, support for the income tax was almost perfectly inversely correlated with its local impact. In a democracy, politicians will always try to follow the dictum usually credited to Senator Russell Long of Louisiana: "Don't tax you, and don't tax me, tax the man behind the tree."

The advocates of an income tax did not make much headway in the prosperous 1880s, but when the great depression of the 1890s struck, and federal revenues plummeted, there were renewed calls for an income tax. With a Democrat, Grover Cleveland, in the White House, and Democratic majorities in both houses of Congress, a new income tax became law in 1894.

It was a very different tax in its impact than the Civil War income tax. The latter had exempted only the poor. The new tax, which called for a 2 percent tax on all incomes more than $4,000, exempted all but the rich. Of the twelve million American households in 1894, only eighty-five thousand had incomes of $4,000 or more. That was well under 1 percent of all households. For the first time in American history, a tax was proposed on a particular class of citizens, one defined by economic success. For that reason, nearly all Republicans opposed it, including Senator Sherman. For that reason as well, Cleveland allowed the bill to become law without his signature.

Not surprisingly, a lawsuit was immediately brought. The plaintiffs argued that an income tax violated the clause in the Constitution that required that direct taxes be apportioned among the several states according to population, something that was obviously impossible with an income tax. The Constitution does not define what constitutes a "direct tax" (Rufus King had asked for a definition during debate in the Constitutional Convention and received no reply). In 1796 the Supreme Court had ruled that a direct tax was any tax that *could* be apportioned on the basis of population. As recently as 1881 the Court had ruled that the Civil War income tax was an indirect tax.

Regardless, with one member of the Court absent due to illness, the Court split four to four on the question of whether an income tax was a direct tax and thus whether it was constitutional. The case, *Pollack v. Farmers' Loan and Trust,* generated enormous interest around the country, far more than *Plessy v. Ferguson,* which upheld the separate-but-equal doctrine of segregation, did the next year. Because of the intense public attention, the Court agreed to a rehearing of the case, and Justice Howell Jackson, who was mortally ill—he died less than three months later— attended, undoubtedly intent on being the fifth vote to uphold the tax.

One of the other justices (it is not quite clear which, but most likely Justice George Shiras) switched sides, however, and the income tax was ruled unconstitutional, five to four. The eastern Republican establishment had prevailed, if just barely. But over the next few years, a progressive wing rapidly developed in the Republican Party, centered in the Midwest and West, and far more sympathetic to the interests of middle-class citizens. The progressives backed an income tax.

When Theodore Roosevelt became president in 1901 on the assassination of William McKinley, he moved sharply in the direction of the progressive wing of his party. In 1906 he even advocated a tax on inheritances with the avowedly social-engineering purpose of preventing the "transmission in their entirety of those fortunes swollen beyond all healthy limits." Mainstream Republicans were, to put it mildly, aghast at the idea, but there was no real threat to the status quo until the panic of 1907 and the short recession that followed, which caused government revenues from the tariff to decline sharply.

During the debate on the tariff bill of 1909, Representative Cordell Hull of Tennessee (later Franklin Roosevelt's secretary of state) proposed reenacting the income tax of 1894 and, in effect, daring the Supreme Court (which, thanks to Theodore Roosevelt's appointments, was far less conservative than it had been fourteen years earlier) to nullify it a second time.

Hull's amendment failed to pass the House, but events conspired in the Senate to change matters. A Democratic senator, Joseph W. Bailey of Texas, introduced the income tax amendment in the upper house, supported by such Republican progressives as William E. Borah of Idaho. Leading the forces opposed to the amendment was Senator Nelson W. Aldrich of Rhode Island. Aldrich, who had made a fortune in the wholesale grocery business and was the father-in-law of John D. Rockefeller, Jr., was one of more than twenty millionaires in the Senate at that time.

Aldrich managed to keep the high, protective tariff intact, but the Republicans were hopelessly split on the matter of an income tax, and

Aldrich appealed to the new president, William Howard Taft, to find a solution.

Taft, a far more conservative man than Roosevelt, revered the Supreme Court. Indeed, he would serve as chief justice, an office far more congenial to his nature than the presidency, for most of the 1920s. He was horrified at the idea of defying the ruling of the Court in *Pollack*. He felt that if the Court acquiesced, its status as the arbiter of the Constitution would be gravely impaired, and if it again struck down the income tax, a crisis between the Court and the two popularly elected branches of the government might result.

So Taft, a very gifted lawyer, proposed an alternative. Taft called for a constitutional amendment that would specifically permit a personal income tax, and meanwhile proposed a corporate income tax on profits. At that time the stock of corporations was owned almost entirely by the very affluent, so a tax on corporate profits was, in effect, a tax on the incomes of the rich. Further, he argued, the tax would not run afoul of the Constitution because it was not actually an income tax at all, but an indirect tax, measured by income, on the privilege of doing business as a corporation. In other words, it was an excise tax. In 1911 the Supreme Court agreed unanimously.

The Sixteenth Amendment, meanwhile, passed the Senate 77–0 and passed the House 318–14. The amendment was ratified by the required number of state legislatures and was declared effective on February 3, 1913.

By that time the Republican Party had split between the conservative Taft Republicans and the progressive Roosevelt Republicans, who stormed out of the 1912 convention to form their own party under the symbol of the bull moose. As a result, the Democrat Woodrow Wilson was elected president with less than 42 percent of the popular vote but with almost 82 percent of the electoral votes. Further, the Republican split had produced solid Democratic majorities in both houses of Congress. Among the first acts of the new Wilson administration was the passage of a personal income tax law.

Although almost comically short by later standards—the law was only fourteen pages long—it contained the seeds of the vast complexity that was to come. Income more than $3,000 was taxed, on a progressive scale from 1 to 7 percent (on incomes more than $500,000, then a vast sum indeed). But there were many exemptions, such as interest on state and local bonds and corporate dividends (up to $20,000). Interest on all debts, depreciation of property, and many other things were deductible from taxable income.

And the corporate income tax, originally intended as only a stopgap, was not merged with the personal tax but remained a completely separate tax. The financial exigencies of the twentieth century's great wars would send income tax rates soaring to heights undreamed of by even its most passionate advocates. As the tax rates bit more and more, accountants and lawyers would begin to find endless ways to shelter income by exploiting the lack of coordination between the two tax systems.

———

IN THE EARLY YEARS of the twentieth century the United States enjoyed a prosperity beyond anything previously experienced. In the ten years between 1897 and 1907, American exports doubled, as did its imports. The amount of money in circulation—national banknotes and gold and silver coins—increased from $1.5 billion to $2.7 billion, while bank deposits soared from $1.6 billion to $4.3 billion—a figure larger than the gross domestic product in 1860. The assets of banks, brokerage houses, and insurance companies increased from $9.1 billion in 1897 to $21 billion ten years later. The rest of the developed world was also enjoying great prosperity.

But there was a problem. With the world on the gold standard, national economies could grow, at least in the long term, only as fast as the gold supply that backed the world's currencies. Gold production had stagnated in the 1880s, but as new finds in the Yukon and, especially, South Africa, came into production in the next decade, the supply grew rapidly. Only $157 million in gold was mined in 1893, but $287 million was pulled from the ground five years later. Production grew to more than $400 million in the first years of the twentieth century, but then stagnated at that level, while the world economy continued to grow rapidly.

Demand for capital to finance consolidations of industries grew (as did the need for governments to finance wars such as the Boer War and the Russo-Japanese War). By 1907 money markets were very tight and growing tighter. James J. Hill, who controlled the Northern Pacific Railroad, began warning of what he called "commercial paralysis" if capital became too costly. By early 1907, even one-year gilt-edged bonds could only be sold if they carried coupons paying 5 to 7 percent, very high rates by the standards of the day. Long-term bonds could not be sold at all.

In March the stock market briefly collapsed but then recovered. Stock markets in other countries, especially Egypt and Japan, also shuddered that spring. Gold began to flow out of the United States as the Banks of England and France sought to shore up their positions and prevent runs on their currencies. The United States, of course, without a central bank since Andrew Jackson, had little if any control over its own money supply.

On October 10, in the wake of an attempted corner in copper stocks, panic hit Wall Street. It soon spread to the banks that had been involved in financing the corner, especially the Knickerbocker Trust. A run began on the Knickerbocker, and soon other banks were under siege as well, as depositors sought to convert their assets to cash while they could. The line outside the Knickerbocker's handsome new headquarters on Fifth Avenue was soon two blocks long. On Wednesday, October 23, the Lincoln Trust bled $14 million in deposits in just a few hours. Other banks were also near to being forced to close their doors.

There was little that the federal government could do. The secretary of the treasury, George B. Cortelyou, came to New York and deposited $6 million in New York banks to add to their liquidity, but the law forbade federal deposits in any but national banks. It was the trust companies and the state banks that were in the most trouble.

Cortelyou did the only thing he could do under the circumstances: he told J. P. Morgan that the federal government would do anything it could to stem the panic and, in effect, relied upon him once again to find a solution to a national financial crisis.

Morgan knew exactly what the problem was. On Thursday, October 24, he told reporters that "if people will keep their money in the banks, everything will be alright." But getting the people, seized by fear, to cooperate was the hard part.

Morgan realized that the Knickerbocker Trust was beyond salvation but that another trust company, the Trust Company of America, while under attack, was basically solvent. "Then this is the place to stop the trouble," he decided. He had Cortelyou deposit $35 million in national banks and told those banks to lend the money to the Trust Company of America. When its depositors had no trouble withdrawing their money, of course, they no longer wanted to withdraw it. The panic at the Trust Company of America ended.

But the situation remained perilous. On Thursday the president of the New York Stock Exchange crossed Broad Street to the Morgan Bank to tell Morgan that the exchange would have to close as call money (lent by brokers to finance margin accounts) simply could not be found. Morgan flatly forbade the exchange to close and raised $27 million among the bankers in five minutes to keep it open. That night he called the bankers to his new and magnificent library on Thirty-sixth Street, and they devised a plan to maintain liquidity, shore up solvent banks that were under attack, and provide more call money to the brokers. He let it be known that anyone selling short to take advantage of the panic would be "properly attended to." Not many brokers wanted to find out exactly what that might mean.

It was a very near-run thing, but the New York financial market managed to get through the rest of the week without another failure. Having summoned the bankers during the week, Morgan now summoned the city's clergy and urged them to give encouraging sermons that Sunday.

Slowly the panic began to abate, and in another week the worst was over. Many of New York's bankers, such as James Stillman and George F. Baker, had contributed materially as well as financially to staving off disaster. But it was generally acknowledged that only Morgan, by then the most powerful banker in the world, perhaps the most powerful banker who had ever lived, had been capable of holding the entire Wall Street community together and getting it to act for the common good.

Even Theodore Roosevelt, so fond of railing against the "malefactors of great wealth," praised "those influential and splendid business men . . . who acted with such wisdom and public spirit."

Thanks to Morgan and the other New York bankers, the crash of 1907 did not mark the onset of a period of severe depression, as the crashes of 1873 and 1893 had. But it did make clear that the country simply could no longer do without a central bank. A man of the stature and probity of J. P. Morgan might be able to avert financial calamity in the future, but there was no guarantee that there would be such a man available. Morgan himself was already over seventy. Still, it took six long years of intricate negotiations to get political agreement on the creation of the Federal Reserve System.

All national banks were required to be members of the new central banking system, and state banks that could meet the capital requirements of national banks (very few of them could) were eligible for membership as well. The advantage of membership, of course, was that in a panic, member banks could use their loan portfolios as collateral to obtain cash in a hurry from the Federal Reserve and thus abort any run. The downside of membership was a new layer of regulation added to, rather than replacing, the older regulatory bodies, such as the Comptroller of the Currency.

And the practical effect was that the banks that most needed discipline and protection against runs were the very banks that did not join the system: the small, stand-alone rural banks. By 1920 these fragile vessels holding the liquid assets of millions of American families and businesses would number almost thirty thousand. They were a disaster waiting to happen.

The new Federal Reserve System came into existence in 1913, and the United States had a central bank, if a flawed one, for the first time since Andrew Jackson had been president. It is one of the great coincidences of American economic history that J. P. Morgan, the country's de facto central banker in much of the post–Civil War era, had been born in 1836, the same year as the charter of the Second Bank of the United States expired. And he died the same year the Federal Reserve, its long-needed replacement, was born.

THE FIRST YEARS of the twentieth century seemed to those alive at the time as the dawn of a new era of progress and prosperity in this country. The country was advancing economically and socially as never before. The United States had one-third of the world's railroad mileage and 40 percent of its steel production. It was the world's greatest exporter of agricultural products. Its per capita income was far ahead of the next richest nation, Great Britain, which had dominated the world economy in the nineteenth century.

The greatest engineering project in history, the Panama Canal, had linked the two oceans on which the nation fronted. The Wright brothers had conquered the air. Automobiles were beginning to replace the horse as the chief means of local transportation.

Ninety percent of the American population was literate and supported more than twenty-two hundred newspapers. The country had a thousand colleges and universities and more high school students than any other on earth.

And this bounteous land, while connected as never before by steamer and telegraph cable, was still far from Europe and its increasingly rancorous international politics and dangerous arms race. The United States Army was among the smallest of those fielded by the major powers, but with the wide Atlantic and the world's third largest navy to insulate it, the country felt secure from whatever might happen in the Old World.

In fact, however, those years proved to be not the dawn of a new era but the golden twilight of an age that was ending. The Victorian age, with its Edwardian coda, had been characterized by a profound belief in the possibility of progress and was, more than any other in history, an age of optimism. That optimism would be among the major casualties of what the diplomat and historian George Kennan correctly called "the seminal catastrophe of the twentieth century," the First World War.

Another casualty would be American innocence and the belief that the New World could remain remote from the troubles of the Old. As Europe plunged into fratricidal war in the late summer of 1914, just as the Panama Canal was opening for business, the *New York Times* editorialized smugly that "The European ideal bears its full fruit of ruin and savagery just at the moment when the American ideal lays before the world a great work of peace, goodwill and fair play."

But less than three years later, an American in Paris would tell the Old World that "America has joined forces with the Allied Powers, and what we have of blood and treasure are yours. . . . [W]e pledge our hearts and our honor in carrying this war to a successful issue. Lafayette, we are here."

In all senses except the calendrical, the twentieth century began on August 1, 1914.

PART IV

THE AMERICAN CENTURY BEGINS

———

Mind your business

— Motto on the first U.S. coin

The trouble with capitalism is capitalists. They're too greedy.

— Herbert Hoover

Transition

THE FIRST WORLD WAR

N O GREAT WAR EVER BROKE OUT so unexpectedly or for so trivial a reason as the First World War.

To be sure, the Great Powers of Europe had been in an arms race for several years and a series of crises had threatened the peace. In 1912 the First Balkan War, when Serbia and Bulgaria declared war on Turkey, threatened to bring in Austria and Russia and ignite a general conflagration. The reaction of the world's markets had been immediate and profound: they all declined abruptly. Interest rates rose. Gold began to flow out of the United States as European central banks liquidated overseas investments and repatriated assets. The crisis soon passed, however, when Russia backed down.

But while the Western world was fearful, many thoughtful people held that Great Power wars were a thing of the past. There hadn't been one in almost half a century. Norman Angell, who would win the Nobel Peace Prize, argued in his very influential book *The Grand Illusion,* published in 1910, that the inevitable disruption of international credit would make it impossible to finance such wars, or would at least bring them to a very speedy conclusion. One economist, writing in the *New*

York Times in 1914, wrote that "no modern war has been conducted to which the business world as a whole was unalterably opposed, for war must draw its sinews from the money chests of business."

So the news of the assassination of the Archduke Franz Ferdinand, heir to the throne of Austria-Hungary, on June 28, 1914, did not raise immediate fears. But Austria, the weakest of the Great Powers, was determined to extract political advantage out of the situation and demanded concessions from Serbia, where the assassination had taken place. Russia, the main Slavic power, backed Serbia and threatened to mobilize its forces. Kaiser Wilhelm II of Germany, who could have defused the situation at any time by reining in Austria, backed it instead.

On July 28 Russia began to mobilize, and events quickly spiraled out of control. Mobilizing armies in the age of the railroad required elaborate and detailed planning. And once mobilization was initiated it could not be reversed without effectively rendering a country defenseless. When the German ultimatum to Russia to cease preparations on its frontier expired on August 1, Germany declared war, and within four days all the Great Powers of Europe were at war. It would be four years, three months, and eight million dead soldiers before peace returned to Europe.

And that continent, the center of the Western world for twenty-five hundred years, would lose far more than a generation of its young men. When the war was finally over, the only winner, in geopolitical terms, would be the United States, which would become by far the strongest nation on earth and the new center of the Western world.

———

AS THE THREAT OF WAR became ever greater in the last days of July 1914, stock markets the world over panicked while the demand for gold soared. On Tuesday, July 28, unable to maintain orderly markets, the exchanges in Vienna, Rome, and Berlin closed. The next day volume on the New York Stock Exchange reached 1.3 million shares, its highest since the panic of 1907, and many leading stocks fell more than 20 percent. On Friday, July 31, the London Stock Exchange closed for the first time in its history, and New York was the only major exchange scheduled to open the next day.

It had no real choice except to close as well. With the world's markets now tied together with a cat's cradle of undersea cables, sellers converged on New York, and sell orders began piling up in mountains awaiting the Saturday opening. (The New York Stock Exchange would have a Saturday morning session until after the Second World War.) The governors voted to close the exchange, and the president of the exchange consulted J. P. Morgan, Jr., now head of the House of Morgan, and the secretary of the treasury, William Gibbs McAdoo. They agreed, and the New York Stock Exchange closed and suspended deliveries until further notice.

It would be December before the exchange reopened, and then on a limited basis. On December 30, 1914, the exchange traded a mere 49,937 shares, the lowest volume of the twentieth century. It was the following April before trading was fully back to normal. By then the situation had changed radically. Nearly all the conventional wisdom regarding the financial and economic effects of a major European war on the United States had proved to be wrong.

The banks had remained opened after the stock exchange had closed, and there were heavy withdrawals, mostly in gold. But these ceased by September, when movement on the Western Front also largely ceased. By the end of that month gold was flowing into New York for safekeeping, and much of it has remained there ever since, now stored eighty-five feet below the Federal Reserve bank on Liberty Street in vaults blasted out of the Manhattan bedrock.

At first American commerce was severely disrupted. Cotton exports declined sharply and so did wheat. Germany had imported 2.6 million bushels of wheat in July, but none in August, as the Royal Navy asserted its sea power.

Soon, however, the situation reversed, and American exports of agricultural products rose quickly. Part of the reason was the poor European harvest that year, which would have assured good exports for American farmers regardless of the war. But more important, with Germany controlling the Baltic, and Turkey, soon Germany's ally, closing off commerce through the Black Sea, Russian exports of grain fell to nearly nothing. Russia had been one of the world's leading exporters of grain in the late nineteenth century, but its share of the world market was quickly taken by the United States, Canada, Argentina, and Australia. It has never recovered it, thanks in large part, of course, to the ruination of Russian agriculture by Communism.

Between December 1913 and April 1914 the United States had exported a total of eighteen million bushels of wheat. In the same period a year later, wheat exports amounted to ninety-eight million bushels. As the war dragged on and more and more agricultural workers were called to military service in Europe, American agricultural exports continued to increase. Net farm income in the war years more than doubled, to $10 billion, and the value of agricultural land, buildings, and equipment increased by nearly 30 percent.

American manufacturing also increased rapidly. Markets in Latin America and Asia, which had been served by European companies, were now open to be taken over by American firms. Far more important was the avalanche of orders that began to roll in to American firms from Great Britain and its allies, for steel, vehicles, railroad rolling stock, and rails. An American invention of the 1870s to help tame the new western farmlands at low cost—barbed wire—was ordered by the combatants by the hundreds of thousands of miles to protect the trenches against infantry assault.

Munitions, of course, were in the greatest demand by the British, French, and Italian militaries. Du Pont had been only a midsized manufacturer of gunpowder before the war, but would come to supply the Allies with fully 40 percent of their munitions. In the four years of the war, Du Pont's military business increased by a factor of 276. And it became one of the world's largest chemical companies as well. Germany had dominated the world's chemical industry in the decades before the war, but it lost its export market with the Royal Navy's blockade. Du Pont and other American chemical companies quickly seized that vast market. By the end of the war, Du Pont had annual revenues twenty-six times as large as they had been in 1913.

Bethlehem Steel, a major armor plate manufacturer and shipbuilder, had never had a foreign contract larger than $10 million, but in November 1914 the British Admiralty offered it a contract for $135 million worth of ships, guns, and submarines. The German government, unable to tap into American industrial capacity, tried to deny it to its enemies by buying it. In 1915 it offered Charles Schwab, president and principal stockholder in Bethlehem Steel, twice the market value of his stock for control of the company. Britain, able to read German diplomatic communications, learned of the bid and was prepared to make a counteroffer. But Schwab assured the British that he would fulfill his contracts and turned down the German offer out of hand.

Overall, the gross national product of the United States increased by 21 percent in the four years of the war, while manufacturing increased by 25 percent.

The country had been in recession in 1914, but thanks to the slaughter in Europe, American industry began to prosper as it had not since the Civil War. The effect of this sudden growth was soon felt on Wall Street, and the Dow-Jones Industrial Average had the biggest annual percentage gain in its history, 86 percent, in 1915. General Motors, then the second largest American automobile company, had seen its stock plummet 39 percent in the last day of trading before the New York Stock Exchange closed on August 1, to 39. By the end of the year, with orders for vehicles pouring in, it had rebounded to 81½. A year later GM stock stood at 500. Bethlehem Steel's stock increased tenfold in 1915.

Early in that year the British government signed an agreement with

J. P. Morgan and Company, making the bank the American purchasing agent for the British government. Its first deal was for $12 million worth of horses, which were desperately needed to move artillery and supplies at the front. (The extremely high prices paid for horses in the war years was a prime reason that tractors so quickly replaced them on American farms at this time.) The bank soon signed a similar agreement with the French government.

No one had any idea at the beginning of the war how much the Allies would purchase in the United States for the war effort, but the British secretary of state for war, Lord Kitchener, thought it might be no more than $50 million. In fact, it would amount to more than $3 billion, more than four times total federal government revenues in 1916. The Morgan Bank's influence on American industry in these years, thanks to its immense buying power, was considerable, and Morgan had a staff of 175 working on finding needed supplies and arranging shipping and insurance.

The vast imports of war matériel by the belligerent powers had to be paid for, of course, and that was no easy matter. Britain's annual defense budget in the immediate prewar years had averaged 50 million pounds. It was soon spending 5 million pounds *a day* to fight the war.

Britain, long the major supplier of capital to the developing American economy, now began to liquidate its American investments. It put a special tax on the dividends of American securities but allowed British taxpayers to pay their income taxes with U.S. securities at face value. The British treasury then turned the securities over to the Morgan Bank, which managed to sell them quietly and without adversely impacting stock values. About 70 percent of the American securities held by French and British citizens were liquidated by the end of the war.

Still, liquidating American investments provided nowhere near enough money to prosecute the war, and both the British and French governments sought American loans. The Wilson administration, especially the secretary of state, William Jennings Bryan, at first opposed any loans to the belligerent powers, Bryan calling them "the worst of contrabands." But Bryan, a profound isolationist and intellectually inflexible, did not last long as secretary of state once the war broke out. His successor, Robert Lansing, managed to convince Wilson that loans were necessary to ensure continued economic expansion here at home.

In September 1915 the Morgan Bank arranged a loan of $500 million for the British government, far and away the largest bank loan in history up to that point, but only the beginning, it turned out. By the time the United States entered the war, Morgan had arranged loans equaling $1.5 billion to Britain and more to France. After the United States entered the war, the federal government took over as the major lender to the Allies. Altogether the government would loan the Allies $9.6 billion, a sum equal to eight times the entire American national debt in 1916.

DESPITE THE MASSIVE FINANCIAL and industrial aid, by 1917 it was becoming clear that the Allies were in deep trouble. The sinking of British merchant ships by the German submarines was threatening starvation in Britain, while the French army was near mutiny. With the fall of the czarist regime in Russia in March of that year, the possibility that the Allies might lose the war was becoming very real.

German behavior in the war, such as the violation of Belgian neutrality and the sinking of the unarmed passenger liner *Lusitania*—skillfully exploited by brilliant British propaganda—had turned American public opinion sharply against the Central Powers. The resumption of unrestricted submarine warfare in January 1917 proved the final straw for President Wilson, and he broke off diplomatic relations with Germany at that time.

A month later the so-called Zimmermann telegram, promising Mexico the return of its "lost provinces" in exchange for a declaration of war against the United States should war break out between the United States and Germany, became public. American public opinion was outraged. Wilson, who had run for reelection in 1916 on the slogan "He kept us out of war," asked Congress for a declaration of war less than a month after his second term began.

With America in the fight, the extraordinary capacity of the American nation to respond to military necessity quickly became apparent. The army had numbered only two hundred thousand before the war, but by November 1918 there would be two million American soldiers in Europe, together with forty thousand cars and trucks (plus forty-five thousand horses) and two thousand airplanes.

The war's effect on federal finances was both great and permanent, just as the Civil War's had been. Since 1865 the government had never spent more in one year than the $746 million it had spent in 1915. The national debt that year was a mere $1.191 billion (John D. Rockefeller could have paid it off all by himself and still have been the richest man in the country). After the First World War, however, annual government outlays were never less than $2.9 billion, and the national debt rose to more than $25 billion in 1919.

Bond drives using techniques invented by Jay Cooke during the Civil War were quickly implemented, now with the addition of using movie stars such as Douglas Fairbanks, Mary Pickford, and Charlie Chaplin to entice citizens into buying Liberty Bonds.

And the income tax, which had been a mere social-engineering device to get the rich to share more of the tax burden, began to bite into the middle class. The personal exemption, which had been at $3,000, was dropped to $1,000. The tax rate, a mere 7 percent on incomes more than $500,000 before the war, rose to 77 percent. The income tax thus became the most important source of federal revenues, as it has remained ever since. And this changed the nature of the endless debate over taxes.

When the tariff had been the primary source of federal revenues, the debate was between sections of the country. New England mill owners and their workers both favored high tariffs on cloth. Southern sharecroppers and their landlords favored low tariffs. With the income tax, the debate was now between economic classes.

There was no change in the economic situation of the United States, however, that compared with its international assets and liabilities. As an undeveloped country, the United States had been a major importer of capital. To be sure, as a curious side effect of the periodic booms, panics, and deep depressions that characterized the American economy in the nineteenth century, much of this imported capital eventually ended up in American hands. In boom times, capital would pour into the country to build new railroads and factories. Then, after the inevitable bust, discouraged European holders of stocks and bonds would dump the greatly depreciated securities for whatever they would fetch in the American market. As a result, the United States would end up having both the asset and the ownership of it.

Still, even as late as 1914, the United States remained the world's largest debtor nation, with investments abroad that equaled $3.5 billion and European investments in the United States worth $7.2 billion. By the end of the war, the situation had nearly reversed, with foreigners holding $3.3 billion in American securities and Americans owning foreign investments worth $7 billion. In addition, foreign governments, principally France and Britain, owed the United States government $9.6 billion in war loans. Thus in four years, the United States had gone from being a net debtor, owing $3.7 billion, to being a net creditor that was owed a total of $12.6 billion.

The new reality of world politics was even more dramatic than what the mere figures imply. While the United States had military deaths amounting to 126,000 in the First World War, France, with a population of less than 40 million, had lost 1,357,000 young men, the British Empire, 908,000; Germany, 1,773,000; Austria, 1,200,000; Russia, 1,700,000.

Austria, shorn of its east European empire, ceased to be a Great

Power. Germany, burdened with the reparations imposed by the draconian Treaty of Versailles and the deep psychological wound of seeing its incomparable army nonetheless defeated, would struggle for a decade and more before falling into the grip of the Nazis. Russia would be consumed with establishing a Communist state.

Even the nominal European victors, Britain and France, had been bled white, militarily, economically, and psychologically by the Great War. They would never again be the forces in world affairs that they had been for centuries.

Only the United States emerged from the struggle materially strengthened. It had been the world's leading industrial power for three decades. Now it was the world's leading financial power as well, replacing Britain in that role, and money would circle around the new financial center of gravity, Wall Street, rather than Lombard Street.

Britain, with its worldwide empire and immense foreign trade, had willingly and efficiently functioned as the world's financial center and de facto central bank. Indeed, Britain had welcomed this role as an instrument of considerable power to wield on the world stage. But the United States, new to the world stage, and still deeply wary of the "entangling alliances" President Washington had warned against more than a hundred years earlier, was not so willing. (The United States had insisted on the odd locution of "associated power" rather than "ally," after it declared war.)

It would take two decades and a resumption of world war before the American people would come to accept the fact that they had become, in the words of President John F. Kennedy, "by destiny not by choice, the watchmen on the walls of world freedom."

Chapter Fifteen

GETTING PRICES DOWN
TO THE BUYING POWER

T HE FEDERAL RESERVE, created in 1913, began to function in the early days of the First World War. But it did not come into its own until after peace was reestablished. It then almost immediately made its first serious policy mistake.

As always when huge deficits have to be financed, the First World War caused a serious inflation and the Consumer Price Index nearly doubled between 1915 and 1920. The Federal Reserve had kept interest rates low during the war to facilitate the government's borrowing needs, and maintained those rates until November 1919. Then it moved the rediscount rate—then its major means for influencing interest rates—in a series of abrupt steps from 4 percent to 7 percent over the next eight months.

The economy, in fact, had already been moving toward recession with the end of vast military orders and the revival of European agriculture. The Federal Reserve's actions turned a decline into a near disaster. The money supply contracted by 9 percent, while unemployment shot up from 4 to 12 percent. GNP declined by nearly 10 percent. At least the Federal Reserve's overcorrection broke the back of the wartime inflation, and wholesale prices declined by nearly 40 percent between 1920 and 1921, one of the sharpest price declines in American history.

Fortunately, the depression of 1920–21 proved to be short-lived. New opportunities abounded in the 1920s and produced a decade of immense prosperity. The new economic engines behind that renewed prosperity were the automobile and electricity.

———

THE LATE-NINETEENTH-CENTURY ECONOMY—dominated by steel, petroleum, and railroads—had been the era of heavy industry. The 1920s would see the emergence of a much more consumer-oriented economy, a trend that only accelerated for most of the rest of the century. But the seeds of the new economy, as always, had been created in the old.

The automobile was not an American invention (although a clever patent attorney from Rochester, New York, named George B. Seldon, who never actually built one, managed to get a patent for "an improved road engine," powered by "a liquid-hydrocarbon engine of the compression type" in 1879, even before the word *automobile* entered the English language). Most of the necessary technology was developed in Europe. A German named Nickolaus Otto had built the first practical internal combustion engine in 1876, and Wilhelm Maybach, also German, invented the carburetor in 1893. The carburetor was the last piece of the puzzle needed to build a practical horseless carriage, and tinkerers and entrepreneurs by the hundreds in both Europe and America began to manufacture automobiles in backyards and blacksmith shops. In 1900 four thousand automobiles were manufactured in the United States, by dozens of different firms and individuals.

A classically Darwinian competition ensued, just as in a natural ecosystem when a new body plan evolves or a new territory is reached. Whenever a major new technology becomes practical, there is always a large number of people and firms who will try to exploit it for profit. Most quickly fall by the wayside as they fail to compete successfully.

Then, as the industry matures, the need for economies of scale and the huge capital needs often required to realize them cause the industry to consolidate into a few firms of great size. This was certainly true of the automobile. In 1903 alone, fifty-seven automobile companies came into existence in the United States, and twenty-seven went bankrupt. Today there are no more than two dozen automobile companies, all of them necessarily multibillion-dollar corporations, in the world.

One of the American companies that opened for business in 1903 was the Ford Motor Company. Its principal owner (sole owner after 1915), Henry Ford, wanted to produce a new kind of car, a car for the masses rather than the rich, the market for automobiles up to that point. If the automobile was invented in Europe, the mass-produced automobile, sold at a price the middle class could afford, was a purely American idea, an idea that transformed the American and world economies.

More than any single other economic development, the mass-produced automobile made the twentieth century very different from the nineteenth. Less than thirty years after the Ford Motor Company was founded, the British novelist Aldous Huxley wrote the science fiction classic *Brave New World*. In it he depicted a future world industrialized to the point where even human beings are produced in baby factories. And that world reckoned time not from the birth of Christ, but from the birth of Henry Ford.

The son of a farmer in Dearborn, Michigan, Henry Ford attended public schools there until going to work at the age of sixteen as an apprentice in a machine shop. Neither well educated nor well read, he early exhibited a marked talent for and fascination with mechanics, and he began tinkering with internal combustion engines in the early 1890s. In 1896 he built his first automobile in a carriage house behind where he was living. In the next few years he built several racing automobiles, which broke speed records, and, with the help of several backers, opened the Ford Motor Company in 1903.

The company was moderately successful at first. Then in 1908 Ford introduced the Model T. It was designed to be both rugged, to handle the often ghastly roads then in existence (there were fewer than two hundred miles of paved roads in the entire country in 1900, outside of cities) and cheap to manufacture. Its initial price was $850, a fraction of what most automobiles then cost, and its running expenses were equally modest, by some estimates only a penny a mile. It was an immediate hit with the public, which bought 10,607 Model Ts that year.

Henry Ford, having developed a product that he regarded as perfect, then bent every effort to finding ways to manufacture it more and more cheaply and thus bring it within reach of an ever larger segment of the population. In 1913 he fully introduced the assembly line at a new plant, built for the purpose, at Highland Park, Michigan. (Ford had visited a meat-packing plant and realized that if animals could be disassembled as they moved along a line, automobiles could be assembled in the same way, with huge savings in labor.)

That year it took only ninety-three minutes to assemble a Model T. In 1916 the price had dropped to only $360, and Ford sold 730,041 of them. By the 1920s, despite the inflation caused by the First World War, the price tag of a Model T was only $265, and Ford was still finding ways to lower labor costs by an average of 7.4 percent a year.

The result of Henry Ford's relentless drive to lower costs in manufacturing the Model T was one of the most astonishing economic success stories in world history. Over the nineteen years that the Ford Motor Company produced the Model T, it manufactured fifteen million of them. By the end of the model run, the company had more than $700 million in undistributed profits. In 1920 Ford was producing half the cars built in the world. The runaway success of the Model T helped mightily to cause the entire automobile industry to take off. From 4,000 cars in 1900, the country produced 187,000 in 1910. By 1920 some 1.9 million cars rolled off assembly lines, and 8.1 million vehicles were in registration. By 1929 production was up to 4.5 million cars, and 23.1 million had been registered. The five-thousand-year reign of the horse as the prime mover of humankind was over.

It would be difficult to overestimate the impact of the automobile on the American economy by the 1920s. The automobile industry not only employed hundreds of thousands of workers itself, but greatly stimulated other industries. In the 1920s automobiles were using 20 percent of the steel produced in this country (and almost all of the sheet steel), 80 percent of the rubber, and 75 percent of the plate glass.

By the 1920s the automobile industry had become the largest in the American economy. The seemingly insatiable national appetite for cars produced a decade of great industrial prosperity. GNP increased by 59 percent between 1921 and 1929, reaching $103.1 billion. Meanwhile, GNP per capita rose by 42 percent, and personal income by more than 38 percent.

The automobile also greatly increased road building and paving, which became a major component of the construction industry and greatly stimulated quarrying and cement manufacture. From hardly any paved roads in 1900, by 1920 there were 369,000 miles, and by 1929 there were 662,000 miles. Just as the old turnpikes built in the late eighteenth century had stimulated commerce along them, the new roads did likewise. Gasoline had originally been sold at general stores and blacksmith shops, which, adapting to a changing market, had mostly become garages by the end of the 1920s. In 1905 the first purpose-built gas station opened in St. Louis. A quarter century later there were tens of thousands of them, most of them franchisees of the major oil companies. And gasoline replaced rapidly fading kerosene as the most important petroleum distillate, giving the petroleum industry a new, and larger, niche in the marketplace.

The fact that automobiles moved so much faster than a horse and carriage required a change in advertising. Seen at a speed of thirty or forty miles an hour, a sign had to be grasped instantly or it wouldn't be grasped at all. Corporate logos became important for the first time, and the wordy style of nineteenth-century advertising disappeared, even in other venues, such as newspaper ads, as short and punchy became perceived to be "modern."

And the automobile began to change the country's demographics. The shift from a predominantly rural population to an urban one had been going on almost since the dawn of the Republic and had reached the tipping point in the census of 1920, which was the first to record more urban dwellers than rural ones. But the automobile allowed the emergence of a whole new demographic region: the suburbs.

A nineteenth-century demographic map of a typical American city would have resembled a daddy longlegs, with a dense urban core and long, thin strings of population along the railroad and trolley tracks. In between the tracks was deep country, for once a person disembarked from the train, he was again reduced to the speed of a horse. With the coming of the automobile, however, people could live miles from the railroad tracks and still be able to reach the city easily. More and more people began living in the country and working in the city.

The automobile also affected the rural areas as much as the urban and suburban ones. For one thing, it ended the suffocating isolation of the typical American farm. European farmers usually lived in villages and walked out to the fields (often owned by someone else) to work. But American farmers mostly lived on their own land often a mile or more from the nearest neighbor and several miles from the nearest town. Visiting was difficult and time-consuming.

The automobile allowed the rural consumer to shop farther afield. Before the automobile, what wasn't available at the local general store had to be ordered from catalogs, such as those put out by Sears, Roebuck and Montgomery Ward. But the inexpensive automobile began to change that. The cozy monopolies of the local mercantile and bank were broken when their clients could drive to a larger town and do their business there, taking advantage of the better prices that are always to be found in larger markets. Commerce in the smallest towns began to decline and has been declining ever since.

Because the local banks were usually stand-alone operations dependent on their local economy, many of them began to decline when their customers began to go elsewhere. There were an astonishing number of them, peaking at 29,798 in 1921, almost all of them one-branch outfits with assets under a million dollars, and which were not members of the Federal Reserve. The number of bank failures in the United States in the 1920s, despite the general prosperity, began to increase. By the end of the decade, more than six hundred rural banks a year were failing, often taking the savings of their customers with them.

The automobile also put a far greater stress on the country's rural economy as a whole. In 1900 one-third of the nation's farmland was devoted to fodder crops to feed the vast number of horses and mules that powered the local-transportation and agricultural industries. By 1929 most of that herd was gone, replaced by automobiles, and much of the land that had grown such crops as hay and oats had been switched over to crops for human consumption, causing food supplies to rise much faster than demand and prices to decline sharply. The result was hard times for many farmers, who never saw prices recover from the fall-off in European orders after the First World War. The depression in American agriculture, largely unnoticed by the urban-based media at the time, would slowly, inexorably both deepen and widen.

ELECTRICITY HAD BEEN AN UTTER MYSTERY in the seventeenth century and a parlor trick in the eighteenth as people like Benjamin Franklin began to explore its nature. Although the early nineteenth century would gain a deeper understanding (in 1831 the great British physicist Michael Faraday proved the identity of electricity and magnetism) and produce the first practical use of electricity, the telegraph, it was only at the end of the century that electricity began to impact seriously on everyday life. No one played a greater role in that than Thomas Edison, who proved to be to Yankee ingenuity what Shakespeare had been to drama.

Thomas Edison is today remembered for his almost endless stream of inventions that helped turn the nineteenth century into the twentieth. Every schoolchild knows that Edison invented, or made substantial contributions to, the phonograph, the stock ticker, the telephone (along with important mechanical improvements to Bell's original machine, Edison also coined the word *hello*), movies, and, of course, electric light.

But two of Edison's greatest inventions are seldom mentioned because, by their nature, they couldn't be patented. One was perhaps his greatest invention of all, the industrial research laboratory. Edison established his own laboratory in Menlo Park, New Jersey, in 1876, and it was there that he created the phonograph (1877), the electric light (1879), and hundreds of other inventions. It was, in essence, an invention factory where engineers, chemists, and mechanics turned new technological possibilities into practical—and, most important, commercially viable—products.

When General Electric was formed in 1892 by J. P. Morgan from the Edison General Electric Company and its major competitor, Thomson-Houston Electric Company, the new company almost immediately established a laboratory of its own at its headquarters in Schenectady, New York. It quickly became the model for a number of other corporate research labs that in the twentieth century would turn out an unending stream of inventions and practical applications of new technology. The list of the fruits of Edison's seminal idea to industrialize the process of invention—to industrialize Yankee ingenuity—is nearly endless: cellophane, nylon, synthetic rubber, transistors, Teflon, and the microprocessor being but a few of the more important. In 2003 IBM alone would take out more than thirty-four hundred patents.

Edison's other unsung invention was the electric power system by which his lightbulb could be lit. Once the lightbulb was a working invention, he set about to build a generating plant and lay electric lines in a one-square-mile area of the Manhattan business district. In 1880 he secured from the city the right to "lay tubes, wires, conductors and insulation, and to erect lamp-posts within the lines of the streets and avenues, parks and public places of the City of New York, for conveying and using electricity or electrical currents for purposes of illumination."

Edison built the world's first power plant on Pearl Street and installed six of the largest dynamos yet built, weighing thirty tons each. Working at night so as not to make New York City's traffic any worse than it already was, he dug trenches for his electric mains, which totaled fifteen miles in length, and sent out crews to wire up houses and stores whose owners were willing to sign up for the new service.

As with any new technology, Edison had to devise solutions on the fly to endless problems that had not been thought of until they arose. One problem was that if there was a leakage of current under the pavement, horses would conduct it through their shoes and panic. Many of Edison's on-the-fly solutions were patentable, and he applied for no fewer than 102 patents in 1882, the most in any one year, as he was building his system.

Finally, at 3 PM on September 4, 1882, Edison, standing in J. P. Morgan's office, closed the circuit, and 106 lamps came on in the offices of Drexel, Morgan and Company. More came on at the *New York Times,* which had also signed up as an Edison customer, and in shops along Fulton Street. They didn't make much impression in daylight. But by that evening it was obvious that something important had happened. The next day the *New York Herald* reported that "in stores and business places throughout the lower quarters of the city there was a strange glow last night. The dim flicker of gas, often subdued and debilitated by grim and uncleanly globes, was supplanted by a steady glare, bright and mellow, which illuminated interiors and shone through windows fixed and unwavering."

In the next few years electricity spread through the business districts and fashionable residential areas of the country's cities, but it remained expensive, and most people continued to get along with gaslight or, beyond the reach of urban gasworks, kerosene. It would be Thomas Edison's former secretary, Samuel Insull, who would prove to be the Henry Ford of electricity and make the new technology affordable for the average person and permanently change the American economy thereby.

Insull was born in 1859 in London, to a lower-middle-class family, and he went to work at an auction house at the age of fourteen while continuing to study at night. At eighteen he went to work for Edison's British representative, who was so impressed with his energy and organizational abilities that in 1881 he sent him to the United States to be Edison's personal secretary. Soon he was indispensable to Edison, who was not nearly as good a businessman as he fancied himself to be.

Edison put Insull in charge of the Edison General Electric Company, which had been struggling, and Insull quickly turned it around. To obtain a measure of independence from Edison, he moved the company to Schenectady, explaining that "we never made a dollar until we got the factory a hundred and eighty miles away from Mr. Edison." In a few years he increased business so much that the labor force rose from two hundred to six thousand, and the company became highly profitable.

When the company merged to form General Electric, however, Insull, although being paid the then-princely sum of $36,000 a year, decided to move on. He was more interested in building a power grid than in manufacturing electrical equipment, and he accepted the top job at one of Chicago's electricity-generating companies, Chicago Edison (it had been named in honor of Edison, but he had no financial stake in the company). It had only five thousand customers when Insull arrived in 1892. And it was one of thirty companies then generating electricity in that city.

One severe restraint on the number of customers for electricity was the cost, then about 1 cent an hour to light a single bulb (which produced only about one-fifth as much light as a modern lightbulb of the same wattage). A factory worker at that time was lucky to earn $750 a year, so electric light was a luxury few could afford.

There were two problems that made the cost of electricity so high. One was that electricity was a very capital-intensive business. This meant that economies of scale were crucial. But in the early days of electricity, generators were relatively small. Thus, large users, such as department stores and factories, were often better off building their own generating capacity than buying the electricity from a utility. With only small customers, the price per kilowatt-hour was necessarily high.

Insull moved to do something about this. He built the world's largest generating plant on Harrison Street in Chicago and installed a new design of generators that used only half as much coal as the previous type. He also began buying up the competition to enlarge his market. By 1898 his company owned all the generating capacity in his distribution area and had doubled the size of the Harrison Street plant. But when he began to supply power to Chicago's streetcars and elevated railways, he needed still more power.

Insull decided to gamble on a radical new technology. Steam-powered generating equipment up to this time had used reciprocating engines, where pistons pounded up and down, turning a crank shaft to produce the power. They were noisy and, at maximum output, vibrated alarmingly. They needed constant maintenance. On a trip to England Insull had seen a speedboat that was powered by a new type of steam engine invented by Charles Parsons, called a turbine. Instead of piston rods pounding up and down, the turbine spun smoothly, turned by steam acting on blades of propellors, at much higher speed, producing far more power per unit of fuel and needing less maintenance.

Insull thought the steam turbine was ideally suited to producing electricity, but had to cajole General Electric into producing engines of the size he wanted, far larger than had ever been built. And his own board was so nervous that Insull had to personally guarantee the company against loss if the new turbine-powered plant he planned on Fisk Street didn't work. When the plant was ready to go on line for the first time, the engineer told Insull to step away in case something went wrong and the turbine blew up. "Well," Insull replied, "if it blows up, I blow up with it anyway. I'll stay."

It didn't blow up, of course. It revolutionized the electricity-generating business, greatly lowering costs per kilowatt-hour. The steam turbine immediately became the standard means of producing electricity, which it remains to this day.

The biggest problem with lowering the cost of electricity, however, is the fact that electricity, almost uniquely among major commodities, cannot be stored. Instead, it must be produced at the instant of demand. Thus generating capacity must be large enough to meet peak demand, even though that means there will be very expensive excess capacity 95 percent of the time, the expenses of which must be pro-rated.

Again on a trip to England, Samuel Insull found a partial solution to the problem. The first electric meters had merely measured how much current was used between readings, as most house meters still do. (Thomas Edison had devised a meter wherein a small amount of the current being used melted zinc, which dripped onto a plate below. The meter reader would weigh the plate to determine how much electricity had been used.) But in the resort town of Brighton, on England's south coast, Insull talked to a man who had invented a meter that not only kept track of how much was used, but, crucially, *when* it was used.

Electricity usage over the course of a day varies widely but predictably, peaking in the hours between 4 P.M. and 8 P.M. and reaching its lowest point between 2 A.M. and 5 A.M. Insull realized that any electricity he could persuade users to use at slack times was, in effect, found money, whatever he charged for it, while getting customers to shift away from peak periods lowered his capital costs by reducing the capacity he had to build and maintain.

In the first year the new meter was in use in Chicago, electric rates fell by 32 percent, while demand began to soar. It came first to stores, factories, and advertising, but by 1910 one household in six in Chicago was electrified and the percent climbed steeply thereafter. By the 1920s the gaslight business was nearing extinction.

The ever increasing use of electricity in the United States is one of the wonders of the twentieth century. In 1902 the United States used 6 billion kilowatt-hours of electricity, about 79 kilowatt-hours per person. In 1929 it was 118 billion, and 960 kilowatt-hours per person, well over ten times as much per capita. Today usage is a staggering 3.9 *trillion* kilowatt-hours, more than 13,500 per person, more than 170 times as much electricity as was used per person in 1902.

This astonishing rise in the use of electricity came about not only because more and more people were switching over to electric light, but also because more and more tasks were being powered by electricity rather than by other means. This affected the American economy in many ways. For one thing, it changed the very shape of factories. Steam engines are very inefficient in terms of the amount of energy in the fuel that is converted into work-doing energy. And the smaller they are, the more inefficient. So nineteenth-century factories were tall, with as large a steam engine as possible in the basement. The engine powered a shaft running up the side of the building, from which horizontal shafts on each floor took power. It was important that these shafts be as short as possible.

But small electric motors are just as efficient as large ones (more so in some ways), so it made sense, once the price of electricity dropped sufficiently, to power each machine separately by electricity, eliminating the need for shafts to transmit power. (An electric motor is exactly the same mechanism as an electric generator, only working backward. The first uses electricity to produce power, the latter uses power to produce electricity.) Once freed from the need to connect machinery to the shaft from the steam engine, factories began to spread out horizontally on one level.

The rapidly widening use of electricity also caused productivity to soar in the 1920s, increasing output per worker by 21.8 percent in that decade. This helped to push manufacturing output up by more than 90 percent. Although electricity and the small electric motor had been around for two decades, the full effects of their use on industrial productivity came only in this decade. This is always the case with new technology because of what economists call the installed base problem. The old technology is already in place and paid for. Therefore it makes no economic sense to replace it until it wears out. The Erie Canal, rendered obsolescent by the railroads in the 1850s, was still carrying freight as late as 1970. Today the personal computer is the main engine behind the extraordinary gains in productivity in recent years although the personal computer has been around for nearly a quarter of a century.

More important, small electric motors began to power an ever-increasing number of household appliances in the 1920s, refrigerators, electric irons, vacuum cleaners, hair dryers, washing machines, radios, and phonographs among them. These began replacing servants in large numbers, and the servants moved on to more wealth-creating jobs. Many servants had gone to work in factories to take jobs left by soldiers in the First World War and had not returned. With the new appliances, the need for servants began its long decline. And households that had never had help became much easier to keep clean and supplied with clean clothes. As in the early nineteenth century, the middle class became able to live in a lifestyle that had once been reserved for the rich.

These wondrous new machines, of course, cost money, especially automobiles, in sums that were often beyond the ready means of the average family. In the early twentieth century few households had bank accounts and even fewer had established means of borrowing money. Bankers, who catered to business and the rich, were usually not interested in customers of modest means, although there were notable exceptions, such as A. P. Giannini, who built San Francisco's Bank of America into the largest bank in the country.

So the manufacturers of the new devices began to offer credit terms of so much down and so much a month for what economists call "durable goods," those with a useful life of more than three years. This greatly increased the potential number of buyers, of course, and the larger market brought down prices, further enlarging the market. Thus the country's rapidly increasing number of people with significant disposable income (income over and above what is needed for necessities) came to be a major influence in the American economy in the 1920s.

This was a profoundly democratizing development as huge industrial concerns came more and more to cater to the needs and desires (and also, of course, to foster these needs and desires) of the average citizen. "Why flounder around waiting for good business?" Henry Ford asked, explaining his business philosophy of pursuing the mass market. "Get costs down by better management. Get the prices down to the buying power."

But there was more to this new mass market than just cheap prices, and some companies were better at understanding it than others. Henry Ford, obsessed with the idea that the Model T was perfect, refused to change the design after 1908, concentrating instead on making it cheaper and cheaper. He even refused to add an electric starter after they became available in 1912, because of the weight of the battery. The starter became standard on other cars almost immediately because it was far safer to use than hand cranking (the crank could, and not infrequently did, break the arm of an unlucky user). It also allowed many people, such as women and the elderly, to use automobiles unassisted. Nor would Ford provide credit or even paint his cars any color other than black (which dried faster than other colors and thus, once again, lowered costs).

By the mid 1920s the Model T was out of date both technologically and commercially, but Ford refused to change. His once unassailable position as the world's largest producer of automobiles began to fade as his chief American competition, General Motors, outcompeted him for the attention of the American consumer. Ford's business model regarded the automobile as transportation and nothing but. The cheaper, therefore, the better. But Alfred P. Sloan and his remarkable associates at GM realized that the automobile had become much more than just transportation; it had become a part of how Americans saw themselves and others. It had become both a status symbol and a means of expressing personality, not unlike clothes.

General Motors set up the General Motors Acceptance Corporation to finance purchases of its products, making it possible for customers to move up market. And it provided instead of just one model, a whole series of models and brand names, from Chevrolet to Cadillac, providing what has now been called in business school for generations a "mass-class" business model.

In 1927, shortly after the fifteen millionth Model T rolled off the once revolutionary assembly line, Henry Ford, facing acres of unsold cars, had no choice but to shut down for eighteen months while he retooled his plants to produce an up-to-date car, the Model A. By the time the Ford Motor Company was back in business, General Motors had become the largest automobile company on earth, as it remains to this day.

The lesson was clear: not even the sole owner of a multibillion-dollar industrial concern and one of the half dozen richest men in the world could defy the power of the new American marketplace for long. The consumer was now in control of the American economy.

———

BUT IF CONSUMERS were in control, they could not lead, and the American economy was not well led in the 1920s. One might say it was not led at all. By the end of the First World War the United States had become financially as well as economically the strongest country in the world. It was now the world's greatest creditor nation, and its share of the world's manufacturing increased from 36 percent in 1914 to 42 percent by the end of the 1920s. It was the world's largest exporter and the second largest importer (after Britain), and the greatest supplier of capital to other countries.

But Woodrow Wilson was outmaneuvered at the Versailles Peace Conference and a draconian, merciless peace was imposed on Germany, requiring it to pay vast war reparations to the victors (but not the United States, which demanded no reparations). The peace treaty assured that Germany, intrinsically the greatest power in Europe, would be economically prostrate for the foreseeable future. Britain, France, and Italy, meanwhile, were saddled with huge war debts to the United States, which they had scant means to pay and the United States had scant interest in forgiving. "They hired the money, didn't they?" Calvin Coolidge asked.

Woodrow Wilson's stubborn refusal to make necessary political compromises prevented the United States from joining the League of Nations and effectively leading the international system. Instead American diplomacy largely pursued a quixotic foreign policy with such treaties as the Washington Naval Treaty, which limited the size and number of battleships and total naval tonnage, and the Kellogg-Briand Pact, which outlawed war as an instrument of national policy (both Germany and Japan signed it).

Further, the United States was determined to maintain a high tariff to protect American producers and to have a favorable balance of trade. Meanwhile the Federal Reserve returned to a policy of low interest rates while European central banks maintained high ones to protect the value of their currencies. The result was a largely unnoticed (at the time) cycle. American investment banks aggressively pushed highly profitable loans and underwritings in Europe. Europe used the proceeds to finance imports from the United States, and Germany used them to fund reparations to the Allied Powers. The Allied Powers then used the reparations to repay their war loans to the United States.

Thus the exported American capital quickly returned to the United

States and provided the wherewithal for more loans to Europe. As long as the cycle continued, everything was fine. But, of course, it didn't continue. The result would be a worldwide economic disaster.

In 1928 American investment bankers began increasingly to turn to a market even more lucrative than European loans, the call money market on Wall Street. *Call money* was the term for the funds used to finance stocks held on margin. At that time a speculator could buy stocks on as little as 10 percent margin, borrowing the rest from the broker. As long as the stock, which served as collateral for the loan, was headed upward, all was fine, and the speculator could increase his capital very quickly. But if the stock price declined, he had to put up more money or he would be sold out, often wiping him out.

The call money market was very lucrative in the late 1920s as Wall Street had entered one of its periodic booms, such as had occurred in the 1830s, the 1850s, and the 1870s. Once it recovered from the short-lived depression of 1920–21, Wall Street, as it always does, had reflected the growing American economy and had reached heights that had seemed impossible only a few years earlier. But in fact Wall Street, at least as measured by the Dow-Jones Industrial Average, increased far faster than the American economy. While the GNP increased by 59 percent in the 1920s, the Dow-Jones went up by 400 percent.

In 1928 the Federal Reserve acted to slow down the economy and thus, it hoped, the boom on Wall Street, which was showing signs of getting out of hand. By the fall of 1928, the New York Federal Reserve, headed by Benjamin Strong, had raised its discount rate to 5 percent from 3.5 percent and had begun clamping down on the money supply. The other eleven Federal Reserve banks had done the same. "The problem now," Strong explained, "is to shape our policy as to avoid a calamitous break in the stock market . . . and at the same time accomplish if possible the recovery of Europe."

Benjamin Strong, former president of Bankers Trust, had been governor of the New York Federal Reserve bank since it began in 1914, and he was the unquestioned leader of the system. The other banks and the board in Washington almost always did what he suggested. But Strong suffered from tuberculosis, and in the fall of 1928 he died after an operation on his lungs. The Federal Reserve was now effectively leaderless.

By the spring of 1929 the economy, reacting to the Federal Reserve's policy, had cooled noticeably. Wall Street, however, appeared not to have. Although usually a leading indicator of the economy, declining and rising before the economy does, the market, as measured by the Dow-Jones Industrial Average and the New York Times Index, continued to rise while the economy began to move into recession. But the wider market, including thousands of secondary stocks and those not included in the most widely watched averages, had begun to decline along with the economy.

Such is the economic power of human psychology, however, that neither the market nor the public noticed, and all attention was focused on the most widely held stocks. The market was caught up in a bubble, and bubbles are always recognized by the public only in retrospect.

Call money rates continued to climb as the action became more frantic on the Street, but speculators are notably indifferent to the cost of borrowing money in a rising market. Banks and even industrial corporations flocked to lend money to brokerage firms at 12 percent, while the brokers lent it in turn to their customers at 20 percent. Bethlehem Steel had $150 million in the call money market by the end of summer 1929. Chrysler had $60 million.

Banks began borrowing money from the Federal Reserve through the discount window at 5 percent and lending it to brokers. The Federal Reserve could have stopped this any time it wanted to, and there is little doubt that Benjamin Strong would have done exactly that. But, leaderless, the bank used only what it called "moral suasion" to urge the commercial banks to end the practice. It didn't work. When a bank can borrow at 5 percent and lend at 12, making a 7 percent return on someone else's money, it is going to do so.

By the summer of 1929 Wall Street and its millions of customers were deeply out of touch with the underlying economy, visions of riches dancing in their heads. Broker's boardrooms were full of people watching prices, and even normally sensible and knowledgeable people were caught up in the frenzy. Irving Fisher, a nationally known professor of economics at Yale, who had made a fortune by inventing the Rolodex, opined that "stock prices have reached what looks like a permanently high plateau." The *Saturday Evening Post* printed a poem that summer that caught the mood of the country.

> *Oh, hush thee, my babe, granny's bought some more shares*
> *Daddy's gone out to play with bulls and the bears,*
> *Mother's buying on tips, and she simply can't lose,*
> *And baby shall have some expensive new shoes!*

On September 3, the day after Labor Day, the New York Stock Exchange closed with the Dow at 381.17, a new all-time high. On September 5 a stock market analyst of no great note named Roger Babson addressed a luncheon group in Wellesley, Massachusetts. The frequent predictions of trouble ahead from this perennial pessimist had been ignored by a market that wanted only optimistic forecasts. On this day he didn't say anything different, merely noting that "I repeat what I said at this time last year and the year before, that sooner or later a crash is coming."

It was a slow news day, and the Dow-Jones Financial News Service put Babson's unexceptionable remark on the news ticker at 2 PM. The effect was extraordinary. In the last hour of trading, prices plunged (U.S. Steel fell 9 points, AT&T 6) and volume that final hour was a fantastic two million shares. Known as the "Babson break," it was like a slap across the face of a hysteric, and the mood of Wall Street changed abruptly from "the sky's the limit" to "every man for himself."

Over the next six weeks the market trended downward, with occasional plunges followed by more modest recoveries. Then, on October 23, a wave of selling swept the market on the second highest volume on record. A mountain of margin calls went out that night, and sell orders by the thousands piled up at brokerage houses across the country. The next day, Thursday, October 24, soon known as Black Thursday, was the most frantic in the history of the New York Stock Exchange up to that time, as stocks plunged, generating more margin calls, which caused more stock to be sold at any price, as the averages spiraled downward. Meanwhile, short sellers added to the downward pressure on stocks in bear raids.

A group of the Street's leading bankers met at J. P. Morgan and Company, across Broad Street from the exchange, to decide what to do. They raised a fund of $20 million to steady the market and entrusted it to Richard Whitney, acting president of the exchange. At 1:30 Whitney went to the post where U.S. Steel was traded and asked the price. He was told that the last trade had been at 205 but that it had fallen several points since with no takers. "I bid 205 for 10,000 Steel," Whitney shouted dramatically. He then went to other posts, buying other blue chip stocks in large amounts.

The effect was everything the bankers had hoped for. Short sellers ran for cover and the market steadied. At the end of the day, U.S. Steel closed up slightly. But the volume had been an utterly unprecedented thirteen million shares.

The rally continued on Friday, with modest profit taking at the Saturday morning session. On Monday selling resumed as rumors flew around the Street about major speculators having killed themselves and new bear pools being formed. The next day, Tuesday, October 29—Black Tuesday—there was no stopping the market. It plunged from the opening bell and kept plunging nearly continuously all day. Volume reached sixteen million shares, a record that would stand for forty years, and the ticker ran more than four hours late, not tapping out the last price until nearly eight o'clock that night. Someone calculated that the nation's tickers that day had used up fifteen thousand miles of ticker tape. The Dow-Jones average at the end of the day's carnage stood at 23 percent below where it had closed on Saturday, and nearly 40 percent below its high of early September.

No one could know it at the time, of course, but the greatest economic calamity in the nation's history had begun.

Chapter Sixteen

FEAR ITSELF

HISTORIANS INESCAPABLY BOTH WRITE with the benefit of hindsight and shape the stories they tell. So history always appears much tidier and more dramatic to the reader than the events depicted seemed to those who lived through them day by day. Human beings have to live with a future that is always unknown while enveloped in the fog of mere existence that can be as hard to penetrate as the far better recognized fog of war. Thus the three-and-a-half-year economic slide from the blue-sky prosperity of late summer 1929 to the black pit of depression in the first, winter months of 1933 seems smooth, perhaps even inevitable to most people today, only a few of whom have personal, adult memories of the time.

As a consequence, many view the story of the Great Depression as something like the story of the *Titanic,* with the stock market crash serving as the iceberg, and only a change of captains in the nick of time providing a different ending. In reality, the market crash was an effect of the forces moving the American and world economies into depression, not a cause. And Herbert Hoover did all he could—and far more than had any previous president faced with economic adversity—both to reverse the course the economy took in those years and to ameliorate the suffering it caused. It was only when the depression had destroyed the economic and historical limitations that had bound him that a new president could move to new, previously politically impossible remedies.

———

THE STOCK MARKET RALLIED SHARPLY on Wednesday, October 30, and the exchange announced that the market would open at noon the following day and then stay closed until Monday to give back offices time to cope with the mountains of paperwork that had piled up in the panic. The market also rose in the abbreviated Thursday session, and many thought the worst was over. They were wrong. The market fell sharply on Monday, November 5, and continued falling for two more weeks, by which time the *New York Times* average had given back all its gains since the summer of 1927.

All markets, however, bear and bull alike, finally run out of steam, and by December stocks were moving upward, although on much reduced volume. By year's end, while many of the high-flying stocks had been savaged, some market sectors—such as airplane manufacturers, department stores, and steel companies—actually showed gains for the year. It was widely thought the crash had been only a severe correction for a much overbought market. In January 1930 the *New York Times* thought the biggest news story of 1929 had been Admiral Byrd's flight over the South Pole.

Although the stock market had been a national obsession in 1929, its crash had not directly affected that many families. The stock exchange had reported that twenty million Americans owned stock in 1928, but in truth the number was not one-tenth that, and less than 2.5 percent of the population had brokerage accounts. While many of those who had been speculating and buying on thin margins had been wiped out, many others had resisted panic and still owned their shares.

One reason most thought the crisis had passed was that the banking system had not been showing signs of unusual stress. To be sure, 659 banks had failed in 1929. But that was slightly below the annual average for the decade, and no major banks had collapsed as a result of the crash. One reason was that the New York Federal Reserve bank and the New York brokers had both moved swiftly. The brokers announced that they would continue to carry accounts that were below margin requirements, thus avoiding additional at-the-market selling that would have further depressed prices.

The New York Federal Reserve helped by lowering its rediscount rate to 3.5 percent by March 1930, helping to lower interest rates generally, and by massive buying of federal securities, greatly improving the banks' liquidity. It had been buying no more than $25 million a week, but bought $160 million shortly after the crash and more in weeks to come. George Harrison, who had succeeded Benjamin Strong as governor of the New York Federal Reserve, had followed his predecessor's advice on dealing with such a situation when Strong had noted that "we have the power to deal with such an emergency instantly by flooding the Street with money."

But George Harrison did not have anything like the influence Strong had had with the other Federal Reserve banks, and he was criticized by the chairman of the Federal Reserve Board in Washington, among others, for overstepping his bounds. The Federal Reserve as a whole did not move decisively to add liquidity to the banking system nationally, a serious mistake.

The winter and early spring of 1930 saw the stock market rebound, regaining about 45 percent of what had been lost in the previous fall's debacle. Hoover had called a conference of businessmen in November 1929 and had urged them to invest in new construction, which many promised to do. And he telegraphed state governors—state governments at this time funded about 80 percent of government construction projects—to do the same. In the spring he proposed that, to stimulate the economy further, federal construction spending be increased by $140 million. That was no small sum in a federal budget that amounted to only $3.3 billion, or about 3 percent of GNP. Twenty-five percent of the federal budget went to debt service, and most of the rest to fund the 139,000-man army and the 95,000-man navy. This was, in fact, about as much as the federal government could have spent in that fiscal year anyway, as construction projects necessarily have long lead times before they can hire workers in large numbers.

And by spring 1930 it didn't look as if more would be needed. In May that year President Hoover was able to tell a religious group who had called on him to urge more public works that "You have come sixty days too late. The depression is over." Unfortunately, the president then made a fateful mistake to add to the earlier one of the Federal Reserve. He signed the Smoot-Hawley Tariff Act.

As a presidential candidate in 1928, Hoover had promised the nation's struggling farmers that he would call a special session of Congress to deal with the agricultural depression, which he did in the summer of 1929. Among his proposals was an increase in agricultural tariffs to protect the American market for American farmers. After the slowing of the economy in 1929, however, and the crash, Hoover's proposal was hijacked by special interests as nearly every industry in the country (tombstone makers, for instance) paraded before Congress, demanding protection from "unfair" foreign competition. Congressmen and senators, anxious to bring home the bacon for local industries, obliged them in case after case. The result was by far the highest tariff in American history.

This was economic folly. Tariffs are taxes, and taxes, inescapably, are always a drag on the economy. But, far worse, high tariffs breed retaliatory tariffs in foreign countries. A country cannot wall off its market from other countries and expect those countries to keep their own markets open. Professional economists knew this, of course, and a thousand of them signed a petition to President Hoover asking him to veto the tariff bill that had emerged from Congress. Thomas Lamont, second only to J. P. Morgan, Jr., at the Morgan Bank, wrote, "I almost went down on my knees to beg Herbert Hoover to veto the asinine Hawley-Smoot Tariff. That Act intensified nationalism all over the world."

The economists' arguments proved all too true, and world trade began to collapse. Great Britain, the great champion of free trade since the 1840s, and the world's largest trading nation, established the "imperial preference system"—in other words, a tariff wall—to keep British trade within the British Empire. Other nations adopted similar restrictions. In 1929 total global trade had amounted to $36 billion. In 1932 it was about $12 billion. American exports had been $5.241 billion in 1929. Three years later they were a mere $1.161 billion, a 78 percent drop and below the export level of 1896 if inflation is factored in.

As soon as the Smoot-Hawley tariff had been signed into law, the stock market began to give up its gains of the spring. By the fall, banks were beginning to fail in growing numbers. The rate of failure had been up in 1930 over 1929, but not alarmingly so. But then in the last two months of the year, 600 banks failed, making a yearly total of 1,352, more than twice the number of 1929. To be sure, most were the small, unitary banks that dotted the rural areas of the country and served the poor urban neighborhoods populated by immigrants.

The Bank of United States was another matter, however, being a large, many-branched bank, headquartered in the country's financial capital, New York City. It had deposits of $268 million, held by 450,000 depositors, most of them small merchants and working-class Jews employed in the city's vast garment trade. When it began to show signs of stress, New York State banking authorities and the New York Federal Reserve tried hard to rescue it. They wanted to merge it with three other banks, but needed a $30 million loan from the great Wall Street banking houses to make the deal work. Wall Street refused to help. "I warned them," Joseph A. Broderick, New York State's chief banking regulator, wrote, that "they were making the most colossal mistake in the banking history of New York."

The Bank of United States closed its doors on December 11, 1930, and was the biggest bank failure in the history of the country up to that time. Its collapse sent a shiver of fear throughout the American body politic and through Europe as well, where many thought from its name that it had an official connection with the federal government. If a bank this big could fail, many thought, what bank couldn't?

And it had been unnecessary. The Bank of United States had not been a well-run bank (two of its principals would end up going to jail), and there have always been charges of anti-Semitism leveled against the refusal of the white-shoe Wall Street banks to help. Regardless, if the Smoot-Hawley tariff had been Congress's biggest contribution to causing the Great Depression, Wall Street's refusal to help the Bank of United States was the American financial establishment's contribution.

Still, at the end of 1930, the country, while certainly experiencing hard times, was only in an ordinary depression, not even one as severe as that experienced in 1920–21. Then unemployment had averaged 11.9 percent. In 1930 unemployment did not reach 9 percent. It was in 1931 that the depression would become the Great Depression.

Again, as in 1930, in the early months of the new year it began to look as if a bottom had been reached and, in the phrase of the day (an increasingly ironic one, to be sure) prosperity was "just around the corner."

Then events in Europe intervened. On May 11, 1931, Credit Ansalt, Austria's largest bank and one of the most influential in Europe, failed. This was a far greater collapse than the Bank of United States, and a number of banks in Austria and Germany quickly followed it into oblivion. Germany's economy, already under immense economic stress thanks to the war reparations, began to implode.

Herbert Hoover, at the urging of Thomas Lamont of the Morgan Bank, proposed a one-year suspension of both loan repayments to the United States by the Allied Powers and German reparations to the Allied Powers. It was a very brave act of political leadership. Sophisticated Wall Street bankers like Thomas Lamont could understand what was at stake. The average American saw the suspension of Allied loan repayments simply as a means of transferring more of the cost of the European war to American taxpayers and of safeguarding the interests of Wall Street bankers. More and more, as the 1920s and 1930s progressed, Americans saw Woodrow Wilson's decision to end American isolation from Europe as a mistake. They wanted nothing to do with Europe or Europe's problems.

French opposition delayed acceptance of Hoover's plan (in fact, payments on both the American loans and German reparations would never resume). On July 13, 1931, Danat Bank, Germany's largest, suspended operations. The German government had no choice but to close the Berlin stock exchange and the city's banks. The European financial system was in danger of collapse, and the crisis soon spread to London, the system's heart, and to sterling, the most important currency in the world after the dollar. With the British budget deep in deficit, thanks to the depression, sterling came increasingly under pressure as banks and traders dumped sterling for gold, which flowed alarmingly out of the Bank of England in consequence.

On September 21, despite loans of 25 million pounds from both the New York Federal Reserve and the Bank of France, Britain went off the gold standard, which Britain itself had first established in 1821. Britain's days as a financial Great Power were over. Because so much of the world's trade was conducted in sterling and so many currencies both within and without the British Empire were tied to sterling, the effects were widespread, and more and more central banks had no choice but to abandon the gold standard as well.

In the United States, however, the Federal Reserve moved aggressively to defend the dollar and maintain the gold standard as foreign central banks and investors moved to repatriate gold. It was an utterly disastrous decision, perhaps the greatest of all the mistakes made in these years. Maintaining the gold standard required raising interest rates and cutting the money supply, causing an already severe deflation to become much more severe. Banks called in loans to stay liquid, while customers postponed purchases in expectation of lower prices. In the next year and a half more than half a million mortgages would be foreclosed. Unemployment ballooned. And bank depositors, well aware of the sudden rash of bank failures a year earlier, rushed to withdraw their money from banks while the withdrawing was good. In just the first month after Britain abandoned the gold standard, 522 American banks failed.

By the end of 1931 the United States faced economic circumstances such as it had never faced before. Bank failures had totaled 2,293 that year, each a tragedy for hundreds or thousands of families who saw their one bulwark against the consequences of unemployment vanish. And unemployment spread relentlessly through the American economy. GNP fell by a further 20 percent; automobile production, which had been 4.5 million in 1929, fell to 1.9 million in 1931, causing massive layoffs not only in the automobile companies, but at rubber, glass, and steel companies, auto dealerships, and insurance companies as well.

Unemployment by the end of 1931 rose to 15.9 percent, and the country's roads began to fill up with men in shabby clothes searching in vain for a job, any job. In rural areas it became not uncommon for people to answer a door and find someone asking for work in exchange for a meal or the right to sleep in the barn. In cities breadlines stretched for blocks and ramshackle communities of huts and lean-tos—dubbed Hoovervilles—spread in parks and open spaces as those who had lost their houses and apartments sought shelter. In some ways the whole economic system of the country seemed to be breaking down. Unsaleable crops rotted in the country, while people in the cities picked through garbage cans looking for something to eat in what was supposed to be the richest country on earth.

With tax receipts plummeting by $900 million while expenses rose by $200 million, the federal budget went into deficit by half a billion dollars in fiscal 1931. That does not seem like much in today's fiscal lexicon, but it represented a deficit equal to almost 13 percent of revenues, the worst peacetime deficit since the dark days of the early 1890s.

At that time, it was the nearly unquestioned conventional wisdom that balancing the budget was the first priority of the federal government after the defense of the Republic. The idea that governments *should* spend in deficit in times of economic crisis would not be fully explicated until John Maynard Keynes published his seminal work, *The General Theory of Employment, Interest, and Money,* in 1936.

But while Keynes's intellectual underpinnings were not yet available, the idea of deficit spending to provide economic stimulation was not unknown. Indeed, Hoover, who was perhaps the most economically sophisticated man ever to be president, had argued for exactly that in a cabinet meeting as early as May 1931. "The President likened [the situation] to war times," Henry Stimson, secretary of state, wrote in his diary about one cabinet meeting. "He said in war times no one dreamed of balancing a budget. Fortunately we can borrow."

Unfortunately, Hoover changed his mind in late 1931 and asked Congress for a huge tax increase to balance the budget. The House had gone Democratic in the election of 1930. The balance had been 267 Republicans, 167 Democrats in the first Hoover Congress, but was 220 Democrats, 214 Republicans in the second. The Senate after 1930 was nearly evenly split, with 48 Republicans and 47 Democrats. But both houses passed Hoover's tax bill with little opposition. The Democrats in Congress were largely from the South and West, and a balanced budget for them was nearly holy writ. Speaker of the House John Nance Garner of Texas, in addition to Hoover's income tax increases, had wanted to add a national sales tax as well, a tax that would have fallen most onerously on the poor.

Hoover also proposed a far better idea than raising taxes in the teeth of a growing depression, the federal Home Loan Bank Act, which would have created a number of Home Loan banks empowered to lend money on the mortgage portfolios of commercial banks. The Federal Reserve Act of 1913 had forbidden the Federal Reserve to loan money through its discount window on such collateral. The effect of that prohibition was to freeze hundreds of millions of dollars in banking assets that could otherwise have been used to add liquidity to the banking system. Congress, however, dawdled about passing the legislation until July 1932, and—the ghost of Thomas Jefferson still abroad in its halls—raised the collateral requirements Hoover had wanted.

Hoover, although long opposed to direct federal relief to banks, industrial corporations, or individuals, also proposed a radical new means to ameliorate the situation, the Reconstruction Finance Corporation. The "dole" was widely thought by Americans of all political stripes to be a European aberration that would make citizens wards of the state. But Hoover was intellectually flexible enough to realize that desperate times call for desperate measures. The RFC, capitalized at $500 million by Congress and authorized to issue up to $2 billion in tax-exempt bonds, did not provide direct relief for individuals. But it did provide emergency loans to banks, life insurance companies, farm mortgage associations, and railroads that otherwise might have collapsed. The bill was signed into law by Hoover on February 2, 1932.

Because it necessarily saved the stockholders of the companies aided, many, like Fiorello H. La Guardia, called it "a millionaire's dole." But it also saved the employees and depositors. And once the barrier to direct federal aid to corporations was broken, could direct federal aid to individuals be far behind? It was not, although it was disguised with a fig leaf. In July 1932 the Relief and Reconstruction Act authorized the RFC to finance public works up to $1.5 billion to generate jobs, and to provide up to $300 million to the states—many of whom were up against constitutional limits on spending and borrowing—so that they could provide direct relief.

Within six months, the RFC had loaned out $1.2 billion, equal to fully a quarter of all federal expenditures that year. And the RFC would prove to be the model for much of the early New Deal. But Hoover would get no credit for it. Instead the first historians of the time would freely use him as a foil against which to set off the glory of his successor. John F. Kennedy once famously remarked that "life is not fair." He would have been the first to agree that neither is history.

By that time it was far too late to save the political, let alone historical fortune of Herbert Hoover. At the end of May about a thousand men who had served in the First World War came to Washington to demand that a bonus due to be paid them in 1945 be paid immediately. A bill to issue $2.4 billion in fiat money to do so passed the House but was defeated in the Senate. Meanwhile, the "Bonus Expeditionary Force" grew alarmingly in numbers, reaching seventeen thousand by the middle of June, mostly camped on the Anacostia Flats on the outskirts of the District. Some, however, built shacks on government property near the Capitol and occupied several government buildings on Pennsylvania Avenue.

Congress authorized funds to pay for the veterans' transportation home, and most departed Washington peaceably. But about two thousand refused to go home, and violence erupted when Washington, D.C., police tried to evict them. Two veterans and two policemen were killed. Hoover called on the army, under Chief of Staff Douglas MacArthur, to evict them from government property and limit them to the Anacostia Flats.

On July 28 MacArthur, in full strutting-rooster mode, personally led fully armed and prepared cavalry and infantry, along with six tanks, to clear the government buildings. But then, in flat contradiction of his orders from the president, he also cleared the flats and burned down the shacks that had been erected there, scattering the remnants of the Bonus Army. Hoover should have fired MacArthur on the spot for gross insubordination. Instead he took full responsibility and paid a fearful political price when Americans were profoundly shocked by images of mounted cavalry, sabers unsheathed, chasing the unemployed down the streets of the nation's capital city.

The president was now routinely depicted in editorials and cartoons as a technocrat who was indifferent to the suffering of the American people, a suffering that could be seen on every streetcorner and country lane. And nothing, it seemed, could stop the descent of the American economy into the bottomless pit of the Great Depression.

The government deficit in 1932, despite Hoover's tax increases, was $2.7 billion. Revenues had been a mere $1.9 billion. It was the worst peacetime deficit in the nation's history. Gross national product that year was $58 billion, a mere 56 percent of what it had been three years earlier. Unemployment stood at an entirely unprecedented 23.6 percent. But that did not tell the whole story, for millions more were working part-time or at much reduced wages. The number of hours of labor worked in 1932 was fully 40 percent below the level of 1929. Another 1,453 banks had failed, bringing the depression total to a staggering 5,096. In 1929 Americans had held about $11 in bank deposits for every dollar in currency and coin in circulation. By 1932 the ratio was five to one, because so many banks had failed and so many more were distrusted. The Dow-Jones Industrial Average fell as low as 41.22, down 90 percent from its high of three years earlier and less than a point above where it had stood the first day it had been calculated in 1896.

It is a measure of how desperate the situation had become by the fall of that year that the interest rate of Treasury bills went negative. Treasury bills, which have maturities of less than one year, are sold at a discount and mature at par. But by the fall of 1932, so many people who still had investable assets wanted to invest in what is by definition the safest of all possible investments—the short-term obligations of a sovereign government—that the price rose above par.

———

THE MAN WHO OPPOSED HOOVER for the presidency that year, Franklin Roosevelt, was his opposite in more ways than politics. Hoover had been born into a family of modest means and had been soon orphaned. Roosevelt had been born rich into a family that lavished affection on their only child. Hoover was shy, sometimes dour, at his best in small groups; Roosevelt was outgoing to a fault, chronically ebullient, and utterly certain of himself. Hoover was both highly intelligent and an intellectual (by no means necessarily the same thing), with an engineer's logical and tidy mind. Roosevelt seldom read a book as an adult, and was highly intuitive.

But if Roosevelt had only a second-class intellect, as Justice Oliver Wendell Holmes is supposed to have observed, he had a first-class temperament. His charm was both undeniable and irresistible. And his unfeigned optimism was highly infectious.

The election of 1932 was not determined by the issues—the two men actually differed little in what they called for to meet the emergency, and Roosevelt hammered Hoover for having failed to balance the budget. It was determined instead by personality, and Hoover, burdened by four years of ever-deepening economic disaster, didn't stand a chance. In 1928 he had carried forty states against Al Smith. In 1932 he carried six against Roosevelt. The nation, in its agony, had placed all its fast-dwindling store of hope upon one man, whose greatest asset was the fact that he never once doubted that he was the right man for the job.

In the long interregnum between the election on November 8, and the inauguration on March 4 (the last before the date changed to January 20), the economic situation continued to deteriorate alarmingly. Raymond Moley, one of Roosevelt's closest advisers, wondered if they might be facing revolution by the time the new administration was sworn in.

But the American people, especially compared with Europeans, were surprisingly unprotesting of the situation in which they found themselves. Socialism—with its promises of equality and security—had long been an important political force in all the major European countries.

But even in November 1932, as the American economy seemed to be on the verge of collapse, Norman Thomas, the Socialist candidate, could not get more than 2.2 percent of the vote.

As they lived through the most desperate winter the country had known since the Continental Army had camped at Valley Forge, the American people waited for the end of the Hoover presidency and the accession to the White House of the Hudson River aristocrat they had elected overwhelmingly. Things only got worse while they waited.

The index of industrial production between December and March dropped 12.5 percent, from 64 to 56, an all-time low. Gold continued to move abroad in large amounts, sometimes at the rate of $100 million a week. Farm mortgages were being foreclosed at the rate of twenty thousand a month. On February 14, 1933, the governor of Michigan ordered all his state's banks closed for eight days to prevent a fast-spreading panic there from destroying what was left of the state banking system.

The next day even the last, best hope of the American people was very nearly dashed. Roosevelt had spent several days cruising in Florida waters on the vast yacht of his Dutchess County neighbor Vincent Astor, and on his return he drove in an open touring car to a Miami park. There he made a short speech from the backseat of the car. When he had finished, Anton Cermak, the mayor of Chicago, came up to talk with him for a few seconds. Shots rang out and Mayor Cermak crumpled over from the bullets that had been meant for the president-elect. Roosevelt, who counted great personal courage and coolness among his attributes, had the mortally wounded mayor put in the seat beside him and sped to a hospital. He himself was unharmed.

The banking panic that had started in Michigan, one of the most important industrial states, meanwhile spread like wildfire throughout the country. Banks were besieged everywhere by frantic depositors shouting for their money. State after state followed Michigan in ordering its banks to close. By March 4 they were entirely closed in thirty-two states and nearly all closed in six others. Governors had sharply limited withdrawals in the ten states where banks continued to function. In Texas no more than $10 a day could be withdrawn. On inauguration day itself, the New York Stock Exchange announced that it would not open that morning and did not state when it would open again.

With most of the nation's banks and its premier stock exchange closed, the very heart of the American capitalist economy had nearly ceased to beat.

Chapter Seventeen

CONVERTING RETREAT
INTO ADVANCE

I F THERE WERE ANY DOUBT as to the primacy of human psychology as a force in the economic universe, one need only look at the first few days of the Franklin Roosevelt administration for proof. On Saturday, March 4, as millions listened on the radio, Roosevelt gave one of the handful of inaugural addresses that have been remembered beyond the day it was given. Its very first paragraph gave us a phrase that instantly became part of the American political fabric. "So, first of all, let me assert my firm belief that the only thing we have to fear is fear itself—nameless, unreasoning, unjustified terror which paralyzes needed efforts to convert retreat into advance." The speech had a near magical effect on the American people. In the first week the White House received 450,000 letter and cards. Hoover had needed one clerk to handle the White House mail; Roosevelt would need seventy.

The next day he summoned Congress into special session for the following Thursday, issued an executive order—under the dubious authority of the Trading with the Enemy Act passed during the First World War—closing all the nation's banks until after Congress met, and convened an emergency meeting of major bankers. The Treasury Department, still largely staffed by Hoover's men, and the bankers worked fever-ishly in the next few days to prepare the Emergency Banking Relief Act.

On Wednesday, March 8, Roosevelt called his first press conference, and 125 reporters crowded into the Oval Office. At the end of his pre-pared remarks the reporters—for perhaps the only time in history—broke into spontaneous applause. The American people, including reporters, wanted FDR to succeed, and in no small degree because of that, he did.

The banking bill was presented to the House at one o'clock on Thursday and passed, unread, by acclamation thirty-eight minutes later. The Senate passed it with only seven dissenting votes (all from rural states), and the president signed it into law at 8:36 that evening.

The act authorized what Roosevelt had already done and conferred on him vast new powers to regulate the banking system and foreign exchange in the future. It set Monday, March 13, as the day for banks declared sound to reopen. On Sunday, March 12, he gave the first of his fireside chats. In a voice that was both aristocratic and avuncular, didac-tic and soothing, he told his vast audience that when the banks began to reopen the next day, it would be "safer to keep your money in a reopened bank than under the mattress."

The people believed him, and the next day money and gold began to flow back into the banking system. The heart of the American economy began to beat again. Raymond Moley, with justifiable pride, reported that "capitalism was saved in eight days."

THE CONSTITUTION of the Roman Republic, its citizens fearful of exec-utive authority after the overthrow of the kings, had divided it between two consuls, who served for a year and alternated daily in command of both the government and the army. But the Romans realized that, in an emergency, such a system wouldn't work. And so the constitution allowed, when necessary, for one man to hold absolute power for six months. The term for this temporary official was *dictator*.

For the first three months after his inauguration—the so-called Hundred Days—Franklin Roosevelt was the American dictator, in the very best sense of that term.

The record of legislation passed and signed into law is simply astonishing.

- March 9. Roosevelt signed the Emergency Banking Relief Act.

- March 20. Roosevelt signed the Economy Act, reorganizing the government and cutting salaries and the pensions of veterans—perhaps the most potent lobby in Washington at that time—to reduce expenses by $500 million.

- March 21. Roosevelt signed the Civilian Conservation Corps Reforestation Relief Act, to employ up to 250,000 young men in construction and environmental projects.

- March 22. Roosevelt signed the Beer-Wine Revenue Act, legalizing beer and wine with less than 4 percent alcohol and taxing it heavily to increase government revenue.

- April 19. Roosevelt took the country off the gold standard, demonetized gold by making gold coins no longer legal tender and recalling them to the Treasury, and forbidding citizens to hold bullion. The next year he devalued the dollar from $20.66 to an ounce of gold to $35.00.

- May 12. Roosevelt signed the Federal Emergency Relief Act to provide grants totaling $500 million to states to fund relief for the unemployed.

- May 12. Roosevelt also signed the Agricultural Adjustment Act to relieve farmers with measures to raise farm prices, limit production, and refinance farm mortgages.

- May 18. Roosevelt signed the bill authorizing the establishment of the Tennessee Valley Authority to develop the Tennessee River Valley by building dams that would provide electric power in seven states.

- May 27. Roosevelt signed the Federal Securities Act, which required full disclosure of pertinent information to investors, the first federal regulation of the securities business.

- June 5. Congress by joint resolution canceled clauses in contracts requiring payment in gold.

- June 6. Roosevelt signed the National Employment Act establishing the U.S. Employment Service to work with state employment agencies to help the unemployed find jobs.

- June 13. Roosevelt signed the Home Owners Refinancing Act establishing the Home Owners Loan Corporation, which was empowered to issue $2 billion in bonds to help nonfarm home owners keep their properties.

- June 16. Roosevelt signed the Banking Act of 1933, usually known as the Glass-Steagall Act after its congressional sponsors. It revolutionized American banking.

- June 16. Roosevelt also signed the Farm Credit Act to help refinance farm mortgages.

- June 16. Roosevelt also signed the Emergency Railroad Transportation Act to increase federal regulation of railroads and railroad holding companies.

- June 16. Roosevelt also signed the National Industrial Recovery Act to create the National Recovery Administration (NRA, in the alphabet soup for which the New Deal became famous), which established a series of industrial cartels to limit "excessive competition."

By definition, a depression is a period of economic contraction, and Roosevelt's Hundred Days brought the contraction that had begun in early 1929 to a close. As the American people, thanks to the force of the president's personality, his swift moves, and his unquenchable optimism, began to have renewed faith in the economy and in the country, recovery began. The year 1933 would prove to be one of the best years of the twentieth century on Wall Street, although, of course, rebounding from a disastrously low base. The Dow that year rose almost 60 percent, and some brokerage firms even began hiring again.

The country was a very long way from prosperity, but, still, the upward path was soon unmistakable. The GNP was a mere $55.6 billion in 1933, its lowest since 1916, without taking inflation into account. The next year it was $65.1 billion. By 1937 it was $90.5 billion. The money supply and wholesale prices also increased at the rate of 10 to 12 percent annually in the four years following early 1933. Unemployment, however, stayed stubbornly high, dropping only to 14.3 percent in 1937. Much of the renewed economic activity was performed by employees who had been working part-time in what were supposed to be full-time jobs, delaying the need to hire new workers.

(One often overlooked, but highly favorable consequence of the persistent unemployment in the 1930s was the fact that children tended to stay in school longer. The number of people receiving high school diplomas almost doubled in the 1930s, while those receiving college degrees increased by 50 percent. In 1940 some 8.1 percent of twenty-three-year-olds received bachelor's degrees.)

NOWHERE DID THE EARLY NEW DEAL have a greater effect on the American economy than in banking, the one sector that had come closest to utter destruction as the depression had deepened. The Glass-Steagall Act greatly strengthened the Federal Reserve's control over the nation's banking system, making many more banks, such as savings banks, eligible for membership. It also gave the Federal Reserve the power to control speculation on Wall Street by setting margin requirements.

In 1935 the Federal Reserve Act further increased and centralized the powers of the Federal Reserve. The heads of the regional Federal Reserve banks, who had been called "governor"—the title of power in central banking—were restyled presidents, and the Federal Reserve Board in Washington became the Board of Governors, where the main power has resided ever since. Each of its members was appointed to a fourteen-year term and its chairman appointed from among them for four years. To ensure independence from politics—which Alexander Hamilton had recognized as necessary 150 years earlier—the members of the Board of Governors could not be removed except for cause.

Open market operations, a major tool for regulating the banking system and interest rates, were centralized in the Federal Open Market Committee in Washington rather than being handled by the twelve regional banks. This committee consisted of the seven members of the Board of Governors and five members appointed by the regional Federal Reserve banks, one being always appointed by the New York Federal Reserve. The Federal Reserve was also empowered to set reserve requirements for member banks, another powerful tool to control interest rates and the money supply.

For the first time in ninety-nine years, since President Andrew Jackson had destroyed the Second Bank of the United States, the country had a fully functioning and empowered central bank. The country had paid dearly for the lack of one in those years, not least because the country had been unable to develop much expertise in the arcane specialty of central banking.

The Glass-Steagall Act also established the Federal Bank Deposit Insurance Corporation (FDIC), which guaranteed the deposits of banks that joined the system (only banks that were members of the Federal Reserve were required to join) up to $5,000 per account. At a stroke, the bank run, a recurring nightmare in the American economy since the first one in 1809, became a thing of the past. Roosevelt had worried about the "moral hazard" created by a system that relieved bankers of the worry that their depositors' assets would be wiped out. But he decided that eliminating bank runs was worth it. There has not been a significant American bank run since, but events long after his death would prove that Roosevelt had been right to worry.

Glass-Steagall greatly strengthened national banks by permitting them to branch within the states where they were headquartered, if that state permitted branch banking. This allowed these banks to diversify over a larger area and thus not be so subject to purely local economic fluctuations, such as a major layoff by a large local employer. Unfortunately the reform did not go far enough, and interstate branching was still not permitted, limiting the size and resources of banks.

But Glass-Steagall also greatly weakened the largest and strongest banks by forcing those that had both a deposit and an investment business to choose one or the other. J. P. Morgan and Company, for instance, remained a depository bank and spun off Morgan, Stanley, and Company.

As corporations had grown in size and profitability in the early decades of the twentieth century, they became much less dependent on the Wall Street banks to finance expansion and acquisitions, using instead their own retained profits and the commercial paper market. So Morgan no longer possessed the overwhelming influence in the American economy that it had had at the turn of the twentieth century. But Glass-Steagall weakened Morgan and similar banks much further.

The separation of deposit and investment functions into noninterlocking companies was required because having the two businesses under the same management was thought to create an inevitable conflict of interest that had exacerbated the banking crisis of the early 1930s. In fact, the evidence for this was slight, and within a couple of years even

Senator Carter Glass, the father of the bill (and, indeed, the father of the Federal Reserve System itself twenty years earlier), recognized that this separation had been a mistake.

The iron law that it is far easier to pass a bill than to repeal one, however, held, and it would be more than sixty years before this part of Glass-Steagall would finally be reversed in an utterly different economic universe than that which had brought it into existence. And only then would banks finally be allowed to have branches in more than one state.

After the reforms of the early 1930s the American banking system would prove to be stable and able to meet the needs of the country. But it would remain the most Byzantine in the world, with many overlapping regulatory agencies at both the state and federal levels. It would also continue to have more banks. Despite thousands of mergers, there would still be more than seven thousand independently chartered banks in the United States in the year 2003, more than the rest of the developed world put together.

WALL STREET CHANGED PROFOUNDLY as well in these years, although it put up far more resistance than did the banking industry. The resistance was led by Richard Whitney, the hero of Black Thursday, who became president of the New York Stock Exchange in his own right in 1930. A series of congressional hearings, remembered by the name of the lead counsel, Ferdinand J. Pecora, had revealed much that was indefensible with the way the Street had operated.

The New York Stock Exchange was a private organization owned by its seat holders. Although the exchange had also long been a vital mechanism in the country's financial system, it still operated solely for the benefit of those seat holders. Specialists, who were exchange members who maintained orderly markets in the various listed stocks, were in a privileged position to know what those stocks were likely to do in the near future. Floor traders were also members, but ones who traded only for their own accounts. Because they had access to the trading floor, they too had what amounted nearly to insider information. The specialists and the floor traders throughout the boom years of the 1920s manipulated stocks by forming pools—"took them in hand," in the phrase of the day—to fleece the outsiders.

As long as stocks boomed in general, there was little sentiment to reform the ways of the exchange. But after the crash, the pressure to reform grew exponentially as more and more abuses came to light. Many on Wall Street welcomed and even worked for reform, especially the brokers who did a regular retail business and were thus dependent on the goodwill of the public.

Those who benefited from the status quo, such as the specialists and major speculators, of course, resisted it ferociously. The situation in the early 1930s was very similar to the situation in the immediate post–Civil War era. Richard Whitney, the leader of the so-called Old Guard, proclaimed that "the exchange is a perfect institution." But even he tacitly admitted it wasn't a perfect institution when he introduced several reforms, including forbidding specialists from taking options in stocks in which they made a market, and giving insider information to friends.

Roosevelt intended much more thorough reform than that, and in 1934 Congress established the Securities and Exchange Commission to oversee the industry. His first chairman of the SEC was Joseph P. Kennedy, a highly counterintuitive choice, seeing as Kennedy had been one of the most successful and ruthless of the Roaring Twenties speculators. *Newsweek* magazine wrote acidly at the time that "Mr. Kennedy, former speculator and pool operator, will now curb speculation and prohibit pools." The Senate, leery of appointing a fox to safeguard the henhouse, delayed his confirmation for six months while it waited to see how Kennedy would perform.

Kennedy was far too smart—and too rich—to try to profit from his position, and he did such a good job of getting the SEC on its feet that the Senate confirmed him with neither discussion nor dissent. Kennedy, who certainly knew where the bodies were buried regarding Wall Street shenanigans, nevertheless regarded his most important task to be ending the so-called strike of capital, the reluctance of the major underwriting banks, badly shell-shocked, to underwrite anything, regardless of how sound the deal.

Kennedy resigned after sixteen months, and it fell to the third chairman, William O. Douglas (later a justice of the Supreme Court for more than thirty years), to bring fundamental reforms to Wall Street. He would have the entirely unintended help of Richard Whitney. Whitney had retired from the office of president of the New York Stock Exchange in 1935, but he remained on the board of governors and was, by far, the best known broker on Wall Street. "I mean the stock exchange to millions of people," he told his successor as president.

Whitney lived lavishly, with a town house on East Seventy-second Street; a large farm in New Jersey, where he often foxhunted—one of the most expensive sports one can pursue on land—and numerous club memberships. He was spending about $5,000 a month at a time when the average annual per capita income was about $700. Unfortunately, he couldn't afford it. His brokerage firm, while it numbered the Morgan bank among its clients (which conferred great prestige but not much business), actually earned very little, and his investments had all lost money. He maintained himself by borrowing from friends and acquaintances, especially his older brother, George Whitney, who was a partner of J. P. Morgan and Company.

When that didn't suffice, he began dipping into the accounts of his clients, his clubs, even his wife's trust fund. As embezzlements usually do, his fell apart, and on March 7, 1938, the New York Stock Exchange interrupted trading to announce that Richard Whitney and Company had been suspended, for "Conduct contrary to just and equitable principles of trade."

The Wall Street establishment was shocked beyond measure. "Wall Street could hardly have been more embarrassed," wrote *The Nation,* a leftist political journal, gleefully, "if J. P. Morgan had been caught help-ing himself to the collection plate at the Cathedral of St. John the Divine."

Six thousand people turned out in Grand Central Terminal on April 12 to watch Richard Whitney, in handcuffs, board the train that took him to Sing-Sing prison. William O. Douglas, meanwhile, moved swiftly to take advantage of the utter disarray of the Wall Street Old Guard. Before the end of the year, the stock exchange had a new constitution, one that took its public responsibilities into account. The president became a paid employee, not a member. Firms had to submit to much more intrusive audits, and members could not buy stock on margin if they did business with the public. The permissible ratio of debt to capital was lowered, making brokerage firms more stable and able to survive marked downturns in the market.

Even more important, short sales could now only be made on an uptick—that is, at a higher price than the previous sale. This ended one of the principal means by which bear raids had so roiled the market in the 1920s and helped lessen the severity of panics such as the market had known in 1929. By the end of the 1930s Wall Street was ready to take advantage of the revolution in stock trading that would happen in the next decade.

———

THE SUPREME COURT would invalidate much of what came to be called the First New Deal, but a fundamental shift had nonetheless taken place in American politics. For one thing, the Democratic Party, under the leadership of one of the ablest American politicians of his or any day, became the majority party. In the sixty-two years after the election of 1932, the Democrats would control the House of Representatives for all but four years, in 1947–48 and again in 1953–54. After the 1936 election—when Roosevelt, in the greatest landslide in American history up to that time, carried forty-six states—the Republican Party would have only sixteen senators and eighty-nine representatives in Congress.

The amount of GNP that flowed annually through the federal government began a steady rise. Government outlays in 1929 had been about 3 percent of GNP. In 1940 they were more than 9 percent. The national debt, meanwhile, had swelled from 16 percent of GNP to more than 50 percent, by far the greatest peacetime increase up to that point. The effect has been to make the federal government's budget a sort of fiscal flywheel, providing economic energy to the system in times of slack demand and preventing a self-reinforcing downward spiral such as occurred in the early 1930s. Unemployment since the Second World War has only once, and then briefly, risen above 10 percent.

And national fiscal priorities had changed both fundamentally and permanently. Before the onslaught of the Great Depression, balancing the budget and paying down the debt had been the most important fiscal responsibilities of the federal government. After 1933, preventing a new Great Depression became the most important responsibility, in some ways the only one. Further, government came to be perceived as, at least, the provider of last resort of the essentials of an adequate standard of living.

This meant, inevitably, that government influence on the economy would increase as well. The national debt would never again be substantially reduced in dollar terms (although it would vary greatly as a percentage of GNP). And the annual budget, which had been in surplus two-thirds of the time before 1933, would be in surplus only 16 percent of the time after that date.

Most of all, while many of the New Deal programs were unsuccessful and many of its economic principles shortsighted, in its totality it was an enormous success. The country since the New Deal has been a far richer, far more economically secure, far more just society. It has been one that has proved to offer far more opportunity for all and produce far more wealth as a consequence. There has never been a serious political effort to reverse the New Deal, although its worst ideas—such as the cartelization of much of the American economy—were discarded and much of it has been reformed, for democracy is a process of endless reform.

In a very real sense—to paraphrase Richard Nixon—we are all New Dealers now.

When the Supreme Court threw out most of the early New Deal (some of whose programs, notably the NRA, had not been working well anyway) the Roosevelt administration, ever pragmatic, tried new programs. In 1935 it initiated what came to be called the Second New Deal. This called for Social Security, further banking reform, more extensive works programs (including the WPA), higher taxes on high incomes and inheritances (including a so-called wealth tax on estates more than $50 million—a threshold only a handful of families met), and strong protections for organized labor to make it easier for unions to organize workers.

This revolutionized the American workplace, especially in manufacturing, which was then the heart of the American economy. The reform of labor relations in this country had begun, as so much else attributed exclusively to the New Deal, under Herbert Hoover. The Norris–La Guardia Act of 1932 made "yellow dog" contracts—in which workers had to agree not to join a union—unenforceable and forbade injunctions against strikes and picketing. But organized labor had been badly hit by the depression. In 1933 the membership in the AFL was only 2.3 million, about what it had been at the turn of the century.

In 1935 the National Labor Relations Act (usually known as the Wagner Act, after its chief congressional sponsor, Senator Robert F. Wagner of New York) was passed. It has often been called the Magna Carta of American organized labor. It guaranteed the right of workers to join a union of their choice and to bargain collectively with their employers. Further, it included a long list of "unfair labor practices" that companies were forbidden to engage in (but, significantly, did not name any practices that unions were forbidden to engage in). It established the National Labor Relations Board to police the labor marketplace and supervise elections. Most major industrial states soon passed laws modeled on the Wagner Act.

American government, long almost instinctively on the side of management, had now swung decisively behind labor. While there was violence as plant after plant was organized, it was far less deadly than the labor violence of the late nineteenth century. Even the aged Henry Ford, essentially a man of the nineteenth century and long adamantly opposed to unionization, had no choice but to bargain with Walter Reuther and the Automobile Workers Union in 1939.

In the six years following passage of the act, union membership more than doubled. By the early 1950s unions would represent about 35 percent of the American workforce. Membership in the craft unions of the AFL advanced greatly, but the greatest gains were among unskilled and semiskilled workers who most needed unions to protect their interests. The Congress of Industrial Organizations (CIO), established as an independent labor organization under the leadership of John L. Lewis of the Mineworkers Union in 1937, had 2.7 million members by 1941.

Another crucial part of the Second New Deal was the move to bring electricity to vast areas of the country that did not have it. The cost of providing electricity to a given area—in terms of building generating capacity and running lines—is nearly the same whether that area is densely or scarcely populated. Thus if the population density is below a certain level, it is simply prohibitively expensive on a per-capita basis for utilities to provide service. The spread of cable television in more recent decades had exactly the same spotty distribution for precisely the same economic reasons.

Because private enterprise could not bring electricity to much of the countryside, Roosevelt established the Rural Electrification Administration to do so. It worked to form publicly owned, nonprofit, electrical cooperatives to provide power to areas that did not have it. When the REA was formed in 1935, only two farms in ten in the United States had electricity. Just a little more than a decade later, eight in ten did. This not only greatly increased the economic productivity of these farms (and increased the flow of workers from farming to other sectors of the economy), but equally increased the quality of life in rural areas and brought the inhabitants into much closer contact with the nation as a whole through such means as radio and the telephone, whose wires could be strung on the same poles as the electric wires.

———

IN 1937 THE FEDERAL RESERVE, using its newly granted power, began to increase bank reserve requirements sharply for various technical reasons. At the same time, the Roosevelt administration began to cut public works spending to help bring the budget closer into balance. The result was a new depression. Unemployment soared back up to 19 percent the following year, while GNP dropped 6.3 percent. It was the first time in the history of the American economy, and the last time, so far, that the peak of the business cycle was lower than the previous peak had been, as the height in 1937 was well below the peak in 1929.

While technically the economy had been in recovery for four years, in popular parlance the word *depression* was applied to the whole decade of the 1930s. So economists dubbed this new depression within a depression a "recession." This has been the term for economic downturns ever since, and the word *depression* is now usually capitalized and refers exclusively to the uniquely dark days of the 1930s.

Recovery began again in 1938, but unemployment remained stubbornly high, being at 14.6 percent as late as 1940. In the end it was not the New Deal that cured what ailed the American economy. It was war. The dreadful peace devised at Versailles in 1919 turned out not to have been a peace at all, but merely a twenty-year truce, an interlude between the worst war in human history and one that would be far, far worse in terms of lives lost and treasure squandered.

And as with the first great war of the blood-drenched twentieth century, when the bombs stopped falling, geopolitical power would have been radically redistributed in favor of the United States.

PART V

A NEW ECONOMIC REVOLUTION

Tyranny, like hell, is not easily conquered.

— Thomas Paine
The American Crisis, 1776

Since the New Deal and the 1930's there has been a revolutionary development in the technology of industry and in the fiscal policy and social doctrine of governments. The assumption of reformers from Theodore Roosevelt through Woodrow Wilson to Franklin Roosevelt was that the poor could be raised up only by a redistribution of wealth. The basic assumption of the pre-war reformers is being dissolved. We have come into an era when the class struggle, as Marx described it a hundred years ago, has been overtaken by events.

— Walter Lippmann, 1964

The hope is that, in not too many years, human brains and computing machines will be coupled together very tightly, and that the resulting partnership will think as no human brain has ever thought.

— J. C. R. Licklider, 1960

Transition

THE SECOND WORLD WAR

I N A FIRESIDE CHAT on December 29, 1940, Franklin Roosevelt first used a phrase that would prove enduring when he called upon the United States to become "the great arsenal of democracy."

War had broken out in Europe on September 1, 1939, after German troops invaded Poland, and France and Great Britain stood by their pledges to come to Poland's aid. Few Americans thought the Nazis anything but despicable, but public opinion in the United States was overwhelmingly to stay out of the conflict. Many newspapers, from the communist *Daily Worker* to those owned by William Randolph Hearst, were strongly isolationist. This isolationism had manifested itself in law in several ways. In 1934 Senator Hiram Johnson of California had pushed through a bill forbidding the Treasury to make loans to any country that had failed to pay back earlier loans. That, of course, included Britain and France. On November 4, 1939, Congress had passed the Neutrality Act, which allowed purchases of war matériel only on a "cash and carry" basis.

Seven months later France fell to the Nazi onslaught, and Britain stood alone. In the summer of 1940 Germany proved unable to defeat the Royal Air Force in the Battle of Britain and thus gain the air superiority necessary to mount an invasion across the English Channel. It tried instead to bludgeon Britain into submission with the blitz and to force Britain into submission by cutting off its trade lifelines across the Atlantic. It nearly worked. As Rudyard Kipling had explained decades earlier:

> *For the bread that you eat and the biscuits you nibble,*
> *The sweets that you suck and the joints that you carve,*
> *They are brought to you daily by all us Big Steamers—*
> *And if any one hinders our coming you'll starve!*

While the Royal Navy was far larger than the Kriegsmarine in total tonnage, it was critically short in escort vessels to protect convoys against U-boat attack. As early as May 15, 1940, the new prime minister, Winston Churchill, was undiplomatically frank regarding what Britain needed from the United States to survive. "Immediate needs," he wrote, "are: first of all, the loan of forty or fifty of your older destroyers. . . . Secondly, we want several hundred of the latest types of aircraft. . . . Thirdly, anti-aircraft equipment and ammunition. . . . Fourthly, [we need] to purchase steel in the United States. This also applies to other materials. We shall go on paying dollars for as long as we can, but I should like to feel reasonably sure that when we can pay no more, you will give us the stuff all the same."

Roosevelt realized what was at stake in terms of America's own security. He knew that Germany was no immediate threat to the United States, safe behind the vast Atlantic. But he also knew that a triumphant Nazi state, lord of the Old World and all its economic and manpower resources, would be a mortal threat to the peace and liberty of the New World in the not-too-distant future. The president felt that Britain must survive long enough to hold the Nazis at bay while the United States rearmed, and he was able to bring the American people around to see where their true interests lay.

The day after receiving Churchill's letter, he appeared before a joint session of Congress and asked for a supplemental defense appropriation of $1.3 billion, a very considerable increase in the total federal budget for that year. He also asked for the production of "at least 50,000 planes a year."

At the time American military forces were puny. The army had about three hundred thousand soldiers—fewer than Yugoslavia—and was so short of weapons that new recruits often had to drill with broomsticks instead of rifles. The equipment it did have was often so antiquated that the chief of staff, General George C. Marshall, thought the army no better than "that of a third-rate power." The navy, while equal to Britain's in size, lacked ammunition to sustain action, and much of its equipment was old or unreliable.

On September 16, 1940, Congress approved the first peacetime draft in America history, and 16.4 million men between the ages of twenty and thirty-five registered. The act called for the training of 1.2 million soldiers and 800,000 reserves in the next year. But it specified that none was to serve outside the Western Hemisphere and that their terms of service were not to exceed twelve months.

Getting congressional and public approval for measures to increase the country's military preparedness was much easier than getting approval for aiding Britain. Roosevelt, at no small political risk to himself, for the election of 1940 was only months away, agreed to transfer fifty destroyers to Great Britain in exchange for fifty-year leases on bases in British New World possessions. And, more important, he began to formulate an all-aid-short-of-war strategy that he presented to the people as the best way to avoid war itself.

Even before Roosevelt's "arsenal of democracy" talk, Britain's financial situation was getting desperate. Britain's dollar and gold reserves were approaching exhaustion. On his return from Britain on November 23, the British ambassador, Lord Lothian, was blunt, to put it mildly, when he told the reporters who met his plane, "Well, boys, Britain's broke. It's your money we want."

In his State of the Union speech to Congress on January 6, 1941, Roosevelt declared that his policies were aimed at protecting the essential human freedoms of speech and religion and freedom from fear and want. He proposed what soon became known as Lend-Lease. Churchill, a month later, would memorably if disingenuously describe Lend-Lease as a matter of "Give us the tools and we will finish the job." Roosevelt, a few days earlier, had described it a bit more prosaically, but no less disingenuously, as the equivalent of lending a neighbor whose house was on fire a garden hose, expecting to get it back when the fire was out.

Congress, after a great national debate, approved Lend-Lease on March 11, appropriating $7 billion. By the end of the war, Lend-Lease aid to the Allies would amount to $50,226,845,387. Churchill called it the "most unsordid act in the history of any nation." It was also, of course, an act of singularly enlightened self-interest. It not only helped the Allies battle Germany and Japan effectively, it also did *not* create a vast and unpayable debt that would be an impediment to American action in the postwar world as the First World War debts had been in the previous decades.

Britain was still very short of escort vessels, and the Battle of the Atlantic showed signs of going Germany's way in 1941. Hemmed in by public opinion and congressional restrictions, Roosevelt maneuvered around them. While American warships did not attack German U-boats, American sightings, by both plane and ship, were passed on to the British. In a move of deft politics if dubious geography, Roosevelt got around the restriction against posting American military personnel outside the Western Hemisphere by simply declaring Greenland and Iceland to be part of the Western Hemisphere, and he stationed patrol aircraft there.

Slowly the United States became more and more involved in safeguarding the Atlantic sea lanes, and the USS *Greer* was attacked in Icelandic waters on September 4. The USS *Reuben James* was sunk by a German torpedo on October 30. But isolationism was still a potent force in American politics. On August 18 the extension of the Selective Service Act had passed the House by a single vote, 203–202.

———

ISOLATIONISM VANISHED from the American political landscape on the morning of Sunday, December 7, 1941, when Japanese carrier aircraft attacked and gravely damaged the U.S. Pacific Fleet at Pearl Harbor. The next day Congress declared war on Japan, and on Thursday, December 11, Germany and Italy declared war on the United States.

The Japanese armed forces ran riot in the Pacific for the next six months, taking Hong Kong, the Philippines, Malaya, Singapore, the Solomon Islands, the Dutch East Indies, and Burma, while threatening Australia and India. In the Atlantic, German U-boats sank numerous American ships within sight of the East Coast (indeed, they often used the lights along the shore to silhouette their victims until blackout regulations were implemented). In North Africa, German troops pushed British forces back toward the Suez Canal, vital both to holding the Mediterranean and to denying Middle East oil to the Nazis. In the Soviet Union, the Wehrmacht plunged ever deeper into Russia.

The huge manpower reserves of the United States, Russia, and the British Empire supplied all the military personnel that was necessary. But if the United States and its hard-pressed Allies were to win the war, this country would have to become indeed the arsenal of democracy.

It did, in one of the most astonishing feats in all economic history. In the first six months of 1942, the government gave out more than $100 billion in military contracts, more than the entire gross national product of 1940. In the war years, American industry turned out 6,500 naval vessels; 296,400 airplanes; 86,330 tanks; 64,546 landing craft; 3.5 million jeeps, trucks, and personnel carriers; 53 million deadweight tons of cargo vessels; 12 million rifles, carbines, and machine guns; and 47 million tons of artillery shells, together with millions of tons of uniforms, boots, medical supplies, tents, and a thousand other items needed to fight a modern war.

The Ford Motor Company alone produced more war matériel than the entire Italian economy. By 1944 its Willow Run plant was turning out B-24 bombers at the rate of one every sixty-three minutes. Henry J. Kaiser, who at first knew so little about ships that he referred to the front and the back instead of the bow and stern, brought the techniques of automobile mass production to shipbuilding. He reduced the time needed to build a liberty ship—the standardized freighter of seventy-two hundred tons and thirty thousand parts—from 244 days to 42. A total of 2,710 were produced during the war, each, in Roosevelt's words, "a blow for the liberty of the free peoples of the world."

At the Tehran conference in 1943, Joseph Stalin, of all people, offered a toast "to American production, without which this war would have been lost."

The United States accomplished this awesome feat of industry by turning the world's largest capitalist economy into a centrally planned one, virtually overnight. Central planning has always proved dismally inefficient at producing the goods and services needed by a consumer economy (largely because the consumers have so little say in what is produced). But central planning has done far better at producing the instruments of war.

When the president first moved to put the American economy on a wartime footing, he relied on the alphabet soup for which the New Deal had become famous. The NDAC (National Defense Advisory Commission), the OPM (Office of Production Management), and the SPAB (Supplies, Priorities and Allocations Board) all came into existence during 1941 but coordinated poorly if at all with one another. In addition, the navy and the army (of which the air force was a part until 1947) continued to operate their separate supply administrations, which often worked at cross purposes with each other and furiously resisted any outside interference from other parts of the government. And with the American economy finally booming again (unemployment fell below 10 percent in 1941 for the first time since 1931, and continued to fall rapidly all year), American companies were not interested in dancing to any tune called by Washington.

With Pearl Harbor, the president quickly realized that a different approach was needed. In early January 1942 he called in Donald Nelson, who was the OPM's director of priorities. Nelson had been the executive vice president of Sears Roebuck, earning $70,000 a year when he went to work for the government at $15,000. Roosevelt told Nelson he wanted him to take over the job of organizing war production.

"I will if I can boss it," Nelson replied.

"You can write your own ticket," the president promised him.

Nelson, Vice President Henry Wallace, and the president discussed the shape of the new agency that would take over the functions of the earlier ones. Nelson suggested calling it the War Production Administration, but Roosevelt suddenly realized that its initials would then be WPA, and decided that War Production Board would have to do instead.

Nelson went back to his office and drew up an executive order creating the WPB and giving it the powers he thought necessary to make the American economy into a war machine and himself the powers as chairman needed to make it an effective and efficient bureaucracy. The president signed the order, and Donald Nelson became, with some exaggeration, the CEO of the American economy.

He was perfectly suited to the job. Born in 1888 in Hannibal, Missouri, he had taken a degree in chemical engineering and planned to get his PhD in the subject, but he went to work for Sears Roebuck as a chemist and stayed for the next thirty years. He soon moved over into management and rose steadily.

During the 1930s Sears stocked in its stores and sold by catalog more than one hundred thousand items, from hat pins to prefabricated houses. (Franklin Roosevelt had once joked that the way to convince the Soviet Union of the superiority of the capitalist system would be to bomb it with Sears Roebuck catalogs.) For years it was Nelson's job at Sears to learn what items were needed by the retail and catalog operations, find out who would sell them or make them at the best price, and see that the merchandise got to where it was needed, when it was needed. It was the perfect training for his new job as head of the WPB, for Nelson had developed a familiarity with the width, depth, and breadth of American industry that was second to none.

At the WPB, Nelson had three overwhelming priorities. First, he had to find out from the services and the Allies what was needed to win the war. Second, he had to inventory the raw materials the country had on hand, together with the country's industrial resources. Finally, he had to find ways to fill any gaps between supply and demand.

The most acute shortage throughout the war was rubber. Most of the supply had come from plantations in British possessions in Southeast Asia. These had been largely overrun by the Japanese advance. The collection of wild rubber from its native habitat in the Amazon rain forest—an industry destroyed early in the century by the development of the plantations—was revived and several latex-producing plants that would grow in the United States were cultivated, but it was synthetic rubber that saved the day. In 1939 the United States had produced little if any. In 1945 Du Pont and other companies turned out 820,000 tons. While this sufficed for critical war needs, there was very little available for civilian use, and rubber was nearly unavailable. The war years would long be remembered as the golden age of flat tires and endlessly patched inner tubes.

Tires were the first product to be rationed, only three weeks after Pearl Harbor, and many rubber products were simply unavailable in the marketplace throughout the war. This was true as well of most industrial goods, such as refrigerators and automobiles. Indeed, between 1943 and 1945, the American automobile industry produced exactly thirty-seven automobiles. By the end of the war, thirteen rationing programs were in effect, covering such scarce commodities as gasoline, sugar, coffee, butter, fats and oils, red meat, and shoes. But rationing was nowhere near as severe as it was in Britain, where so much more had to be imported through the Atlantic lifeline.

The most politically difficult job facing Donald Nelson was deciding what was to be produced first and what could wait. The army air force wanted one sort of plane, the navy needed another, and both wanted them *now*. But there wasn't enough aluminum available in the early days of the war to produce all the aircraft needed, and it was Nelson who had to decide who waited.

The WPB was divided into several "industrial branches," each responsible for a particular industry and charged with knowing exactly what every plant in that industry could produce, what it was producing at the moment, what it was already committed to produce in the future, and what inventory it possessed. These data were sent up the line to WPB divisions in charge of overall materials, allocations, production, and procurement decisions. It was at this level that individual orders for equipment and matériel were weighed against one another, approved, given a priority, and sent to the plant that was to produce them, along with requisitions for the necessary raw materials. By the end of 1942 the WPB was the largest of the wartime bureaucracies in Washington, with twenty-five thousand employees. It used as much paper every day as a good-sized newspaper.

———

IN 1940 THE GROSS NATIONAL PRODUCT was $99.7 billion. In 1945 it was $211.9 billion. Even taking into account the 25 percent wartime inflation (kept in check by stringent wartime wage and price controls), GNP increased by 56.3 percent.

Unemployment, meanwhile, became essentially nonexistent. With 20 percent of the male population in uniform, millions of industrial jobs were filled by women, soon known collectively as "Rosie the riveter." By the end of the war they constituted fully one-third of the American labor force, and there was hardly a job they did not fill, from cowhands to lumberjacks (nicknamed, inevitably, limberjills). Under some government pressure at first, industrial firms offered on-the-job training in such previously men-only specialties as welding and crane operating.

The war also greatly increased the migration of poor black families from the South to northern industrial cities, which had begun during the First World War. Both of these consequences of the economic requirements of the war would have a great impact on the American economy and take decades to play out. But they both changed the country profoundly and, once again, increased economic opportunities for large segments of the population.

As with all of the country's major wars, government revenues and outlays rose both sharply and permanently. The government had never spent more than $18.5 billion in a single year (1919) before the First World War. Since the war it has never spent less than $33 billion (and, for all but five years in the late 1940s and early 1950s, never less than $60 billion).

The national debt soared along with expenses. It had stood at $43 billion just before the war. By 1946 it was at $269.4 billion, equal to 130 percent of GNP, by far the highest it has ever been before or since. Once again, bond drives were instituted on a mammoth scale to help fund the war. But only about a quarter of these bonds were taken by individuals. The rest were taken by banks, insurance companies, and other financial institutions. Commercial banks had held less than a billion in Treasury securities in 1941. By 1945 they held some $24 billion.

The United States was able to pay for about 45 percent of the war through taxation, far more than in the First World War or the Civil War. The Revenue Act of 1942 transformed the federal tax system. Before

1942 the income tax largely affected the middle class and the rich. In that year only about four million Americans paid any income tax at all. By the following year, after the personal exemption had been lowered from $1,231 to $624, seventeen million owed income taxes, bringing those taxes, "from the country club . . . district down to the railroad tracks and then over to the other side of the tracks." By the end of the war, as more and more Americans took wage-paying jobs, 42.6 million Americans were paying income taxes. Meanwhile rates on high incomes were raised sharply to as much as 94 percent. For the first time, the personal income tax yielded more revenue than the corporate income tax, twice as much by the end of the war.

The Revenue Act of 1942 brought another large and permanent change to the income tax: withholding. Until then, people who owed income taxes simply paid them every year. Now estimated taxes were taken out of each paycheck, which helped greatly to smooth out the Treasury's cash flow and, once withholding became familiar, also helped to hide the tax bite from the citizenry.

Despite the greatly increased federal taxes and strict wage and price controls, that citizenry was prospering economically as never before. But because so much of the economy was absorbed in war production, there was little to buy over and above necessities. The surplus income went into savings. In 1940 personal savings had amounted to $4.2 billion, the same as in 1929. Over the next five years, personal savings totaled an astonishing $137.5 billion, which went into savings accounts, insurance policies, and government bonds, and to paying down debt. Where little of it went was into Wall Street, despite the enormous growth of corporate profits. The memories of 1929 and the early 1930s were still too vivid.

HAD IT NOT BEEN for wage and price controls, corporations, desperate for workers, would have competed for the available labor by raising wages sharply. Unable to do so, they competed using other, nonmonetary forms of compensation. The most important of these in the long term was hospitalization insurance.

Medical insurance was something very new. The first hospitalization plan had been introduced only in 1929 when Baylor University Hospital in Dallas, seeking to smooth out its cash flow, had agreed to provide up to twenty-one days of hospital care each to a group of fifteen hundred schoolteachers in return for a premium of $6 a year. This type of insurance quickly spread, soon evolving into the Blue Cross and Blue Shield plans that would dominate medical insurance for decades.

But this model contained the seeds of great problems. For one thing, it was what is called "front loaded"; in other words, it paid for the first dollar of medical or hospital costs, not the last dollar. It covered the costs of a short illness, but not a protracted one. It was a bit like house insurance that paid for a broken window but not for the roof being torn off in a storm. In 1929, however, twenty-one days was a very long hospital stay, and given the level of medical technology at that time, the daily cost of hospital care was about the same, whatever ailed the patient. Also, these policies paid to cure illness but not to prevent it.

And unlike ordinary insurance, it paid the bill, whatever it was, rather than giving the insurance holder a check and letting him decide how best to correct the situation. This made consumers (that is, insured patients) indifferent to the costs involved, a vital element of a well-functioning free market. Further, these policies paid off only if the patient was treated in a hospital, the most expensive form of medical treatment. Thus doctors found themselves under pressure to admit patients to hospitals who might have been treated just as well, and far more cheaply, in the office or at home.

Medical insurance was still quite uncommon on the eve of the war, but it was perk of employment for millions by the end of the war. After the war, the IRS tried to tax this considerable benefit, but Congress quickly stepped in and ordered the IRS to treat it as a nontaxable, deductible business expense of corporations (but medical insurance purchased by individuals had no such tax advantage, an increasingly serious distortion of the labor market as more and more people became self-employed). In 1948 the National Labor Relations Board ruled that health benefits were subject to collective bargaining, and medical insurance spread quickly through the American economy. By 1950 some 54.5 million people were covered by employer-paid health plans, more than a third of the total population.

By removing the threat of economic catastrophe, medical insurance greatly improved the quality of life for those covered. But, because of its peculiar history, when American medical insurance interacted with the revolution in health care that had been gathering force since the 1930s and exploded in the decades after the Second World War, the results would be an economic problem of enormous size for the country.

———

ON SEPTEMBER 1, 1945, the USS *Missouri* entered Tokyo Bay, and Japan formally surrendered the next day to the Allied forces, ending the war. More than fifty million people had died, not only soldiers but, for the first time in the history of modern warfare, even greater numbers of civilians. The costs had been immeasurable. But those costs had not been evenly shared by the major combatants.

Germany and Japan lay in ruins. Roosevelt's assistant Harry Hopkins, flying over Berlin shortly after the war, called it, "a modern Carthage." The Soviet Union, although greatly strengthened geopolitically, had lost a greater percentage of its population than any other country, and had had much of its most productive areas laid waste by combat. Britain and France were militarily and financially exhausted (Britain, by one estimate, expended about one-quarter of its total national wealth on the war). Their great colonial empires would soon melt away.

The vast territory of the United States and its industrial base, however, had been untouched by the war. Its productive capacity had been hugely increased and its population enriched. Its economy, by far the largest on earth before the war, now produced fully 50 percent of the world's gross product. Eighty percent of the world's monetary gold belonged to the United States; most of the rest was stored in the vaults beneath the New York Federal Reserve bank. Under the Bretton Woods Agreement of 1944, the dollar, convertible into gold by central banks, would be the world's primary reserve currency and the basis of world trade in the future.

The American army was the best equipped in the world and second in size only to that of the Soviet Union; the navy and air force were larger than the navies and air forces of the rest of the world combined. It had a monopoly on the most fearsome weapon of war ever conceived, the atomic bomb. No country in history had possessed such a preponderance of military and economic power.

Most of all, the American people had been greatly energized by the battle, just as they had been by the fearful struggle of the Civil War eighty years earlier. At the cost of $300 billion and four hundred thousand lives, they had not only maintained their own liberty and saved that of countless others, they had brought liberty—for the first time—to many millions more.

Gone from the American body politic was the soul-destroying fear that had gripped the nation in 1933. Gone too was the nostalgic pull of isolationism. The United States now accepted the inevitable—that it must lead the world—because no one else could, and American security would soon depend on it.

Chapter Eighteen

THE GREAT POSTWAR BOOM

I T WAS WIDELY PREDICTED by both economists and business leaders that the postwar American economy would be characterized by renewed depression. Federal government expenditures would be drastically scaled back (and in fact they fell by nearly two-thirds over three years) while most of the twelve million men and women in the armed services would pour into the job market, forcing down wages and driving up unemployment. Proving that economists have uniquely clouded crystal balls, the longest sustained boom in American history was in the offing.

So worried was the government about renewed depression that it moved in 1944 to prevent it. On June 22 that year President Roosevelt signed the GI Bill of Rights (formally the Serviceman's Readjustment Act), passed unanimously by Congress. Ostensibly it was intended to reward veterans for their bravery and sacrifice in defeating Germany and Japan. In fact, a major purpose was to slow down the return of veterans to the job market. The GI Bill of Rights provided generous assistance to all honorably discharged veterans in paying for education and housing while in school and assistance in buying houses and businesses after school was completed.

The law of unintended consequences is usually invoked to explain the pernicious results stemming from well-intentioned legislative action, such as Prohibition. But the GI Bill, while perhaps moderately effective in smoothing the flow of GIs into the American economy, had in fact almost nothing but unintended consequences, and almost all of them were profoundly good for the country.

It allowed no fewer than eight million veterans to obtain more education both in college and in technical schools than they otherwise would have. It greatly enlarged the percentage of the population that had college degrees. In 1950 some 496,000 college degrees were awarded, twice the number of a decade earlier.

Between 1945 and 1952 the federal government spent $14 billion on GI educational benefits but added far more than that to the human capital that would power the postwar economy. Considered as a "public work," the GI Bill proved to be the Erie Canal of the new, postindustrial economy that was then, quite unrecognized, coming into being.

The GI Bill also powered a social revolution. It opened up high-level jobs to many segments of the population that had rarely known such jobs before, thus greatly enlarging and diversifying the country's economic elite, which had long been dominated by people with British or northwest European names. Because children in this country have historically received on average two years more schooling than did their parents, these benefits have continued generation after generation. Further, because the benefits of the GI Bill have been extended to veterans who served in subsequent wars, including the cold war, it has been a continuing engine of human capital creation and technological capacity for the last sixty years, allowing the country to dominate the new information economy as much as it had dominated the industrial economy of decades past.

The GI Bill of Rights also revolutionized housing in this country.

Housing had been a growing problem in the 1920s and 1930s, when new construction did not equal the number of new families needing housing by six hundred thousand units. The war had brought housing construction virtually to a halt. With the end of the war, veterans returning in the millions, getting married and initiating the baby boom, the pressure for new housing became intense.

Many New Dealers envisioned government-built or -sponsored housing in apartment complexes to be built on areas of cleared slums, such as Parkchester in New York's borough of the Bronx, or housing built privately with government subsidies, such as Stuyvesant Town in Manhattan, owned and built by Metropolitan Life Insurance Company. Many of these would be built in cities across the country in the years after the Second World War, but they were almost always economic and social failures, simply and quickly evolving into high-rise slums that were often far worse than the slums they replaced.

The GI Bill of Rights provided for Veterans Administration mortgages, by which the Veterans Administration guaranteed, at first, half the mortgage up to $2,000. This was soon amended to allow guarantees of up to $25,000 or 60 percent of the loan, whichever was less. With no fear of loss through a default, many banks were willing to make loans with no money down to veterans. All that was needed was the housing to buy.

Entrepreneurs such as William Levitt supplied it. Levitt had been a contractor before the war, erecting houses one by one, as single-family houses had always been built. With the VA mortgages, he saw opportunity. "The market was there," he said years later, "and the government was providing the financing. How could we lose?"

He acquired 7.3 square miles of what had been Long Island potato fields in suburban Nassau County. His brother Alfred designed two basic models of houses, "ranch" and "Cape Cod," and in four years he built fully 17,500 units of single-family housing by industrializing the process of construction. "What it amounted to," Levitt explained, "was a reversal of the Detroit assembly line. There, the car moved while the workers stayed at their stations. In the case of our houses, it was the workers who moved, doing the same job at different locations."

The first houses could be rented for $65 a month or purchased for $6,990, a figure soon raised to $7,990. By 1949 they could only be purchased. What the family got on a sixty-by-one-hundred-foot lot was a two-bedroom house on a cement slab, with living room, kitchen, and bathroom. An attic could be converted into two more rooms, and a bathroom. The neighborhood consisted of hundreds or thousands of nearly identical houses. At first, they were nearly devoid of trees.

Intellectuals, with characteristic snobbery, were appalled. The social critic John Keats wrote a best-selling book called *The Crack in the Picture Window,* in which he mourned the fact that the inhabitants of these new suburbs, which began springing up around every American city in imitation of Levittown, "were not, and are not, to know the gracious dignity of living that their parents knew in the big two- and three-story family houses set well back on grassy lawns off shady streets."

Of course, most of the people who moved into Levittown and its thousands of imitations had never known any such thing. Instead they had grown up in crowded, walk-up apartments in urban neighborhoods where parks were few and far between. To these people, the new suburbs were an affordable paradise.

Far more important economically, this new type of housing allowed millions of new families to have something their parents had never known: home ownership. Instead of paying rent, they were building equity. As family income rose with age and experience, they could trade up, using the equity in their old house to serve as the down payment on the new house. The GI Bill thus helped millions of families acquire not only better housing than their parents had ever dreamed of, but something else: capital, the financial assets that are the defining characteristic of the middle class in this country.

Further, once a family had some financial assets, it became much easier for it to obtain credit. Bank loans and charge accounts had been attributes of the rich before the Second World War; now they rapidly became an aspect of everyday life. In 1951 a banker named William Boyle, who worked for the Franklin National Bank headquartered in the middle of the burgeoning Long Island suburbs, came up with the idea of the credit card. It relieved merchants of the trouble and expense of maintaining their own charge accounts, allowed ordinary people to charge at numerous businesses, and provided the issuing banks with handsome profits by charging interest on unpaid balances.

The idea, as good ideas always do, spread rapidly. By the 1960s credit cards were common. In the early 1970s MasterCard and Visa gave credit cards national and soon international scope. Today they and their offspring, debit cards, are replacing cash in most transactions. The ubiquity of credit has become so intense and credit so important in everyday life that maintaining a good credit rating is now a great concern of most Americans. Having one's credit cut off, and thus losing one's access to the marketplace, is not altogether dissimilar today to what being excommunicated meant in the Middle Ages.

As the new suburbs grew explosively, many of the cities around which they grew declined in population. Except for New York, all the cities with major league baseball teams in 1950—which is to say the major cities of the northeastern quarter of the country—lost population, sometimes by as much as 50 percent in the decades after the Second World War. The population that remained was mostly poor and minority, often needing more services than the cities could provide. Within a few decades, the population of the suburbs exceeded that of the nation's cities, and had become the linchpin of American politics.

———

THE RETURNING VETERANS and their families, besides investing in real estate, also began, slowly at first, to invest in securities.

Wall Street had seen business seriously decline from the glory days of the late 1920s. Volume in 1929 had averaged 2.5 million shares a day. In

1939 it dropped below 1 million. There was no panic on Wall Street when war broke out in September that year, but rather, as elsewhere, a sullen acceptance. However, a three-year decline in both prices and volume began. In 1942 daily volume averaged a dismal 455,000 shares, while the Dow-Jones Industrial Average fell below 100 for what turned out to be the last time in April that year, even while corporate profits were soaring thanks to war orders.

Even after the war, prices on Wall Street lagged behind the fast-growing economy. On December 31, 1949, the Dow stood at 200, only twice where it had been in 1940, although the economy had nearly tripled and corporate profits increased even more. Some blue-chip stocks were selling for a mere four times earning and paying dividends over 8 percent. But a revolution was already under way on Wall Street that would remake the brokerage business and the American economy as well.

Charles Merrill, a southerner, arrived on Wall Street at the age of twenty-two, just in time to experience the panic of 1907. He opened his own firm in 1914 and two years later merged it with the firm owned by Edmund Lynch, to form Merrill Lynch and Company. (The partnership papers accidentally left out the comma between the two names, and it has been left out intentionally ever since.) In the 1920s he helped underwrite the stock issues of several chain stores. Early in 1929 he saw what was happening on the Street and urged his customers to get out and was himself largely out of the market when the crash came.

Correctly forecasting that the depression that began to develop in 1930 would be a long one, he sold his seat on the exchange and sold his firm to another brokerage house, E. A. Pierce and Company, becoming a limited partner there but not active in its management. He spent most of the 1930s consulting with the various chain stores he had helped underwrite, such as Western Auto and Safeway, and began to think about applying chain store techniques to the business of brokerage.

Most Wall Street brokerage firms at this time were small, family-owned, and uninterested in customers who had only small accounts.

Research, such as it was, was casual at best, mere rumor gathering at worst. In 1940 Merrill took over as the senior partner at E. A. Pierce and Company, and the Merrill Lynch name reemerged on Wall Street. Merrill immediately began to create an entirely new kind of brokerage business. His customers men (soon renamed registered representatives) were thoroughly trained and provided with information gathered by a large research department.

In 1948 he began to advertise—unprecedented on Wall Street—to acquaint the average person with Wall Street and the investment opportunities to be found there. The ads discussed the mechanics of how stocks are bought and sold and the risks involved. They often as well made a subtle political point. When President Truman, running for another term in 1948, made a rabble-rousing reference to "the money changers," Merrill replied in an ad. "One campaign tactic did get us a little riled," he admitted. "That was when the moth-eaten bogey of a Wall Street tycoon was trotted out. . . . Mr. Truman knows as well as anybody that there isn't any Wall Street. That's just legend. Wall Street is Montgomery Street in San Francisco. Seventeenth Street in Denver. Marietta Street in Atlanta. Federal Street in Boston. Main Street in Waco, Texas. And it's any spot in Independence, Missouri [Truman's hometown], where thrifty people go to invest their money, to buy and sell securities."

Merrill's idea of bringing Wall Street to Main Street worked. By 1950 Merrill Lynch was the largest brokerage house in the country. By 1960 it was four times the size of its nearest competitor, with 540,000 accounts, and was already known on the Street—with a mixture awe and envy—as "the thundering herd." Other brokerage firms had no choice but to imitate Merrill Lynch's business model, and the family firm, catering to a few rich clients, began to disappear on Wall Street.

While those investing directly in stocks remained a relatively small group, those who had indirect investments grew very rapidly. Pension programs for hourly wage employees had been nearly nonexistent in the 1920s (Sears Roebuck was a notable exception). But the Wagner Act made it possible for labor unions to insist on negotiating for them, and an increasing number of people in corporate management favored the idea and it began to spread rapidly through corporate American in the 1940s.

Charles E. Wilson, the president of General Motors in the 1940s (and later Eisenhower's secretary of defense), was told that if the money in such programs was invested in the stock market, workers would be the owners of American business in a few decades. "Exactly what they should be," he replied.

By the 1950s pension funds, controlled by both corporations and unions, had become major players on Wall Street. In 1961, when the federal budget was less than $100 billion, noninsured pension funds held stock worth $17.4 billion and were making new investments at the rate of $1 billion a year. Mutual funds, which had first appeared in 1924, as well began to play a larger and larger role on the Street, as people sought to invest in common stocks without having to make the actual decisions themselves about which ones to buy. Only $500 million was invested in mutual funds in 1940. Investment was five times as high a decade later, and $17 billion, nearly five times higher still, in 1960.

Finally, in 1954, the market, deeply undervalued, began to move up. It reached a new postdepression high on February 13, when it closed at 294.03, its highest since April 1930. By June it stood at 330, and in December it finally broke through its September 3, 1929, high of 381.17 after more than twenty-five years, the longest period between highs in the Dow-Jones in its 108-year history. The Great Depression was finally over, psychologically as well as economically.

———

THE BIGGEST PROBLEM of the postwar economy turned out to be not unemployment but inflation. Although GNP dipped slightly in 1946, as military orders fell from an annual rate of $100 billion in early 1945 to $35 billion a year later, GNP had recovered by year's end and grew strongly thereafter.

The reason, clear in retrospect, was the vast pent-up demand for durable goods created by the war. Virtually no cars, no housing, and no appliances had been manufactured during the war. Those in use were nearing the end of their productive utility, and many were already far beyond that point. Further, the huge pool of personal savings that had accumulated during the war was there to pay for the goods demanded.

But it required time for the country's industry to shift back from war production to consumer goods, while irresistible political pressure brought wage and price controls to a premature end in 1946. The result was a roaring inflation, the greatest in the peacetime history of the country up to that point, as nongovernmental spending rose by 40 percent while the supply of goods did not rise nearly so quickly. Farm prices rose 12 percent in a single month and were 30 percent higher by the end of that year. Automobile production, virtually nonexistent since 1942, reached 2,148,600 in 1946, but wouldn't top 1929's production until 1949.

Corporate profits in this intense sellers' market rose by 20 percent, and labor unions demanded large increases in hourly wages and benefits. Strikes multiplied alarmingly once the wartime prohibition against them lapsed. In January 1946 fully 3 percent of the labor force, including workers in the automobile, steel, electrical, and meat-packing industries, was on strike. Never before (or since) had so many workers been on the picket lines. Many people thought labor had become too powerful and that the Wagner Act had pushed the pendulum too far in labor's direction.

The new president, Harry Truman, took much of the blame for the economic disruption of the immediate postwar era, and "to err is Truman," became a national joke. In the off-year campaign of 1946 the Republicans ran on the slogan of "Had enough?" and for the first time since 1928 won a majority in both houses of Congress. Truman would famously label it the "do-nothing Eightieth Congress," but it produced at least one piece of major legislation, the Taft-Hartley Act. Taft-Hartley, unlike the Wagner Act, allowed employers to fully inform their workers on the company's position regarding the issues in an election to certify a union, as long as they used no threats. It also allowed management to call an election on its own if it chose to do so and forbade unions to coerce workers or to refuse to bargain, just as the Wagner Act had forbidden management to do so.

Secondary boycotts, a powerful weapon in labor's arsenal, were outlawed, as was the closed shop (where workers have to be a member of the union before they can be hired). Union shops (where workers must join the union upon being hired) required a vote of the workers, and states were allowed to outlaw the union shop. Most visibly, Taft-Hartley gave the president the power to interrupt a strike by calling an eighty-day cooling-off period while government mediators sought a settlement.

Labor, of course, fought Taft-Hartley tooth and nail and President Truman vetoed it, calling it "shocking—bad for labor, bad for management, bad for the country." Congress overrode the president's veto. But as so often happens, the problem the legislation was meant to address, the great upsurge in strikes immediately following the war, was already correcting itself. In 1946 there had been a total of 125 million man-days lost to strikes. In each of the next three years there was an average of only 40 million.

And Truman was wrong; Taft-Hartley proved good for the country. With a more level playing field and the great prosperity of the times, labor and management learned how to be less confrontational and to work better toward achieving a just division of the wealth created by corporations and their workers. By 1992, in a vastly larger economy and workforce, fewer than four million man-days were lost to strikes. And while Taft-Hartley was often invoked to end strikes in the two decades after the law was first passed, and by both Democratic and Republican presidents, it has been invoked only once in the last quarter century.

By 1992, of course, the economic universe that had brought the modern labor movement into being was rapidly vanishing. The percentage of the workforce that was unionized peaked in 1945 at 35.6 percent and has been declining ever since. By 1960 it was only 27.4 percent of the nonfarm workforce. Today it stands at only 14 percent and would be lower than it had been in 1900 were it not for the spread of unionization among government workers, which began only in the 1960s.

The heart of the old union movement had been in manufacturing, among the everybody-does-the-same-job assembly line workers. But just as agriculture, the country's first great economic sector, has continually increased output while using an ever-declining percentage of the workforce, so has the country's second great economic sector, manufacturing. The age-old American drive to increase productivity and thus minimize labor continues unabated.

———

ALTHOUGH MANY NEW ENGLAND cotton mills had moved to the Piedmont area after the turn of the twentieth century, to take advantage of the cheap labor available in the area, the South was still overwhelmingly agricultural at mid-century, and unions had never been strong there. Permitted by Taft-Hartley to do so, many southern states outlawed the union shop, ensuring that the labor movement would remain relatively weak.

Corporations, always looking to minimize labor costs, began building more factories in these states. But economic growth in the South was still impeded by two major factors: its summer climate, which nonnatives found very difficult to deal with, and the enduring and bitter legacy of racism.

The first factor was solved with air-conditioning. A primitive but working air-conditioning system had been built as early as 1842, in Apalachicola, Florida, to cool a hospital. Early in the twentieth century, Stuart Cramer and Willis H. Carrier, working independently, developed practical air-conditioning systems that could be manufactured on an industrial scale. Their use in large commercial buildings, such as theaters and department stores, began in the early 1920s. The development of

Freon, a highly stable and efficient refrigerant, in 1930 brought the running cost of air-conditioning down substantially and allowed it to spread rapidly.

After the Second World War most public and office buildings were designed with air-conditioning systems, and small air-conditioning units developed for railroad cars were adapted for home and automobile use. By the 1960s air-conditioning was standard equipment in middle-class houses in the South and was spreading rapidly in all areas of the country with hot summers.

Racism, needless to say, was a far more difficult problem to deal with. Franklin Roosevelt, who needed southern votes to enact his economic and foreign policy programs, did little if anything to end the rule of Jim Crow. It was his successor, Harry Truman, who began the battle for equal rights in 1946 by ordering the integration of the armed forces. In 1954 the Supreme Court unanimously overturned the separate-but-equal doctrine that allowed segregation, and ordered the integration of schools and other public facilities "with all deliberate speed."

Meanwhile, one of the most remarkable political movements in American history, epitomized by the Reverend Martin Luther King, Jr., began to demonstrate peacefully for the end of racial discrimination in the marketplace as well as in the law. It took ten years of ever larger demonstrations and courage in the face of sometimes brutal repression, but finally, in 1964, the Twenty-fourth Amendment outlawing the poll tax and the Civil Rights Act were passed. The following year the Voting Rights Act was also passed by Congress. They put the federal government, and the nation as a whole, squarely behind the doctrine of racial equality.

Although at the time it seemed that the fight for equal rights had just begun, in fact, the battle was won. With the rapidly increasing political power of blacks in the South, politicians began attending to their interests, including even die-hard segregationists such as George Wallace and Strom Thurman. Within a decade, Jim Crow was largely an ugly memory, and the most divisive and shameful aspect of American life was expunged from the body politic and increasingly from the hearts and minds of Americans as well.

With its great economic advantages, especially low land and labor costs, the South began to grow and modernize rapidly. As its economy grew and developed increasingly into a First World economy, it population grew likewise, as did its national political influence. In 1940 the eleven states of the old Confederacy had had 25 percent of the electoral votes. Today they have 35 percent. The Civil War was over at last.

THE INTERNATIONAL EUPHORIA of V-E Day (May 8, 1945) and V-J Day (August 15) that greeted the return of peace in Europe and Asia did not continue long. Even before the war was over, it was becoming obvious that the Soviet Union did not intend to live up to its agreements regarding postwar Europe. On April 30, 1945, the day that Hitler committed suicide, President Truman, not yet even moved into the White House, summoned the Soviet foreign minister, Vlacheslav Molotov, who was in Washington, to a meeting. He told him in no uncertain terms that the Soviet Union must carry out its agreements regarding Poland, including free elections. Molotov was stunned by the tongue lashing, but it did no good. By the end of the year Soviet forces were in firm control of most of eastern Europe and clearly had no intentions of withdrawing.

The Soviet Union soon began to pressure Turkey to make concessions to its interests, and it supported Communist guerrilla movements in Greece and elsewhere.

By early spring 1946 the wartime alliance between the Western democracies and the Soviet Union had been shown to be, at best, a matter of "the enemy of my enemy is my friend." With the collapse of Nazi Germany, the alliance collapsed as well. On March 15, 1946, in Fulton, Missouri, Winston Churchill delivered his famous "Iron Curtain" speech with President Truman in the audience. It was clear that a new confrontation between Great Powers had begun and that while the period between the first and second world wars had lasted twenty years, the new peace had hardly lasted one. Worse, the possibility of atomic war—out of which there could come no winners—was frighteningly real.

The Communist tyranny in the Soviet Union was every bit as odious as that of the Nazis in Germany, and, as with all tyrannies, every bit as aggressive. How to confront it was the question. The United States was overwhelmingly powerful militarily, but it had been demobilizing as quickly as possible since the end of the war. There had been more than twelve million people in the armed forces in 1945. By 1947 there were fewer than two million. Thousands of ships and planes had been scrapped or sold to other countries. And while the United States still had a monopoly on the atomic bomb (the Soviets would not explode their first one until September 1949), the country as yet had very few of them.

In early 1947 Britain, which had been aiding Greece against the Communist insurgency, as well as Turkey, told the United States that it could no longer carry the burden financially. Truman felt that the United States had no option but to take over Britain's role in this area, for otherwise these countries were almost certain to fall under Soviet domination.

The United States decided to fight this third Great Power confrontation of the twentieth century, soon dubbed the cold war, in a new way: with money instead of bullets. It would contain the Soviet Union with alliances and with sufficient forces to deter an attack, but would place most of the emphasis on reviving and enlarging the economies of potential victims of Soviet aggression.

On March 12 Truman addressed a joint session of Congress and announced what quickly came to be called the Truman Doctrine. "I believe it must be the policy of the United States," he told Congress, "to support free peoples who are resisting attempted subjugation by armed minorities or by outside pressure." And he said that this assistance would be "primarily through economic and financial aid."

There was little opposition. The former arch-isolationist senator from Michigan, Arthur Vandenberg, now chairman of the Senate Foreign Relations Committee, told Truman after he had been briefed beforehand, "if you will say that to the Congress and the country, I will support you and I believe most of its members will do the same."

In June, at a commencement address at Harvard, General George C. Marshall, Army chief of staff during the war, now the secretary of state, proposed what would be called the Marshall Plan, one of the most extraordinary, and extraordinarily successful, acts of statesmanship in world history. It called for European nations, including the Soviet Union, to cooperate in the economic recovery of the Continent, with the United States providing the capital necessary. Stalin quickly rejected the idea, as did the countries in eastern Europe under his control, but he inadvertently helped sell it to other European countries by engineering a coup in Czechoslovakia in early 1948.

In the next several years the Marshall Plan provided Europe with $13 billion and helped especially to get the economies of West Germany, France, and Italy back on their feet. However, the Marshall Plan aid was but a small fraction of American foreign aid in these years. Between 1946 and the early 1970s, when the foreign aid program began to wind down, the United States spent about $150 billion on economic aid to foreign countries. About one-third of this went to Europe, the rest in Asia, Latin America, and elsewhere.

Like Lend-Lease, it was both extraordinarily generous (indeed, it was entirely unprecedented in world history for a dominant power to help its potential economic rivals build their economies) and a perfect example of enlightened self-interest on a massive scale. That free trader Adam Smith would have approved. With half the world's GNP, the United States was running large export surpluses. But foreign countries, many of them economically devastated, could pay for imports only by exporting in turn to the United States.

With the help of such international organizations founded as a result of the Second World War as the World Bank, the International Monetary Fund, and the General Agreement on Tariffs and Trade (GATT, now the World Trade Organization), the United States worked to establish a new world trading system and to lower tariffs on a worldwide basis to increase world trade, to the mutual benefit of all. The result was, once again, a spectacular success. World trade increased by a factor of six in the fifteen years after the war, greatly strengthening the economies of all involved. And the trend has continued unabated. In 2000 total world trade was 125 times the level of 1950, equaling an astounding $7.5 trillion, including both manufactures and services. Free trade has proved the greatest engine of economic growth the world has ever known.

———

IN 1946 CONGRESS PASSED the Employment Act. It established the president's Council of Economic Advisors as well as the Joint Economic Committee of Congress. Most importantly, it declared it the policy of the federal government to maximize employment, production, and purchasing power. Largely forgotten today—because its purposes are now so taken for granted—the Employment Act of 1946 was revolutionary in its time.

Until the 1930s governments were thought to have only very limited economic responsibilities. It was, certainly, the business of government to maintain a money supply that kept its value and to help in the enforcement of contracts. But government was not thought to have any more responsibility for the business cycle as a whole than for the weather. The Great Depression and the most influential economist since Adam Smith changed that.

John Maynard Keynes (Lord Keynes after 1944) was born in Cambridge, England, and would be associated with the university there all his life. He studied under Alfred Marshall, the leading economist of the previous generation, whose *Principles of Economics,* first published in 1890, was immensely influential in the profession. Marshall had trained as a mathematician and physicist before switching to economics. His conception of a national economy was what Keynes called "a whole Copernican system, by which all the elements of the economic universe are kept in their places by mutual counterpoise and interaction."

Keynes first rose to fame when he attended the Versailles Peace Conference in 1919 with the delegation from the British treasury and, outraged with the peace treaty that emerged, wrote a book called *The Economic Consequences of the Peace,* which proved to be all too prescient. In 1936 he published his masterpiece, *The General Theory of Employment, Interest, and Money,* to explain the origins of the Great Depression and why it persisted.

Before Keynes, economists had been mostly interested in what is today called microeconomics, the myriad allocation of resources that determine prices and affect supply and demand in the marketplace. Keynes was interested in macroeconomics, how aggregate supply and demand affect national economies. Keynes argued that supply and demand must equal in the long term, but as he noted in his most famous aphorism, "in the long term we are all dead." In the short term, supply and demand can be out of balance. If there is too much supply, depression results; too much demand and inflation breaks out.

Keynes felt that government, by running a deliberate deficit in times of slack demand (or cutting taxes) and expanding the money supply, could prevent depression. Equally, by doing the opposite—running surpluses and raising interest rates—governments could keep booms in check. Economists took to Keynesian ideas immediately. For one thing, these ideas were immensely exciting intellectually, seemingly a powerful tool for preventing such economic disasters as they had all just lived through.

Of course, Keynesianism also made economists important as they had never been before. Politicians, before Keynes, had not needed economists to help them govern, any more than they had needed astronomers. After Keynes, they were vital, hence the President's Council of Economic Advisors. The quality and consistency of the advice, however, was equivocal. Truman, the first president to have such formal economic advice, joked that what he needed was a one-armed economist because the ones he had were always saying, "on the one hand . . . but on the other hand."

The first postwar presidents, Truman and Eisenhower, had been born in the nineteenth century and educated in pre-Keynesian economics. They remained skeptical. Eisenhower's secretary of the treasury, George Humphrey, expressed the attitude of the pre-Keynesian generation perfectly when he said, "I do not think you can spend yourself rich." While neither president paid down the national debt to any appreciable degree (in fact it increased under their tenures), neither did they allow it to rise significantly. Half the budgets of Eisenhower and Truman were in surplus, including two during the Korean War. Because the economy grew strongly in the fifteen years after the war, the debt as a percentage of gross national product declined precipitously. It had been nearly 130 percent of GNP in 1946. By 1960 it was a mere 57.75 percent.

When Kennedy became president, however, in 1961, full-blown Keynesianism was adopted. Walter Heller, a professor of economics at the University of Minnesota before becoming Kennedy's chairman of the Council of Economic Advisors, talked about being able to "fine tune" the national economy. In other words, he wanted government to be the "engineer" driving Alfred Marshall's economic machine. He proposed budgeting based on what he called a "full-employment budget." In other words, the budget should spend what revenues the government would be receiving if the economy was operating at an optimum level. If the economy was at that level, the budget would be in balance; if below it, the budget deficit would automatically stimulate the economy, driving it toward the optimum.

Money, of course, is "the mother's milk of politics." Once politicians had intellectually defensible reasons for spending in deficit, the political pressure to do so began to increase relentlessly. As a result, the national debt rose by a third in the decade of the 1960s, the first time the debt had risen significantly during a time of peace and prosperity. But because of that prosperity (and a gathering inflation), the debt-to-GNP ratio continued to decline, dropping below 40 percent by 1969.

Kennedy, relatively conservative, moved to stimulate the economy by means of tax cuts rather than increased spending, especially in the highest brackets, cutting them from 91 percent to 70 percent. When enacted shortly after his death, they worked as intended. Between 1963 and 1966 the economy grew between 5 and 6 percent a year, while unemployment dropped below 4 percent from 5.7 percent.

In the twenty years between the end of the Second World War and the mid-1960s, the American economy almost doubled in real terms, GNP growing from $313 billion to $618 billion (both figures in 1958 dollars), while inflation remained low. Many foresaw an extended period of prosperity with continued "fiscal dividends" as government continued to cut taxes (or increase spending) as revenues rose. The stock market, mired below its 1929 high for a quarter century, rose steadily after finally breaking through that high in 1954 and was approaching 1,000 on the Dow by the mid-sixties. After two decades of continuing economic boom (interrupted only by three recessions so short and so shallow they would have gone unnoticed in a less statistics-drenched age), it seemed the dream of permanent prosperity, dashed by the Great Depression, was at hand once more.

It was not to be.

Chapter Nineteen

THE CRISIS OF THE
NEW DEAL ORDER

W HEN LYNDON JOHNSON SUCCEEDED to the presidency on the assassination of John F. Kennedy, he proved to be a very different president. A decade older than Kennedy, Johnson was fully a son of the New Deal, one with deep faith that government could solve social and economic problems. He also possessed what were perhaps the most surpassing legislative skills of any American president. These skills had made him the most effective majority leader in the history of the Senate, and he was determined to use them to achieve what he saw as the completion of the New Deal of his political hero, FDR. "In your time," he told an audience at the University of Michigan on May 2, 1964, "we have the opportunity to move not only towards the rich society and the powerful society, but upward to the Great Society."

With the help of an overwhelming electoral victory in November that year, Johnson prodded Congress to pass bill after bill. The Equal Opportunity Act (1964), Mass Transit Act (1964), Medicare and Medicaid (1965), Older Americans Act (1965), Appalachian Regional Development Act (1965), Head Start (1965), the Demonstration Cities and Metropolitan Development Act (1966), Higher Education Act (1967). Along with many other, smaller, programs that involved the federal government in areas of national life it had never before been concerned with, these caused a breathtaking rise in federal expenditures. Nondefense government expenditures rose by a third in just three years, from 1965 to 1968, from $75 billion to $100 billion. Two years later they were $127 billion. Meanwhile, the Vietnam War escalated quickly. In 1965 the defense budget had been $50 billion. In 1968 it was $82 billion.

Had the economy been underperforming, as it had been in the 1930s, the result of all this new spending would have been stimulating. But the economy in the mid-1960s was near full employment, so the inevitable result was that inflation began to increase. A vicious circle quickly developed. Increased inflation caused interest rates to rise as lenders wanted protection from the inflation. But the Federal Reserve, operating on a Keynesian model, was afraid that increased interest rates would cause economic growth to end, and so it expanded the money supply to keep interest rates low. An increased money supply, relative to the goods and services the money could buy, ineluctably caused further inflation.

The American economy began to deteriorate sharply. Unemployment, which had been at only 3.5 percent in 1968, rose to 4.9 percent in 1970 and to 5.9 percent in 1971. According to Keynesian theory, increasing inflation and increasing unemployment could not happen at the same time, and a new word was coined in 1970 to denote the unprecedented situation: *stagflation*.

The stock market, which had been rising steadily since the early 1950s, crossed the 1,000 mark on the Dow, intraday, for the first time in 1966. Then it stalled at that point. Four times in the next six years the Dow crossed 1,000 intraday before the market finally closed above 1,000, reaching a high of 1,051.70 on January 11, 1973. But it soon sank back in the worst bear market since the early 1930s. By the end of 1974 it was down to 577.60. The by-then-raging inflation masked how deep the fall of the stock market actually had been. Measured in constant dollars, the Dow-Jones Industrial Average was below its level of the early 1950s, when the great bull market had begun.

With the inflation, which was unprecedented in the peacetime history of the American economy, and the easy money policy of the Federal Reserve, the dollar became overvalued in terms of other currencies. This made American goods seem expensive to those using other currencies and foreign goods look cheap to Americans. The trade balance, which had been strongly in favor of the United States in the early years after the Second World War, had inevitably shrunk as foreign economies recovered. In 1959 it had shown a small deficit. But it began to deteriorate rapidly in the late 1960s, and in 1971 it fell into deficit once more and continued to deteriorate.

Because the dollar was the currency of world trade and, under the Bretton Woods Agreement, was convertible into gold at a fixed price of $35 to the ounce, dollars accumulated in foreign central banks and financial institutions. They often circulated abroad without ever returning to the United States. As American inflation increased, these "eurodollar" holdings began to seem precarious, and gold began to flow abroad in quantity for the first time since the early 1930s. The international currency market, a growing force in the international economy, began betting against the dollar.

On August 15, 1971, President Nixon acted decisively, if not necessarily wisely, to solve the increasing economic problems that confronted the country. First, he renounced the Bretton Woods Agreement and severed the link between the dollar and gold. The dollar would now float in value, and the gold standard, after 150 years, was dead. Second, he froze all wages, rents, and prices for a period of ninety days, to be followed by strict wage and price controls.

Wage and price controls have a long history, almost all of it bad. In a free market it is prices that signal, in their uncountable millions, where resources should be allocated and where opportunity lies, what is becoming scarce and what plentiful, allowing people to adjust their economic behavior accordingly. When prices are fixed, however, shortages and surpluses inevitably and quickly develop. That is why there is a permanent shortage of housing wherever there is rent control. Price controls also transfer power from free markets—in other words, the people—to politicians. Politicians, of course, are always tempted to use this power to benefit favored groups, while the disfavored continue to pursue their self-interests through black markets.

Price controls were first tried on a grand scale by Diocletian, who became emperor of Rome in the late third century. Earlier emperors had progressively debased the coinage with base metals, setting off a rampant inflation. Diocletian attempted serious reform of both the coinage and taxation, but lacked enough precious metal to create an adequate money supply. So he was forced to issue base-metal coins as well, with a wholly artificial value that he sought to enforce by law.

It didn't work, of course; Gresham's law makes it impossible. So Diocletian, unable to contain inflation by setting the price of money, tried to contain it by setting the price of everything else. His edict to that effect, which survives, is an invaluable historical window into the economy of the late Roman world, listing legal prices for all sorts of goods and services. But, as with all subsequent attempts to control by law prices that would otherwise be set by millions of people each pursuing his or her self-interest, it was a dismal failure, despite liberal use of the death penalty as a means of enforcement. Goods simply went into hiding, were bartered, or were traded illegally by parties who had a mutual self-interest in not informing the state.

Nixon's price and wage controls fared no better. Within two years they were abandoned, and inflation, now unchecked by any link to gold, continued unabated around the globe. As a result, interest rates soared as lenders demanded protection against the rapidly falling value of the dollar and other currencies. William Zeckendorf, a famously risk-taking New York City real estate developer, often said, "I'd rather be alive at twenty percent than dead at the prime rate." Four years after his death in 1976, however, the prime rate itself stood at 20 percent. The inflation in 1980 reached 13.5 percent, by far the largest peacetime inflation in the nation's history.

Meanwhile, the American trade balance continued to deteriorate. As foreign countries recovered from the war and rebuilt their economic infrastructure, they often became the low-cost producers with their new plants. As transportation costs and tariffs fell relentlessly in the postwar years, these countries were more and more able to compete successfully in the American market with American companies.

This was also true with some raw materials, especially petroleum. The petroleum industry had been born in the United States, and the country remained a net exporter of oil until the 1950s. But by the 1970s, as rich American fields were increasingly depleted and new ones became ever more expensive to exploit, cheaper foreign oil began to flow into the country in larger and larger amounts. Naturally, it wasn't long before the oil-exporting countries sought to take advantage of this situation, forming a cartel called OPEC (Organization of Petroleum Exporting Countries) to raise prices.

As a result of the 1973 Yom Kippur War between Israel and its Arab neighbors, many oil-exporting countries refused to export to the United States. Long lines formed at gas stations in what had always been the quintessential "land of plenty," while prices for oil products rose steeply. It came as a profound shock to most Americans and to the American economy, as the cost of petroleum affects the price of nearly every other product.

It also came as a shock to what had been the linchpin of the American economy for sixty years, the automobile industry. The industry had had the American market almost entirely to itself since the war. It had evolved into a loose cartel, enforced by antitrust laws that kept the big three automobile companies, General Motors, Ford, and Chrysler, from aggressively seeking market share from one another.

With no need to take the risks and expenses of innovation, the industry had stagnated technologically. The last major technological advance had been automatic transmission, first introduced in 1948. Instead the automobile companies concentrated on appearance, size, and power. American cars in the postwar years became larger and larger and often sported such nonfunctional features as tail fins and much chrome. The evolution of American automobiles in these years is strikingly analogous to the tendency of living things isolated from competition on lush islands to evolve into giant and often grotesque forms. Just as markets are ecosystems, ecosystems are markets.

With the oil embargo of 1973, the isolation of the American automobile market ended abruptly. With gas in short supply, far more fuel-efficient European cars saw a sharp rise in demand. The Volkswagen Beetle, which had had only a niche market among college students and families with second cars, became an icon of the new automotive era and would have a longer production run than even the Ford Model T. Japanese cars also began to invade the American market, revealing starkly how poorly many American cars were manufactured and how inefficient American manufacturing had often become.

With the long period needed to redesign and retool, the American automobile industry would struggle for more than a decade to regain its footing. By the time it did, the automobile business had become one of the first heavy manufacturing industries to be thoroughly globalized. No car today is manufactured entirely in one country, and, increasingly, such words as "American," "German," and "Japanese" refer only to the location of corporate headquarters and the largest concentration of stockholders.

Many American factories closed as they became obsolete, and the term *rust belt* entered the American lexicon. But new factories, many built by foreign companies, were also opening, many in other areas of the country, especially the South and West. These new factories were often able to produce the same quantity of goods with less labor, due to increased productivity.

Because the media have a natural tendency to concentrate on the bad news, this was often perceived as a decline in American economic power, with the rust belt as its symbol. In fact, it was the beginning of a profound restructuring. Manufacturing had been declining as a percentage of the economy since the end of the Second World War. Then half the jobs in the country were manufacturing jobs. By the mid-1970s more than two-thirds of the jobs in the American economy were in services.

But this restructuring, which is continuing after more than two decades, produced wrenching change and much individual and local economic pain. The official poverty rate, which had fallen from 22 percent in 1959 to 11 percent in 1973, rose to 15 percent by 1983. The amount of steel produced in the country remained constant at about 100 million tons a year, but the number of steel workers declined from 2.4 million in 1974 to less than a million in 1998.

Also undergoing wrenching change were government fiscal and regulatory policies. The policies, largely in place since the New Deal, were characterized by a progressive income tax with high marginal rates on large incomes, and considerable regulation of major sectors of the economy, both to prevent the formation of monopolies and to limit "excess competition." They had become the accepted wisdom for how to manage a modern economy, but it was becoming increasingly clear that policies developed in the 1930s simply did not work in the different economic universe of the 1970s.

THE FOUNDING FATHERS had perceived the executive as the major threat to financial prudence, not Congress. After all, Parliament had come into existence in the Middle Ages precisely to be a check on royal extravagance and to limit the taxing power of the king. It remained such a check in the late eighteenth century, as only men of property—taxpayers, in other words—held the franchise. The Founding Fathers expected Congress, also largely chosen by property owners in its earliest years, to fulfill the same function. But the coming of true democracy with universal male suffrage in the age of Andrew Jackson began to change Congress's attitude toward spending.

While Congress as a whole has a collective obligation to control spending, each individual member has a personal self-interest in obtaining the most government spending in his or her own district or state. The phrase *pork barrel* entered the American political lexicon as early as 1904. With the coming of the New Deal and safety net programs such as Social Security, there also developed a pressure to vote for popular new benefits.

Presidents have always been at a disadvantage in budget negotiations with Congress, which has the sole power to appropriate money from the Treasury. While presidents can veto money bills just as they can any other bill, the veto is a very blunt instrument at best, as the entire bill must be accepted or rejected, not just the objectionable spending. Often the one means that the president—the only official in Washington elected by the entire country other than the powerless vice president—had to limit this spending was impoundment, first used by Thomas Jefferson and then by every president since. With impoundment, the president simply refused to spend the appropriated money.

As inflation began to heat up in the later Johnson years, Johnson attempted to control spending by impounding more and more money. In 1966 he impounded no less than $5.3 billion out of a total budget of $134 billion, including $1.86 billion in such popular programs as highways and education. While the Democrat-controlled Congress complained loudly, it did not relish a showdown with a Democrat president. Nixon did not fare so well. When he vetoed the Federal Water Pollution Act in 1972 because it was, in his view, too expensive, Congress passed the bill over his veto. Nixon then impounded the $6 billion the bill had appropriated. Congress saw this, not surprisingly, as a direct threat to its power of the purse.

As Nixon's political leverage began to vanish in the Watergate scandal, Congress passed the Budget Control Act of 1974, perhaps the most misnamed major piece of legislation in American history. It removed the president's power of impoundment, which had never had statutory authority, and created the Congressional Budget Office, which gave Congress much the same budgetary apparatus that the Office of Management and Budget gave the president and thus the power to dispute his estimates with estimates more congenial to Congress's purposes.

With the Budget Control Act, the federal budget went out of control. The deficit in 1974 was $53 billion, the largest deficit in dollar terms since the middle of the Second World War. It increased the national debt by nearly 10 percent in a single year. By the end of the decade, the national debt was two and a half times as large as it had been in 1970, although, thanks to the accelerating inflation, it continued to fall as a percentage of GNP.

What did not fall was the percentage of GNP that passed through the government's fiscal machine every year. Because, under a progressive income tax, higher incomes are taxed at higher rates, inflation had been pushing people into higher and higher brackets. Thus income taxes increased in real terms while incomes in real terms were often stagnant. To many in Washington, this was a highly satisfactory situation, as government revenues increased without Congress having to take the onus of voting to raise taxes.

The Democratic Party, which had dominated the country's politics since 1932, had become increasingly out of touch with the electorate and failed to heed the increasingly clear signals of popular discontent as the American economy floundered in the 1970s. In 1978 the people of California began a "tax revolt" by sharply limiting by referendum how much local property taxes could be raised. This sparked tax revolts elsewhere and calls for reform in the increasingly complex and arbitrary federal tax code.

To help end the stagflation that was plaguing the American economy in the 1970s, Congressman Jack Kemp and Senator William Roth proposed cutting the marginal rates on personal income taxes, just as President Kennedy had done more than a decade earlier with great success, and to index tax rates to inflation so that people were not pushed into higher brackets when their incomes were not rising in real terms.

The Kemp-Roth tax proposal was ridiculed by the Democrats. President Jimmy Carter, running for reelection, tried to tie his opponent in 1980, Ronald Reagan, to the proposal by calling it the Reagan-Kemp-Roth proposal, a move that his opponent shrewdly welcomed.

Gas shortages reemerged in the late 1970s while inflation only increased. Stock prices, which had recovered from their disastrous 1974 low, began to decline again. American industry was having more and more trouble competing with other countries. New York City, which had carried the redistributionist model of social welfare much further than other cities and which was heavily dependent on taxes on financial services, went broke. With the city unable to borrow, its quality of life dropped alarmingly, with unkempt parks; graffiti-ridden, unreliable subways and buses; and a relentlessly rising crime rate.

And the American military, starved for funds by the now adamantly antiwar Democrats, who were shell-shocked by the results of the Vietnam War, was unable to respond effectively to the seizure of our embassy and more than four hundred hostages in Iran in 1979.

The Soviet Union, whose ambitions to world domination had been contained for three decades by American economic and military power, was flexing its muscles as it had never done before. In late 1979 it invaded Afghanistan to secure a shaky puppet regime. There seemed little the United States could do in response. The most powerful nation in the world seemed to be becoming a helpless giant. Many wondered if the American century was coming to a premature end.

As a result, for the first time since Herbert Hoover, an elected incumbent president was turned out of office in a landslide. The American people voted decisively for change, and they got it. Ronald Reagan would prove to have the most consequential presidency of the twentieth century, save only for that of Franklin Roosevelt, a man and a president he greatly admired.

———

WHILE RONALD REAGAN is often given the credit for deregulation and lower taxes, they were, in fact, already under way when he took office, although he helped powerfully to continue and increase the restructuring of the American political economy. Much of the federal regulatory apparatus established as early as 1887, with the creation of the Interstate Commerce Commission, and greatly expanded under the New Deal, had evolved into cartels that protected the interests of the industries they regulated more than the interests of the economy as a whole.

The Civil Aeronautics Board (CAB), which regulated routes and fares for interstate air travel, kept fares on these routes far higher than those on comparable noninterstate routes. In 1978 its power to set rates and routes was taken away by Congress, despite the ferocious opposition of both the airline companies and the airline unions (when both management and labor oppose a change in regulation, it's a sure sign that a cartel is in operation). The airline business, regulated since its infancy, underwent a painful period of adjustment as airlines began to compete by means of fares. A hub-and-spoke route system soon evolved, and air fares changed frequently as price wars broke out among the airlines. Many old airlines, such as Pan American, Eastern, and Braniff went bankrupt, and many new ones, such as Southwest Airlines, entered the business. Air fares fell drastically on average and air travel increased rapidly.

The Motor Carrier Act of 1980 freed the trucking business to compete, and the Staggers Act the same year freed the railroads to do the same. The railroad business, in decline for most of the century, began to revive, and pointless inefficiencies—most trucks had to return to their place of origin empty, for instance—were quickly drained out of the transportation business. In 1980 transportation amounted to about 15 percent of GDP. By the 1990s it had dropped to 10 percent. Since transportation is what economists call a transaction cost—necessary expenses that do not add to the intrinsic value of the product, such as advertising and packaging—this was a pure gain for the economy as a whole.

The most important deregulation of the 1970s was on Wall Street. The New York Stock Exchange had begun as an agreement among brokers to set minimum prices for stock trading, and commissions had been fixed ever since. But on May 1, 1975, under orders from the SEC, most commissions were allowed to be set by competition for the first time in 183 years.

Fixed commissions, which were calculated as a percentage of the share price, had been under pressure for years as the number of large trades had increased substantially. There had been only an average of 35 trades a day involving more than ten thousand shares in 1965. By 1975 the average was a 135 such trades a day. (Today there is an average of more than 5,000.) These trades cost little more to execute than hundred-share trades, and the big institutions, such as mutual funds, that traded in large blocks had been demanding change. Opposing it were the smaller brokerage houses that could not match the efficiency of the major houses.

With the end of fixed commissions, the price of stock trading plunged by about 40 percent overnight and has been falling ever since. As a result, Wall Street underwent a great consolidation as the smaller firms, unable to compete, merged with larger firms. Meanwhile new firms, discount brokers, opened, offering minimal services and minimal prices as well. The most important result of Mayday, as it was inevitably called, however, was the growth in volume. In the next seventeen years it increased 800 percent and has continued to grow exponentially since. Wall Street had had its first billion-share year in 1929; by the end of the twentieth century billion-share days were the norm. As the cost of stock ownership plunged, the percentage of Americans owning stocks directly increased rapidly as well, and the role of capitalist was played by an ever-wider part of the population, with growing consequences for American politics.

Even the tax code began to change in the late 1970s. In 1969 the outgoing secretary of the treasury in the Johnson administration had testified before Congress that in 1967 there had been 155 tax returns showing incomes more than $200,000 and 21 showing incomes more than a million that had no income tax liability, owing to various provisions of the tax code such as the exemption from taxes for municipal bonds. Congress responded by passing laws requiring minimum tax payments and then an entire supplementary tax code known as the Alternative Minimum Tax. This had the effect of pushing up rates on high incomes and could raise the tax on capital gains (the increase in value in assets that are sold over the purchase price) to as high as 50 percent.

This adversely impacted risk taking by lowering the potential reward without also lowering the risk involved. And investing in new technological possibilities is always very risky, for far more such ventures fail than succeed. It is an inescapable law of economics that if the rewards of success do not match the risks of failure, new ideas will not be tried.

It had also lowered total receipts from capital gains taxes, evidence that excessively high tax rates cause tax receipts to fall, not rise—the so-called Laffer Curve after its developer, the economist Arthur Laffer. In 1968, when capital gains taxes had been no higher than 25 percent, receipts from the tax amounted to $33 billion. By 1977 receipts in inflation-adjusted terms were down to $24 billion. And while there had been three hundred start-up technology companies in 1968, there were none at all in 1976. For an economy that had held the technological lead for more than a century, this was an ominous trend indeed.

Congressman William Steiger, a Republican from Wisconsin, decided to fight to change that. The Republican Party in the late 1970s was a party that seemed to many to be headed for political oblivion.

Stained by the Nixon Watergate scandal, and largely out of power for more than four decades, it was perceived as the party of the past. In the election of 1976, Republicans had won only 158 seats in the House while the Democrats dominated with 277.

In fact, the Republican Party was beginning to crackle with new ideas to address the new economic realities. The Democrats would largely cling to the New Deal model that had served it so well for forty years, but that the great political commentator Walter Lippmann had noticed as early as 1964 was becoming outdated. As a result, only one Democratic candidate for president since that year, Jimmy Carter in 1976, has won a majority of the popular vote (and Carter won only 50.46 percent). In 1994 the Republicans would take complete control of Congress for the first time in four decades and hold it thereafter.

Steiger, who sat on the tax-writing House Ways and Means Committee, proved himself a persuasive politician and lined up his fellow Republicans in support of lowering the capital gains tax. Soon he had converted many Democrats as well, giving him a two-to-one majority on the committee. The Democratic establishment fought the proposal hard. President Jimmy Carter threatened to veto the Tax Reform Act of 1978, and the *New York Times* argued for eliminating all distinctions between capital gains and regular income, which would have raised the capital gains tax to as high as 77 percent.

Regardless, the bill passed Congress, and President Carter, despite his threat, signed it. The effect was immediate. In 1977 the venture capital industry had raised only $39 million. In 1981 the sum was $1.3 billion. Reagan would build on this tax reform by getting Congress to enact Kemp-Roth, sharply reducing the marginal rates on high incomes—inevitably the source of most new capital—in 1981. In 1986 Reagan struck a remarkable deal with Congressman Dan Rostenkowski, Democrat of Illinois and chairman of the House Ways and Means Committee. Together they agreed to further cuts in the marginal tax rates, the highest being reduced to a mere 28 percent, the lowest since the 1920s. In exchange, thousands of deductions and loopholes were closed, greatly simplifying the tax code and further improving the investment climate.

———

BY THE TIME RONALD REAGAN took office, even inflation was finally being brought under control, thanks to the Federal Reserve and its new chairman, Paul Volcker. Volcker, appointed by Jimmy Carter in the summer of 1979, changed the old Federal Reserve policy of controlling interests to one of seeking to rein in the money supply that had been growing very quickly and fueling the inflation. As a result, interest rates soared to their highest point in U.S. history in the next few years. Even the federal government, by definition the best credit risk in the country, had to pay 15.8 percent to sell twenty-year bonds.

The inevitable result of Volcker's policy, which was bravely endorsed by the new Reagan administration, was a deep recession, the worst since the 1930s. For the first time since the Great Depression, the unemployment rate rose above 10 percent while the stock market fell below 800 on the Dow. The monetary medicine was bitter indeed, but with a broad safety net now in place, including such programs as unemployment insurance and the widespread provision of layoff benefits in union contracts, there was no widespread profound distress.

And the benefits were not long in coming. Inflation began to break. Inflation had raged at 13.5 percent in 1980. The next year it was at 10 percent. In 1982 it was 6.2 percent, the lowest since the early 1970s. In 1983 it was 4.1 percent. It averaged less than that for the rest of the decade.

With inflation under control, interest rates began to decline, although not nearly as quickly, as lenders still demanded protection from a feared resurgence. With lowering interest rates, borrowing and investment picked up, and the recession came to an end. The end was heralded, as usual, by a resurgent stock market, which began rebounding with a classic buyers' panic in August 1982. By the end of that year the Dow had crossed 1,000 for the last time. The greatest bull market in world history was under way.

Part of that bull market was a new wave of mergers and acquisitions, the fourth to work its way through the American economy and in many ways strikingly similar to the first one in the 1890s. Low stock prices in terms of corporate assets, falling interest rates, and new capital-raising techniques such as "junk bonds"—bonds that paid high interest rates and financed risky, often untried ideas, such as CNN, the first all-news cable network—fueled the movement. By the end of the decade, more than one-third of the Fortune 500 companies would be taken over or merged. Just as in the 1890s, some of these mergers produced greatly improved economic performance and leaner, more flexible organizations. Others were misbegotten and failed. And some were tainted by fraud and shady dealing. But there is no doubt that the American economy was far stronger at the end of this merger wave than it had been before it.

By 1987 the Dow-Jones Industrial Average had reached 2,500, three times higher than where it had been a mere five years earlier, and the underlying economy was seen as basically sound. Regardless, in October of that year the market suffered the worst crash since 1929 and the worst one-day decline in percentage terms—22.8 percent—in history. The volume was a then utterly unprecedented 604 million shares, more than twice the previous volume record.

Many thought that this signaled the start of a new Great Depression. In fact, the market recovered 104 points the next day (on even higher volume, 608 million shares) and reached a new high on the Dow within fifteen months. The reason, principally, was that the Federal Reserve acted immediately and decisively to stem the panic and to protect the economic institutions of the country from harm. It "flooded the street with money," in the words of Benjamin Strong, as it pumped massive liquidity into the economic system.

For the first time since Alexander Hamilton had stemmed the panic of 1792, federal monetary authorities had performed as they should in a moment of financial crisis. As a result, there was little long-term damage to the system as a whole, and the 1987 crash is today hardly remembered at all. The ghost of Thomas Jefferson's hatred of getting and spending, it seemed, had at last been laid to rest. Unfortunately, that ghost was to have one more turn upon the stage of the American economy.

———

FRANKLIN ROOSEVELT had been reluctant to accept the idea of deposit insurance, for he feared the "moral hazard" it ineluctably created. "We do not wish to make the United States government liable for the mistakes and errors of individual banks," he said, "and put a premium on unsound banking in the future." But politics is usually a choice between imperfect means to desirable ends, and the Federal Deposit Insurance Corporation functioned well in the banking cartel that developed under the New Deal.

Commercial banks, savings banks, and savings and loan associations divided the world of American deposit banking among them. Commercial banks became full-service banks, offering savings and checking accounts to individuals while concentrating on business loans. Savings banks and S&Ls offered savings accounts at a slightly higher interest than commercial banks (interest rates were set by law), while concentrating on real estate loans. Even this market was allocated, for savings banks specialized in commercial real estate while the S&Ls lent almost exclusively on single-family houses. New charters were limited to prevent "excess competition." While the number of S&Ls remained steady at around six thousand after the debacle of the 1930s, their collective assets rose from $8.7 billion to $110.4 billion between 1945 and 1965.

It was a cozy, low-stress business, what someone called 3–6–3 banking, because S&Ls paid 3 percent on deposits, charged 6 percent on loans, and management hit the golf course by 3 PM. But as the booming

1950s and early 1960s gave way to the gathering inflation of the late 1960s and 1970s, the business model of the S&Ls began to fall apart. Unregulated interest rates soared while the regulated banking rates stayed the same. Wall Street brokerage houses and mutual funds began offering money market funds that paid a far higher rate of interest than savings accounts.

People increasingly withdrew money from savings banks and savings and loan associations and moved it to the new money market funds, a movement referred to by the sonorous economic term "disintermediation." Commercial banks, most of whose deposit base was in noninterestbearing checking accounts, could cope. The other banks could not. Faced with a rapidly declining deposit base and with long-term real estate loans paying low interest, they went to the federal government for help.

Congress was anxious to do so. As Senator David Pryor explained, "You got to remember that each community has a savings and loan; some have two; some have four, and each of them has seven or eight board members. They own the Chevy dealership and the shoe store." In other words, they were exactly the people whose support members of Congress need. As so often happens in a democracy in the short term, politics trumped economic reality, and what followed was a near textbook case on how not to deregulate an industry.

The savings banks and S&Ls should have been forced to merge with stronger institutions or to become commercial banks themselves, with the same capital and reserve requirements. Instead the ceiling on interest rates was removed, allowing banks to pay market rates on deposits, and the federal guarantee on bank deposits was raised to $100,000 from $40,000.

But Wall Street had found a way around even that generous limit with a device called brokered deposits, bundled deposits that exactly matched the federal guarantee. It was a simple means to allow people with large liquid holdings to have as much of their money federally insured as they wished. It was also what came to be called "hot money," money that followed the highest interest rates.

With S&Ls paying higher and higher interest rates on deposits, while still stuck with their low-interest real estate loans, the industry rapidly went broke. The S&Ls had a collective net worth of $32.2 billion in 1980. Two years later it was $3.7 billion. Bank regulators, under pressure from Congress, came up with new quick fixes. Reserve requirements were lowered and accounting rules were relaxed. This had the effect of making the books look better without addressing the problem. It was a bit like a doctor declaring a temperature of 102 degrees to be normal so that a sick patient could be declared healthy.

And they changed the rules on who could own S&Ls. Instead of local people only, almost anyone could own a thrift and even use noncash assets, such as land, the most illiquid of all assets, as reserves. Latter-day Willy Suttons, sensing opportunity, moved into the industry.

In 1982 Congress allowed the S&Ls to write nonresidential mortgages and make consumer loans, just like commercial banks, but without anything like the capital and reserve requirements or accounting strictures of commercial banks.

A disaster was now unavoidable. Congress and the banking regulators had allowed the creation of an economic oxymoron, the high-yield, no-risk investment called brokered deposits, and allowed people with little banking experience and often dubious respect for the law to try to make them profitable. They quickly destroyed the thrift industry, and when the dust had settled, the federal government had to pay out $200 billion to depositors in failed thrifts.

It was the greatest financial scandal in American history. But, as with all scandals, it pointed the way to reform by overcoming entrenched resistance to it. In 1994 the Banking Reform Act finally freed the American banking industry from the last of its Jeffersonian shackles. Banks were allowed to branch over state lines and to become much larger, providing the protection of diversity and setting off a wave of bank mergers that continues to this day. The distinction between investment banks and deposit banks created in the 1930s by Glass-Steagall was repealed, as was much of the distinction between brokers and bankers and insurance firms. At last, the United States had a banking system that matched the American economy in both scale and scope.

BECAUSE OF THE RESTRUCTURING of the American economy in the 1980s, the dollar strengthened markedly against other currencies. It had been worth only 1.8 German marks in 1980; by 1985 it was worth 3 marks. The value against the French franc more than doubled. And foreign investment in the United States rose sharply. By the late 1980s foreigners owned about $400 billion more in assets in the United States than American citizens owned abroad, reversing the situation that had been the case since the First World War.

While some commentators lamented this as a sign of American weakness, it was, in fact, the opposite. Foreign capital poured into the country precisely because the American economy came to be increasingly seen, once again, as one of great opportunity. Foreign immigration also increased sharply in the 1980s, as the poor, like the rich, sought to prosper in the empire of wealth.

The reform of taxes and regulation that marked the true end of what Arthur Schlesinger, Jr., had termed the Age of Roosevelt two decades earlier could not have been better timed. The world economy was undergoing the most profound transformation since the coming of the Industrial Revolution two centuries earlier, perhaps since the coming of agriculture ten millennia ago. And because the United States was the first major country to undergo the wrenching changes required of a political economy based partly on redistribution, it was positioned to exploit first and most fully the boundless opportunities of a new political economy based on opportunity.

Chapter Twenty

A NEW ECONOMY,
A NEW WORLD, A NEW WAR

WARS ARE OFTEN ENGINES of technological development, and none more so than the greatest war in history, the Second World War. The developments would have come in time regardless, but the war, with its bottomless funds for research and pressing demand for results, accelerated the process considerably, in some cases by decades. With further funding provided by the exigencies of the cold war, these new technologies transformed the world economy in only a generation.

The need for bombers capable of carrying heavy loads long distances produced a quantum leap in both airframe design and construction techniques. After the war these were quickly adapted to civilian uses, and the price of air travel declined to the point where usage increased dramatically. When the large airframe was combined with the jet engine, which allowed near sonic speed, and radar, which allowed the far closer spacing of aircraft near airports through air traffic control, the airplane quickly became the dominant form of long-distance passenger transportation.

Within a decade of the introduction of the Boeing 707 in 1958, both the Atlantic passenger steamship and the long-distance passenger train were near extinction and the world had shrunk by an order of magnitude. What had been a three-day trip between New York and Los Angeles was now five hours; the trip between New York and London took seven hours, not nearly a week.

The large rocket, capable of moving a significant payload hundreds of miles, was developed by the Germans and perfected in the V-2, which weighed fourteen tons and could deliver a one-ton warhead to a target two hundred miles away. V-2s began to hit Britain in late 1944, but they came into use far too late to affect the outcome of the war. Their obvious utility both as a terror weapon and as a means of exploiting the potential of outer space, however, caused a scramble at the end of the war between Western and Soviet forces to secure both the remaining stockpiles of rockets and the scientists who had developed them.

Both the Soviet Union and the United States poured vast resources into rocket research to develop larger vehicles, longer range, and greater accuracy. By the end of the 1950s the rocket, mated to the hydrogen bomb—another technology born in the Second World War—had changed the nature of war. ICBMs, capable of utterly destroying whole cities in an instant, made wars between Great Powers irrational because they were now unwinnable in any real sense. So the United States and the Soviet Union, locked in a profound geopolitical struggle, had to find other ways in which to compete.

But ICBMs also engendered a deep fear that events might spin out of control, as they had on the eve of the First World War, and lay the world waste in a nuclear holocaust. Twice, in the Cuban missile crisis of 1962 and the Yom Kippur War of 1973, that came perilously close to happening.

One means for the United States and the Soviet Union to battle for supremacy was via proxy wars, such as those in Korea, Vietnam, and Afghanistan. Another was by using the technology of the rocket to explore outer space. The Soviet Union stunned the world on October 4, 1957, when it launched the world's first earth-orbiting satellite, Sputnik. (Its purpose was purely propagandistic, as its radio signal, heard around the world, transmitted no information, serving merely as a beacon. But it was very effective propaganda.) The United States soon launched a satellite of its own, and hundreds more followed. A "space race" quickly developed that was won in 1969 by the United States when it landed men on the moon.

Many of these satellites were for military purposes, such as spying, but many others had both civilian and military uses such as communications and weather data gathering. By the late 1960s geosynchronous satellites were able to transmit television pictures that could be received simultaneously by anyone, anywhere, who was able to pick up the signal. The global village, first named by the philosopher Marshall McLuhan in 1960, but which had begun with the laying of the Atlantic cable nearly a hundred years earlier, was now at hand.

The economic applications of space technology, especially since the end of the cold war when many of them were declassified, are nearly limitless and increasing every day. Agriculture, transportation, cartography, navigation, and communications are but a few. Geopositioning satellites have made it possible to determine, with the help of a simple device, one's location within a few feet. The devices are now appearing in many automobiles, giving directions through the use of synthetic voices, a technology that would have seemed utterly miraculous only a couple of decades ago.

Communications satellites, together with an ever-increasing number of undersea cables, have helped greatly to lower the cost of long-distance telephony, leading to an astonishing upsurge in its use. In 1950 about a million overseas phone calls were initiated in the United States. By 1970 the number had grown to twenty-three million; by 1980 to two hundred million. By 2001, as the cost plummeted, the number was 6.3 billion and rising fast.

The collapse in the cost of international communications allowed the world's financial markets to become tightly integrated and increasingly to function as one seamless market, operating twenty-four hours a day.

Just as the teenage runners had once bound together the New York financial market by dashing back and forth between the stock exchange, banks, and brokerage houses, keeping all apprised of the latest prices, now a cat's cradle of undersea cables and satellite links bound together the new global markets.

This had profound political as well as economic consequences. As early as 1980, a unified market trading the major world currencies was in place, trading on average a trillion dollars a day at that time. In 1981 France elected a socialist government under François Mitterrand, and he tried to implement a traditional socialist program, raising taxes on high incomes and nationalizing parts of the French economy, including banks. The French franc quickly plunged on the currency market and continued to do so until the French government had no choice but to reverse course.

It was a pivotal moment in world history. For the first time a free market was able to dictate policy to a Great Power. Just as when the newspapers became a mass medium in the mid-nineteenth century, a major new player in the game of national and international politics had come onto the stage. And the world's governments learned that the old gold standard—which had been implemented by the quasi-governmental Bank of England—had been replaced by the global currency market standard. It was a standard more flexible, more exacting, and far more democratic than the one that it replaced. Inflation, the number-one economic concern twenty-five years ago, has largely vanished from the list of the world's financial concerns.

———

NO TECHNOLOGY COMING OUT of the Second World War can begin to compare with the computer for creating a divide between the past and the present.

The word *computer* has been a part of the English language since the middle of the seventeenth century. But until the middle of the twentieth it referred to a person who calculated for a living, compiling such data as actuarial and navigation tables. (They were mostly women, who were thought more reliable for such work.) Humans, however, have two big limitations as computers. They can only do one calculation at a time, and they make mistakes. In the middle of the nineteenth century a mathematician named William Shanks calculated the irrational number pi out to 707 decimal places, an astonishing intellectual feat. It would be more than a hundred years before it was discovered that he made an error at digit 527 and the last 180 digits of his calculation, therefore, are wrong.

The idea of calculating by machine is a very old one. An Englishman named Charles Babbage, after correcting astronomical tables, lamented as early as the 1820s that "I wish to God these calculations had been executed by steam!" He later began to build, from finely machined brass parts, a hand-powered calculating machine, but never finished its construction. He also designed an analytical machine that was a mechanical precursor of a true computer because it was programmable. It too was never finished.

As governments and businesses grew in size and came to rely on ever more statistics, the need to speed up the processing of data became acute. The 1880 United States census, tabulated by hand, required seven years of mind-numbing work to complete. To help with the next census, a young mining engineer and statistician named Herman Hollerith devised a solution based in part on the eighteenth-century Jacquard loom, which had allowed the machine weaving of complicated cloth patterns. Hollerith's device used punch cards with holes. When a needle passed through a hole, it completed an electrical circuit by dipping into a tiny cup of mercury, and a counter ticked upward.

Hollerith's device was able to tabulate the data on punch cards at the rate of a thousand cards an hour, and the sixty-two million cards generated by the 1890 census were processed in only six months. (Ironically, a fire in 1921 destroyed the database of the 1890 census, and while the totals are known, almost all the individual data are lost.) Hollerith formed a company that merged with two other companies and in 1924 changed its name to the International Business Machine Corporation, IBM for short.

Because of the need to calculate the trajectories of artillery shells quickly and to decrypt codes, the American and British governments poured money into developing a true electronic computer during the Second World War. The first successful general purpose one was called ENIAC (an acronym for Electronic Numerical Integrator and Computer) completed at the University of Pennsylvania by Presper Eckert and John Mauchly in 1946, after three years of effort.

It was a monster, the size of a bus. It filled forty filing cabinets, each nine feet high, with eighteen thousand vacuum tubes and thousands of miles of wiring. The vacuum tubes and cooling system used as much electricity as a small town, and it was, by modern standards, glacially slow. Programming was done by physically switching the wiring on switchboardlike grids. People had to stand by constantly to replace the vacuum tubes as they blew out and to remove the occasional errant insect (the origin of the term *debugging*).

Computers shrank rapidly in size and cost, especially after 1947, when Western Electric, the manufacturing arm of American Telephone and Telegraph, developed the transistor. The transistor does exactly what vacuum tubes do, but are much smaller, are much more durable, and cost far less to operate and to manufacture. By the 1960s banks, insurance companies, government agencies, and large corporations depended on computers to do, at a fraction of the cost, the work once done by hundreds of thousands of clerks. IBM dominated this market with such machines as the 7090, introduced in 1959.

But computers remained very large and mysterious, hidden away in special air-conditioned rooms, tended by men wearing white coats who spoke languages no one else understood. These computers intruded not at all into the daily lives of most people. They also remained very, very expensive for a reason that mathematicians call the tyranny of numbers.

The power of a computer is relative not only to the number of transistors but also to the number of connections between them. If there are two transistors, only one connection is needed. If there are three, then three connections are needed to fully interconnect them. Four transistors need six, five need ten, six need fifteen, and so on. As long as these connections had to be made, essentially, by hand, the cost of building more powerful computers escalated far faster than did their power.

The solution to this problem was the integrated circuit, first developed in 1959 by Jack Kilby of Texas Instruments and Robert Noyce of Fairchild Semiconductor. An integrated circuit is simply a series of interconnected transistors laid down on a thin slice of silicon by machine. In other words, the transistors and the connections between them are manufactured at the same time. In 1971 Intel produced the first commercial microprocessor, which is nothing less than a very small computer laid down on a silicon chip.

The tyranny of numbers was broken. For while the cost of designing a microprocessor and building the machines necessary to produce it are extremely high, once that investment is made, the microprocessors themselves can be turned out like so many high-tech cookies, bringing the cost of each one down by orders of magnitude. They quickly increased in complexity and therefore in power and speed.

Gordon Moore, a founder of Intel, predicted in the early days of the company that the number of transistors on a chip, and thus the chip's computing power, would double every eighteen months. He proved correct, and "Moore's law," as it is called, is expected to continue to operate for the foreseeable future. The first Intel microprocessor had only twenty-three hundred transistors. The Pentium 4, the current standard for personal computers, has twenty-four million. As the power increased, the price per calculation collapsed. Computing power that cost a thousand dollars in the 1950s costs a fraction of a cent today. Its use, therefore, began to increase by orders of magnitude, an increase that has not begun to level off.

The computer, like the steam engine, produced an economic revolution, and for precisely the same reason: it caused a collapse in the price of a fundamental input into the economic system, allowing that input to be applied to an infinity of tasks that previously had been too expensive or simply impossible. The steam engine brought down the price of work-doing energy; the computer brought down the price of storing, retrieving, and manipulating information.

Previously, only human beings could do this sort of work; now machines could increasingly be employed to do it far faster, far more accurately, and at far, far lower cost. And just as the steam engine could bring to bear enormous energy on a single task, the computer can bring a seemingly infinite capacity to calculate and manipulate information. A computer model of the early universe designed in the 1980s was estimated to have required more calculations than had been performed by the entire human race prior to the year 1940.

Computers began to invade everyday life with astonishing speed. It was more than sixty years between Watt's first rotary steam engine and the coining of the phrase Industrial Revolution, but it was clear that a computer revolution was under way less than a decade after the first microprocessor was produced. The first commercial products were hand-held calculators that quickly sent the adding machine and the slide rule into oblivion. Word processors began to replace the typewriter in the mid-1970s. And microprocessors, unseen and usually unnoted, began to be used in automobiles, kitchen appliances, television sets, wristwatches and a hundred other everyday items. Many new products—cordless phones, cell phones, DVDs, CDs, VCRs, digital cameras, PDAs—would not be possible without them. By the 1990s they were ubiquitous. The modern world would cease to function in seconds if microprocessors were all to fail.

But while computers had become much smaller and much cheaper, they were still very difficult to use by people without considerable special training. In the early 1970s the Xerox Corporation, at its Palo Alto

Research Center, developed many ways to make computers far easier for nontechnical people to use, including the mouse and the graphical-user interface. But Xerox was unable to develop a marketable product using these new concepts. Steven Jobs and Stephen Wozniak, the founders of Apple Computer, did so. When IBM entered the PC market in 1981, using an operating system developed by Microsoft, the market for personal computers took off and has been increasing exponentially ever since as the price has dropped relentlessly.

Today tens of millions of children and adolescents have on their desks, and use constantly, computing power that would have been beyond the reach of all but national governments thirty years ago. Their developing brains are literally being wired to use computers as an adjunct of their own intellect. Further, they have at their beck and call what is without question the most extraordinary machine ever developed by a species with an abiding genius for machinery.

A personal computer can play chess, or any other game, better than all but grand masters; keep books; store and retrieve vast amounts of data; edit photographs; produce CDs; play movies; create art; and do a thousand other tasks. No one who lived before the last third of the twentieth century would regard a personal computer that costs no more than 5 percent of average annual income as anything but magic, or an elaborate fraud.

But PCs can do still more. They can communicate. Personal computers have become the portal into an entirely new but already major part of the human universe, the Internet. Just as the railroad proved the most consequential spinoff technology of the steam engine, so the Internet has been for the computer. Once again, it was war, or rather the possibility of war, that brought it into being.

After the launching of Sputnik in 1957, the Defense Department formed the Advanced Research Projects Agency (ARPA) to organize and coordinate science and technology projects with military applications. In 1962 Paul Baran of the RAND Corporation was asked to propose means whereby command and control could be maintained after a nuclear strike. Communications networks had always been either centralized, with all communication through a central hub, or decentralized, with a number of hubs, with subnetworks. Telegraph and telephone networks were structured this way, with switchboards serving as the hubs.

Both were highly vulnerable to a nuclear strike. This not only increased the possibility the system would fail, but also increased the possibility that one side or the other in a confrontation would be tempted to strike first, for fear that it might not be able to respond to a strike.

Baran proposed a "distributed network" with no central hubs, only an infinity of nodes, similar to the crossroads in a street grid. If one or more nodes was wiped out, messages could still travel by other routes. A web of computers that came to be called ARPANET was established in 1968 with a total of four computers connected to it by phone lines, three in California and one at the University of Utah.

In 1972 the first e-mail program was designed, and a protocol named TCP/IP was designed the following year to allow different computer systems, even those using different languages, to connect easily via the net, which by then had twenty-three computers connected to it. One of the designers of TCP/IP, Vinton Cerf, coined the term Internet the following year, because it was beginning to connect not only individual computers but also subnetworks of them. In 1983, by which time there were 563 computers on the net, the University of Wisconsin developed the Domain Name System, which made it much easier for one computer to find another on the net. By 1990 there were more than three hundred thousand computers on the Internet and the number was growing explosively, doubling every year.

But it was still mostly a network connecting government agencies, universities, and corporate research institutions. Then, in 1992, Tim Berners-Lee, an Englishman working for CERN, the European nuclear research consortium, wrote and released without copyright the first Web browser, a program that allows people to easily find and link to different sites set up for the purpose. The World Wide Web (WWW) was born. Individuals and corporations quickly saw the potential of this new means to communicate as well as advertise and sell their products. In 1994, when the number of users of the Internet was still only nearing four million, Pizza Hut began to sell pizza via its Web page.

Internet usage exploded in the mid-1990s, and today, a mere decade later, uncountable millions of computers around the globe are linked together by this system. It is by far the most powerful communications tool ever developed. The Internet has already caused profound restructuring of many businesses.

Indeed, all businesses that are essentially brokers—businesses that bring buyers and sellers together and take a small percentage of any sale that takes place, such as real estate, travel agents, stock and insurance brokers, auction houses—are finding their business changing its nature or disappearing altogether. The Internet, especially together with such search engines as Google, simply makes it far easier for buyers and sellers to find each other without a broker.

Retailers as well began to sell more and more via the Internet, delivering around the country and overseas, often overnight by means of such delivery services as FedEx and UPS. Internet retail sales, which are often cheaper because of lower overhead and no sales tax, have been growing at the rate of better than 30 percent a year for the last seven years. Amazon.com, one of the first of these Internet retailers, now controls about 10 percent of the retail book market in the United States and is expanding quickly into other areas.

The news media as well are undergoing fundamental change as a result of the Internet. Soon after the dawn of the mass media in the 1830s, the cost of entering the news business increased sharply. James Gordon Bennett had begun the *New York Herald* in 1835 with only $500 in capital. The *New York Times,* founded sixteen years later, needed $85,000. Radio and television also required large amounts of capital (and a government license) to reach the public.

But the Internet made it possible for anyone with a PC and a Web page design to enter the news business, and thousands did. In 1998 Matt Drudge of *The Drudge Report* broke one of the biggest news stories of the 1990s, the Monica Lewinsky scandal. Weblogs (called blogs for short, their authors bloggers) sprang up by the tens of thousands as people began to express their opinions through this new medium. The good ones attracted large audiences and quickly developed real influence. The effect of the Internet has been to sharply democratize the media by allowing many more voices to be heard.

Because the Internet needed almost no infrastructure not already in place, it developed spontaneously, with little government help or interference and no government direction. Once the world's most communicative animal discovered this powerful and very inexpensive new means of communication, the species flocked to use it. Those countries whose elites depended on close control of the population and their access to information to maintain their power, found their power slipping and often vanishing altogether.

The computer and its most important creature, the Internet, became the most potent weapon against tyranny since the concept of liberty itself.

———

WHILE WAR HAD BROUGHT the computer into existence, the computer profoundly changed the nature of warfare. In the industrial age, success in war depended more than anything else on which side could field the most men with the most guns, ships, and airplanes. Quantity trumped quality. The Soviet Union, unable to match the West in technological development, had depended on this fact and the ability of its vast intelligence operation to steal Western technology, to maintain its superpower status, and to fight the cold war.

But with computer guidance, bombs became smart, allowing them to be much more accurate and far less lethal to civilians even in crowded urban neighborhoods. More sophisticated radars made possible by the microprocessor changed the nature of air battles. In 1982 the Israeli air force was able to use unmanned surveillance aircraft to mimic fighter planes attacking Syrian radar sites in the Bekaa Valley in Lebanon. When the radars turned on in order to track the aircraft, real fighter planes swept in, honing in on the radars' own beams, and destroyed them. With its battle management radar, supplied by the Soviets, out of action, the Syrian air force was effectively blind and the Israelis shot ninety-six Syrian planes, also Soviet supplied, out of the sky, losing none of their own.

With electronics advancing very rapidly, the Soviets could not keep up on their own, or even steal fast enough to do so. The military advantage of its huge army and massive numbers of ships, tanks, and planes was rapidly eroding. When the United States began supplying the Afghans with handheld Stinger antiaircraft missiles, the Soviet control of the air and thus its military advantage in Afghanistan's rugged terrain vanished, and the war there became unwinnable. The Afghan war quickly became the Soviet Union's Vietnam, and the Soviet government found itself unable to hide the truth from its people.

Ronald Reagan seized the advantage and pushed through Congress a massive rearmament program, increasing defense spending by 50 percent in real terms in the first six years of his administration. He also announced the development of a space-based missile defense system, promptly dubbed "Starwars," that would have cost billions but would have nullified the Soviet's nuclear capacity if it worked. Reagan gambled, correctly, that the Soviets would not be able to take the chance that it would fail.

The president had quite deliberately decided to use the country's strongest weapon—the American economy—to win the cold war, just as Roosevelt had used it to win the Second World War. The United States could afford these massive expenditures. The Soviet Union, it turned out, could not. Its economy, controlled by bureaucracies, not markets, and profoundly corrupt, was in far worse shape than American intelligence estimated.

The top-down Soviet government was paralyzed in the early 1980s by the deaths of three general secretaries in quick succession. But when Mikhail Gorbachev took over in 1985, he tried both to negotiate with the United States to reduce the Soviet Union's defense expenditures, and to loosen controls on Soviet economy and society so that the country could become more productive and use the new possibilities created by the microprocessor more effectively.

But once the people sensed that the hand of tyranny was lifting, the Soviet government quickly lost control over events. First the Communist governments in the East European satellites collapsed and then the Soviet Union itself fell apart. The non-Russian republics declared independence, and the Soviet Union ceased to exist in 1991, when the hammer and sickle was lowered over the Kremlin and the old flag of Russia replaced it.

The Soviet Union, which had presented itself to the world for its entire existence as the wave of the future, a claim accepted—remarkably, in retrospect—by a large part of the Western intelligentsia, was revealed to have been nothing more than a Russian and very old-fashioned empire, the last empire on earth to have been based on military power. The third Great Power conflict of the twentieth century, a global one like the first two, was over after nearly fifty years.

The United States now stood alone, unchallenged as the most powerful country in the world, and with no challenger in sight. But the United States and its allies were not the only victors in the cold war. Capitalism and democracy were also victors, as socialism in all its many forms had been shown to be universally a failure as an economic system. It simply could not produce the goods and services that the United States and other capitalist countries had in such abundance and which the new communications media displayed to the world.

The so-called Second World, the Communist bloc, disappeared with the end of the cold war, leaving a world of modern, developed countries, such as the United States, Western Europe, and Japan; countries that were rapidly developing along modern lines and growing explosively, such as South Korea, Taiwan, mainland China, India, and Brazil; what used to be called the Third World, countries that had yet to cast off the old ways of top-down government and economies controlled by economic oligarchies, such as the Arab world and much of Latin America and Africa; and failed states such as Rwanda, Haiti, and Liberia, hellholes of poverty and chaos.

WHILE CONGRESS HAD BEEN WILLING to fund Reagan's military buildup, it had not been willing to enact his cuts in domestic social programs. As a result, the annual federal deficit mounted sharply. Just as in the 1970s, it more than tripled in dollar terms, from $909 billion in 1980 to $3.2 trillion in 1990. But because the rampant inflation of the 1970s had been stopped, the size of the debt relative to GDP—the true measure of a national debt—had increased rapidly. Only 34.5 percent of GDP in 1980, by 1990 it was 58.15 percent and climbing rapidly. It was the first time in American history that the national debt had increased in these terms in peacetime.

And the American economy, which had grown so robustly in the 1980s that it added productive capacity equal to that of the entire West German economy—the largest in Europe—to what it already had, began to stall shortly after the end of the Reagan presidency in 1989. The rising national debt, however, did not stall. By 1994 it had reached $4.6 trillion and equaled 68.9 percent of GDP.

The recession of 1990–91 was the mildest of the twentieth century, and the economy soon began to grow again, at first fitfully, and then more and more robustly as the possibilities of the Internet and the seemingly limitless potential application of the microprocessor became manifest. The debt, if it did not shrink, grew far more slowly, thanks in part to the sale of assets of failed S&Ls that had been taken over by the government, to rapidly declining defense expenditures, and to rising tax revenues that increased even faster than Congress's prodigious predilection for spending. By 1998 the federal operating budget was in surplus for the first time in thirty years and stayed in surplus for the next three years.

About the only perceived weakness in the American economy was the unfavorable trade balance, which deepened relentlessly during this time. But that was to a large extent an artifact not of American weakness, but of the weakness of foreign economies. Japan's once wondrous economy had peaked in 1989 and then sank into protracted recession from which it has yet to fully recover, its major stock market index falling by three-quarters from its high. Europe, the other of the globe's major economic centers, was also not growing anywhere near as quickly as the United States, and many countries there had unemployment rates that stayed stubbornly above 10 percent.

Wall Street boomed as never before. While the Dow-Jones Industrial Average had tripled in the 1980s, it nearly quintupled in the 1990s, reaching more than 11,000 by the end of the decade. The NASDAQ index, heavily weighted with tech stocks, did even better. Under 500 in 1990, it rose to 5,700 in early 2000.

The late 1990s in the United States were the greatest period of wealth creation in the history of the world. The numbers are almost beyond imagination. In 1988 the richest man in the United States was Sam Walton, age seventy, founder of Wal-Mart, a retail chain that by that time was the third largest in the country. The secret of its success was its revolutionary use of the computer to track and control inventory and squeeze costs out of the operation. He was worth that year $6.7 billion.

Bill Gates, the founder of Microsoft and only thirty-three, was worth $1.1 billion, one of only forty-four Americans with a net worth over a billion dollars. That year it took $225 million to make the Forbes 400 list (that was up from $92 million in the list's first year, 1982).

In 2000 the minimum wealth required to make the Forbes list was

$725 million, and the average worth was $3 billion, with three-quarters of the people on the list worth more than $1 billion. The richest of all was now Bill Gates, worth $63 billion, almost ten times the wealth of the richest American only twelve years earlier. The Walton fortune, now in the hands of the heirs of Sam Walton, had grown to $85 billion, and Wal-Mart had become the largest retailer on earth with four thousand stores and $165 billion in annual sales, equal to the gross domestic product of Poland, a country with a population of nearly forty million.

As always in the American economy, most of richest were self-made. Indeed, 263 of the 400 richest Americans in 2000, almost two-thirds, created their own fortunes from scratch; only 19 percent of the people on the Forbes list in 2000 inherited enough money to qualify for it.

The enormous rise in the stock markets in the late 1990s was bound to end in a correction, and the bubble began to burst in March 2000. But there was no crash. Instead the averages declined, sometimes sharply, sometimes gently, although many individual stocks, especially those that had been publicly offered toward the end of the great bull market, lost much of their value. There was no reason to believe that anything more than a normal, if considerable stock market correction with a very mild recession was taking place.

―――――

THEN, ON THE MORNING of September 11, 2001, an almost lyrically beautiful late summer day, a hijacked airliner slammed into the north tower of the World Trade Center in New York. A few minutes later a second hit the south tower. In less than two hours both buildings collapsed, killing thousands of innocent people. A pall of smoke and dust spread across the nation's largest city and its financial district that had been the very beating heart of world capitalism for three generations. It was a direct attack on the financial capital of the empire of wealth.

A third plane struck the Pentagon, the symbol of American military might, and a fourth crashed in a Pennsylvania field as its passengers gave their lives to prevent it from striking elsewhere. For the first time since Pearl Harbor, the United States had been attacked on American soil. It was the first time since British troops landed in Louisiana in December 1814 that the mainland itself had been under assault.

For the fourth time in less than a century, the United States was at war with forces who sought to prevent the spread of modernity, whose hallmarks are democracy and capitalism. But this time it was not a nation-state that had attacked, but rather a shadowy conspiracy of fanatics. It was an enemy far weaker by all ordinary measures of geopolitical power than the enemies of earlier wars, but also an enemy whose war-making potential would be far harder to destroy. No one thought the war would be easy or short or cheap.

But all, except perhaps for a few of its enemies, blinded by ideology, thought that the United States would prevail in this new struggle. As Cicero had explained in the final days of the Roman Republic two thousand years earlier, "the sinews of war are infinite money." The American economy at the dawn of the twenty-first century was more nearly capable of producing those sinews than any other economy the world has ever known.

NOTES

CHAPTER ONE

THE LAND, THE PEOPLE, AND THE LAW

Page

10 *to settle and plant:* Taylor, *American Colonies,* p. 129.

13 *There was no talke:* Ibid., p. 131.

15 *the marketplace, and streets:* Gately, *Tobacco,* p. 72.

18 *rights of Englishmen:* Paul Johnson, *History,* p. 26.

20 *unhappy effect of owning:* Taylor, *American Colonies,* p. 155.

CHAPTER TWO

IN THE NAME OF GOD AND PROFIT

24 *extraordinary scheme:* McCrady, *South Carolina,* vol. 1, p. 94.

24 *If to convert all things:* Taylor, *American Colonies,* p. 93.

25 *Charles-Town Traded:* Taylor, *American Colonies,* p. 229.

27 *in the name of God and profit:* Ratner, Soltow, and Sylla, *Evolution of the American Economy,* p. 15.

35 *our Iron works: First Iron Works,* p. 19.

35 *Every new undertaking:* Ibid.

35 *the furnace runnes:* Ibid.
35 *engines for mills:* Ibid.

CHAPTER THREE
THE ATLANTIC EMPIRE

45 *Though I desire:* Taylor, *American Colonies,* p. 226.
45 *Tobacco, our money:* Gately, *Tobacco,* p. 107.
50 *as soon as the time:* Taylor, *American Colonies,* p. 441.

TRANSITION
THE AMERICAN REVOLUTION

64 *The touch of a feather:* Bailyn, *Great Republic,* p. 327.
64 *how far a uniform system:* Middlekauff, *Glorious Cause,* p. 600.
64 *to take into consideration:* Ibid.
65 *sedition itself:* Bailyn, *Great Republic,* p. 329.

CHAPTER FOUR
THE HAMILTONIAN CREATION

70 *in all cases:* Malone, *Jefferson and His Time,* vol. 1, p. 417.
70 *monetary unit of the United States:* Goodwin, *Greenback,* p. 84.
72 *damned sharp:* John Miller, *Alexander Hamilton,* p. 225.
72 *were not well acquainted:* Ibid.
74 *on end as if the Indians:* John Miller, *Alexander Hamilton,* p. 246.
75 *safe and free from reverses:* Ibid., p. 253.
77 *To attach full confidence:* Hamilton Papers, p. 83.
78 *My zeal against those institutions:* Galbraith, *Money,* p. 73.
78 *Little less than a prohibitry clause:* John Miller, Alexander Hamilton, p. 265.
79 *The national government like every other:* John Miller, *Alexander Hamilton,* p. 266.
79 *ourselves to each other:* Sobel, *Big Board,* p. 21.
80 *'Tis time:* John Miller, *Alexander Hamilton,* p. 305.
80 *At length our paper bubble is burst:* Sobel, *Big Board,* 28.
81 *No calamity truly public:* John Miller, *Alexander Hamilton,* p. 305.

CHAPTER FIVE
A TERRIBLE SYNERGY

84 *heard much said:* Norman, *Inventing of America,* p. 19.

89 *Let any great social:* http://www.hnoc.org./fabric.htm.

89 *What would happen if no cotton:* http://www.sewanee.edu/faculty/willis/ CivilWar/documents/Hammondcotton.html.

91 *We are destitute:* http:www.uh.edu/engines/epi384.htm.

92 *Mills and machines:* Cameron, *Samuel Slater,* p. 60.

CHAPTER SIX
LABOR IMPROBUS OMNIA VINCIT

99 *the wreck of a coach:* Larkin, *Reshaping of Everyday Life,* p. 212.

101 *endless procession:* Larkin, ibid. p. 219.

105 *our minds are not yet enlarged:* Page Smith, *Shaping of America,* p. 770.

106 *and you talk of making:* Tobin, *Great Projects,* p. 170.

106 Labor improbus: Ibid.

106 *promote agriculture:* Cornog, *Birth of Empire,* p. 117.

108 *to commemorate the navigable:* Page Smith, *Shaping of America,* p. 772.

108 *the longest canal:* Ibid.

108 *Having taken your position:* Ibid., p. 774.

109 *London of the New World:* Milton Klein, ed., *Empire State,* p. 276.

110 *that tongue:* Gordon, *Scarlet Woman,* p. 30.

111 *the subscribers have undertaken:* Fox, *Transatlantic* p. 4.

111 *Such steadiness:* Ibid., p. 5.

CHAPTER SEVEN
THE JEFFERSONIAN DESTRUCTION

114 *Every dollar of a bank bill:* Goodwin, *Greenback,* p. 189.

115 *Every banker knows:* Cloyd Norris, "A Savings Bank Bucks the Trend," *New York Times,* July 5, 1990.

117 *a solemn question:* Page Smith, *Shaping of America,* p. 593.

119 *We have hardly enough money:* Wilson, *Stephen Girard,* p. 266.

119 *The Treasury was so far exhausted:* Wilson, *Stephen Girard,* p. 270.

120 *to the residue of the said loan:* Ibid. p. 274.

121 *on demand:* Page Smith, *Shaping of America,* p. 657.

123 *the last great name:* Burstein, *Passions of Andrew Jackson,* p. xv.

124 *He acquired large tracts:* Gordon, *Hamilton Blessing,* p. 58.

125 *My vow shall be to pay:* Remini, *Andrew Jackson,* vol. 2, p. 34.

125 *nothing less than a voice:* Ibid. p. 253.

125 *New Orleans and the National Debt:* Remini, ibid. vol. 3, p. 223.

126 *the first time in the history:* Ibid., p. 219.

129 *The banks let out their notes:* Sobel, *Panic on Wall Street,* p. 40.

130 *The immense fortunes:* Nevins, *Hone,* p. 1181.

C H A P T E R   E I G H T

<h2 style="text-align:center">NEW JERSEY MUST BE FREE!</h2>

138 *Our road to market must:* Hunter, *Steamboats on the Western Rivers,* p. 11.

139 *a sudden wrench:* Tobin, *Great Projects,* p. 9.

139 *Capt. Shreve:* Ibid., p. 12.

141 *with a gentle motion:* Ferguson, *Oliver Evans,* p. 30.

142 *New Jersey Must Be Free!:* Lane, *Commodore Vanderbilt,* p. 34.

143 *I shall stick to Mr. Gibbons:* Parton, *Famous Americans,* p. 384.

144 *Great anxiety is manifested:* Warren, *Supreme Court,* pp. 607–8.

144 *low, feeble voice:* Ibid., p. 609.

144 *Commerce undoubtedly is traffic:* Ibid., p. 609.

144 *The opinion of the Court:* Ibid., p. 610.

144 *Some of the New Yorkers:* Ibid., p. 615.

145 *The steamboat* United States: Ibid.

145 *cries of "down with all monopolies":* Ibid.

145 *released every creek and river:* Ibid., p. 602 (footnote 3).

145 *emancipation Proclamation:* Ibid., p. 616.

146 *when people will travel:* Norman, *Inventing of America,* p. 54.

148 *I consider what I have just now done:* Martin, *Railroads Triumphant,* p. 5.

148 *Two generations ago:* Hadley, *Railroad Transportation,* p. 65.

150 *I'll knock together an engine:* Dictionary of American Biography, s.v. "Cooper, Peter."

151 *It's a great sight:* Nevins and Halsey, eds., *Diary of George Templeton Strong,* vol. 1, p. 108.

151 *This world is going too fast:* Nevins, *Hone,* p. 722.

CHAPTER NINE

CHAINING THE LIGHTNING OF HEAVEN

155 *If the presence of electricity:* Dictionary of American Biography, s.v. "Morse, Samuel."

158 *Money always has a tendency:* Medbery, *Men and Mysteries,* p. 9.

158 *furnished once a moneth:* Encyclopedia Britannica (1966), s.v. "Newspapers."

159 *looked at me with one eye:* Croffut, *American Procession,* p. 10.

161 *The daily newspaper: North American Review,* April 1866.

163 *Gas is now considered:* Gordon, "Ancestors."

164 *The method of heating:* Ibid.

165 *nothing is talked of:* Nevins, *Hone,* p. 624.

165 *I've led a rather amphibious life:* Nevins and Halsey, eds., *Diary of George Templeton Strong,* vol. 1., pp. 203–4.

CHAPTER TEN

WHALES, WOOD, ICE, AND GOLD

169 *the impestuous energy:* Yergin, *The Prize,* p. 23.

170 *Gentlemen, it appears to me:* Ibid., p. 22.

172 *I never beheld a scene:* Norman, *Inventing of America,* p. 49.

175 *A cross between:* Ibid., p. 53.

176 *The reaping machine:* Ibid., p. 53.

177 *No joke:* Weightman, *Frozen-Water Trade,* p. 37.

180 *It made my heart thump:* Marks, *Precious Dust,* p. 27.

185 *Every beat:* Sobel, *Panic on Wall Street,* p. 92.

185 *What can be the end:* Ibid.

TRANSITION

THE CIVIL WAR

193 *in the face of the people:* Dictionary of American Biography, s. v. "Cooke, Jay."

196 *The bottom is out of the tub:* McPherson, *Battle Cry of Freedom,* p. 444.

198 *General Lee's left wing:* Oberholtzer, *Jay Cooke,* vol. 2, p. 141.

200 *Many brokers earned:* Medbery, *Men and Mysteries,* p. 247.

201 *Where three years ago:* Thomas, *Confederate Nation,* p. 211.

204 *We began without capital:* Richardson, *Greatest Nation of the Earth,* p. 64.

204 *to-day the most powerful nation:* Ibid.

CHAPTER ELEVEN

CAPITALISM RED IN TOOTH AND CLAW

208 *Heaven be praised:* Nevins and Halsey, eds., *Diary of George Templeton Strong,* vol. 2, p. 331.

208 *that another body so reckless:* Gustavus Meyers, "History of Public Franchises in New York City," *Municipal Affairs* 4 (1900).

208 *No* conviction: Tanner, *"Lobby,"* p. iv.

208 *The Supreme Court is:* Nevins and Halsey, eds., *Diary of George Templeton Strong,* vol. 2, p. 202.

208 *in New York there is a custom: Fraser's Magazine,* May 1869.

210 Daniel says up: Lane, *Commodore Vanderbilt,* p. 236.

211 *the Commodore's word:* Smith, *Twenty Years,* p. 119.

211 *It has been much the fashion: Harper's Weekly,* March 5, 1859.

212 *a Gaetulian lion: Fraser's Magazine,* May 1869.

213 *the brute force:* Adams and Adams, *Chapters of Erie,* p. 6.

213 *a Tammany healot:* Stedman, *New York Stock Exchange,* p. 200.

213 *Since they were forbidden:* Ibid p. 202.

214 *If this printing press:* Croffut, *American Procession,* p. 91.

214 *observed a squad:* Fowler, *Ten Years in Wall Street,* p. 502.

214 *flocked to Albany: Fraser's Magazine,* May 1869.

215 *the letter of the law: Commercial and Financial Chronicle,* Oct. 31, 1868.

216 *It remains for the brokers:* Medbery, *Men and Mysteries,* p. 344.

218 *the best position:* Ambrose, p. 98.

218 *there is more poetry:* Nathan Miller, *Stealing from America,* p. 185.

219 *The members of it:* Ambrose, *Nothing Like It,* p. 320.

220 *THE KING OF FRAUDS:* Nathan Miller, *Stealing from America,* p. 188.

221 *The abused machinery:* Nevins and Halsey, eds., *Diary of George Templeton Strong,* vol. 4, p. 264.

221 *for a lawyer to come:* Barlow, *Facts,* p. 218.

CHAPTER TWELVE

DOING BUSINESS WITH GLASS POCKETS

225 *Tap the wires?:* Harper's Monthly, May 1870.

226 *if a man has a thousand:* House of Representative. Report, p. 113.

226 *Broad Street was thronged:* Harris, *Memories of Manhattan,* p. 48.

226 *for the remainder of the day:* New York Herald, Sept. 25, 1869.

229 *an artificial being:* Moss, *When All Else Fails,* p. 57.

230 *Railroad makes no report:* Previts and Merino, *History of Accounting,* p. 81.

230 *a very blind document:* Ibid.

231 *Alaska has a tropical climate:* New York Tribune, Jan. 21, 1870.

231 *The one condition: Commercial and Financial Chronicle,* May 14, 1870.

233 *Is not commercial credit:* Strouse, *Morgan,* p. 13.

237 *The curious fact remains:* Friedman, *History of American Law,* p. 450.

237 *a bargain in which:* Ibid., p. 453.

237 *a large majority:* Ibid., p. 451.

239 *Congress to the realization:* Warren, *Supreme Court,* vol. 2, p. 729.

239 *glass pockets:* Strouse, *Morgan,* p. 5.

CHAPTER THIRTEEN

WAS THERE EVER SUCH A BUSINESS?

242 *there is a glamor:* Krass, *Carnegie,* pp. 115–16.

243 *This country's far better:* Ibid. pp. 19–20.

246 *a cloud which almost:* Freese, *Coal,* p. 109.

246 *But the city itself:* Ibid.

247 *Ah, Andra:* Wall, *Andrew Carnegie* p. 484.

248 *While your metallurgists:* Kirkland, *Industry Comes of Age,* p. 175.

248 *Was there ever such:* Wall, *Andrew Carnegie* p. 536.

251 *who are now* prominently: *The '40s.* New York: Melville Publishing Co. [1889].

254 *It would seem:* Serrin, *Homestead,* p. 88.

260 *find themselves supplanted:* Kessner, *Capital City,* p. 45.

262 *uneasiness over the magnitude: Wall Street Journal,* Feb. 27, 1901.

262 *God made the world:* Holbrook, *Age of the Moguls,* p. 152.

263 *If we have done:* Allen, *Morgan,* p. 220.

Chapter Fourteen

A CROSS OF GOLD

265 *out of work:* Wall Street Journal, Dec. 3, 2003, p. B1.

265 *So long as we have:* House of Representatives, Report, p. 90.

268 *It will be all over:* Chernow, *Morgan,* p. 75.

268 *Have you anything:* Strouse, *Morgan,* p. 343.

269 *Upon which side:* Safire, ed., *Lend Me Your Ears,* pp. 768–72.

270 *that never-to-be-forgotten speech:* Brands, *Reckless,* p. 276.

270 *We have petitioned:* Safire, ed., *Lend Me Your Ears,* p. 850.

270 *The Jacobins are in full control:* Bailyn, *Great Republic,* p. 885.

271 *The sympathies of the Democratic Party:* Safire, ed., *Lend Me Your Ears,* p. 852.

273 *Here we have in New York:* Paul, *Taxation in the United States,* p. 26.

278 *commercial paralysis:* Sobel, *Panic on Wall Street,* p. 303.

279 *if the people will keep:* Satterlee, *J. Pierpont Morgan,* p. 476.

279 *Then this is the place:* Sobel, *Panic on Wall Street,* p. 313.

279 *properly attended to:* Allen, *Morgan,* p. 257.

280 *those influential and splendid:* Cooper, *Pivotal Decades,* p. 221.

282 *The European ideal:* Ibid.

282 *America has joined forces:* Bartlett's, 12th ed., p. 747.

Transition

THE FIRST WORLD WAR

286 *no modern war:* Sobel, *Big Board,* p. 207.

290 *the worst of contrabands:* Chernow, *Morgan,* p. 186.

Chapter Fifteen

GETTING PRICES DOWN TO THE BUYING POWER

303 *lay tubes, wires, conductors:* Tobin, *Great Projects,* p. 94.

303 *in stores and business places:* Ibid., p. 108.

304 *we never made a dollar:* Ibid., p. 111.

306 *Well, if it blows up:* Ibid., p. 121.

309 *Get costs down:* Brinkley, *Wheels for the World,* p. xxi.

311 *They hired the money:* Oxford Dictionary of Quotations, 3rd ed., p. 162.

312 *The problem now:* Greider, *Secrets of the Temple,* p. 296.

314 *stock prices have reached:* Sobel, *Panic on Wall Street,* p. 368.

314 Oh, hush thee, my babe: Ibid.

314 *I repeat what I said:* Brooks, *Once in Golconda,* p. 111.

315 *I bid 205: New York Times,* Dec. 6, 1974.

C H A P T E R   S I X T E E N

FEAR ITSELF

319 *we have the power:* Greider, *Secrets of the Temple,* p. 298.

320 *You have come sixty days:* Garraty, *Great Depression,* p. 32.

321 *I almost went down:* Kennedy, *Freedom from Fear,* p. 50.

321 *I warned them:* Chernow, *Morgan,* p. 326.

325 *The President likened the situation:* Kennedy, *Freedom from Fear,* p. 79.

326 *a millionaire's dole:* Ibid., p. 85.

C H A P T E R   S E V E N T E E N

CONVERTING RETREAT INTO ADVANCE

332 *So, first of all:* Safire, ed., *Lend Me Your Ears,* p. 859.

333 *safer to keep your money:* Kennedy, *Freedom from Fear,* p. 136.

333 *capitalism was saved:* Ibid.

340 *the exchange is a perfect institution:* Brooks, *Golconda,* p. 198.

340 *Mr. Kennedy, former speculator:* Collier and Horowitz, *Kennedys,* p. 75.

341 *I mean the stock exchange:* Sobel, *NYSE,* p. 49.

341 *Conduct contrary:* Ibid., p. 50.

341 *Wall Street could hardly:* Brooks, *Once in Golconda,* p. 273.

T R A N S I T I O N

THE SECOND WORLD WAR

349 *the great arsenal:* Kennedy, *Freedom from Fear,* p. 465.

350 For the bread that you eat: Adams, *Ocean Steamers,* p. 6.

350 *Immediate needs:* Kennedy, *Freedom from Fear,* p. 445.

351 *at least 50,000 planes:* Ibid., p. 446.

351 *that of a third-rate power:* Phillips, *1940s,* p. 139.

352 *Well, boys, Britain's broke:* Kennedy, *Freedom from Fear,* p. 465.

352 *Give us the tools:* Oxford Dictionary of Quotations, 3rd ed. p. 150.

352 *most unsordid act:* Kennedy, *Freedom from Fear,* p. 475.

354 *a blow for the liberty:* Botting, *U-Boats,* p. 135.

354 *to American production:* Bailey, *Home Front,* p. 77.

355 *I will if I can boss it:* Phillips, *1940s,* p. 142.

359 *from the country club:* Paul, *Taxation in the United States,* p. 318.

CHAPTER EIGHTEEN

THE GREAT POSTWAR BOOM

365 *The market was there:* Sobel, *Great Boom,* p. 89.

365 *What it amounted to:* Ibid., p. 90.

366 *were not, and are not:* Ibid., p. 93.

369 *One campaign tactic:* Ibid., p. 335.

370 *Exactly what they should be:* Drucker, *Adventures,* p. 277.

372 *shocking—bad for labor:* Dulles, *Labor in America,* p. 345.

376 *I believe it must be:* Paul Johnson, *History,* p. 811.

377 *if you will say that:* Ibid., pp. 810–11.

379 *a whole Copernican system:* Rothschild, *Bionomics,* p. 51.

CHAPTER NINETEEN

THE CRISIS OF THE NEW DEAL ORDER

382 *In your time:* Paul Johnson, *History,* p. 873.

385 *I'd rather be alive:* Personal communication from William L. Zeckendorf.

398 *We do not wish:* Davis, "Chronicle of a Debacle Foretold," p. 50.

399 *You got to remember:* Rosenbaum, "Financial Disaster with Many Culprits," *New York Times,* June 6, 1990.

CHAPTER TWENTY
A NEW ECONOMY, A NEW WORLD, A NEW WAR

406 *I wish to God:* Tobin, *Great Projects,* p. 268.

419 *the sinews of war:* Brewer, *Sinews of Power,* p. v.

BIBLIOGRAPHY

The First Iron Works Restoration. New York[?] First Iron Works Association, 1953.

Adams, Charles Francis, and Henry Adams. *Chapters of Erie, and Other Essays.* Boston: James R. Osgood, 1871.

Adams, John. *Ocean Steamers: A History of Ocean-Going Passenger Steamers 1820–1970.* London: New Cavendish Books, 1993.

Allen, Frederick Lewis. *The Great Pierpont Morgan.* New York: Harper & Brothers, 1949.

Ambrose, Stephen E. *Nothing Like It in the World: The Men Who Built the Transcontinental Railroad 1863–1869.* New York: Simon and Schuster, 2000.

Bailey, Ronald H. *The Home Front: U.S.A.* Alexandria, Va.: Time-Life Books, 1978.

Bailyn, Bernard, et al. *The Great Republic: A History of the American People.* Boston: Little, Brown, 1977.

Barlow, Francis C., and David Dudley Field. *Facts for Mr. David Dudley Field.* Albany, New York: Parsons and Company, 1871.

Berlin, Ira. *Generations of Captivity: A History of African-American Slaves.* Cambridge, Mass.: Belknap Press, 2003.

Botting, Douglas. *The U-Boats.* Alexandria, Va.: Time-Life Books, 1979.

Bowden, Witt. *The Industrial History of the United States.* New York: Augustus

M. Kelley, 1967. Reprint of the 1930 ed. published by Adelphi Company.

Brands, H. W. *The First American: The Life and Times of Benjamin Franklin.* New York: Doubleday, 2000.

———. *The Reckless Decade: America in the 1890s.* New York: St. Martin's Press, 1995.

Brewer, John. *The Sinews of Power: War, Money, and the English State, 1688–1783.* New York: Alfred A. Knopf, 1989.

Brinkley, Douglas. *Wheels for the World: Henry Ford, His Company, and a Century of Progress.* New York: Viking, 2003.

Brookhiser, Richard. *Alexander Hamilton, American.* New York: Free Press, 1999.

Brooks, John. *Once in Golconda: A True Drama of Wall Street, 1920–1938.* New York: Harper & Row, 1969.

Bruchey, Stuart. *The Wealth of the Nation: An Economic History of the United States.* New York: Harper & Row, 1988.

Buchanan, James M., and Richard E. Wagner. *Democracy in Deficit: The Political Legacy of Lord Keynes.* New York: Academic Press, 1977.

Buck, James E., ed. *The New York Stock Exchange: The First Two Hundred Years.* Essex, Conn.: Greenwich Publishing, 1992.

Burrows, Edwin C., and Mike Wallace. *Gotham: A History of New York City to 1898.* New York: Oxford University Press, 1999.

Burstein, Andrew. *The Passions of Andrew Jackson.* New York: Alfred A. Knopf, 2003.

Cameron, E. H. *Samuel Slater: Father of American Manufactures.* No city: Bond Wheelright Company, 1960.

Chernow, Ron. *The House of Morgan: An American Banking Dynasty and the Rise of Modern Finance.* New York: Atlantic Monthly Press, 1990.

———. *Titan: The Life of John D. Rockefeller, Sr.* New York: Random House, 1998.

Cohen, Lizabeth. *A Consumers' Republic: The Politics of Mass Communication in Postwar America.* New York: Alfred A. Knopf, 2003.

———. *Making a New Deal: Industrial Workers in Chicago, 1919–1939.* New York: Cambridge University Press, 1990.

Collier, Peter, and David Horowitz. *The Kennedys: An American Drama.* New York: Summit Books, 1984.

Cooper, John Milton, Jr. *Pivotal Decades: The United States 1900–1920*. New York: W. W. Norton, 1990.

Cornog, Evan. *The Birth of Empire: DeWitt Clinton and the American Experience, 1769–1828*. New York: Oxford University Press, 1998.

Croffut, William A. *An American Procession 1855–1914: A Personal Chronicle of Famous Men*. Freeport, N.Y.: Books for Libraries Press, 1968. Reprint of the 1931 ed.

Davis, L. J. "Chronicle of a Debacle Foretold, How Deregulation Begat the S&L Scandal." *Harper's Magazine,* September 1990.

Drucker, Peter. *Adventures of a Bystander*. New York: HarperCollins, 1991.

Dulles, Foster Rhea. *Labor in America: A History*. Arlington Heights, Ill.: Harlan Davidson, 1984.

Ferguson, Eugene S. *Oliver Evans, Inventive Genius of the American Industrial Revolution*. Greenville, Del.: Hagley Museum, 1980.

Fogel, Robert William. *Without Consent or Contract: The Rise and Fall of American Slavery*. New York: W. W. Norton, 1989.

Fowler, William Worthington. *Ten Years in Wall Street*. Hartford, Conn.: Worthington, Dustin, 1870.

Fox, Stephen. *Transatlantic: Samuel Cunard, Isambard Brunel, and the Great Atlantic Steamships*. New York: HarperCollins, 2003.

Freese, Barbara. *Coal: A Human History*. Cambridge, Mass.: Perseus Publishing, 2003.

Friedman, Lawrence M. *A History of American Law*. 2nd ed. New York: Simon and Schuster, 1985.

Galbraith, John Kenneth. *Money, Whence It Came, Where It Went*. Boston: Houghton Mifflin, 1975.

Garraty, John A. *The Great Depression*. San Diego, Calif.: Harcourt Brace, 1986.

Gately, Iain. *Tobacco: A Cultural History of How an Exotic Plant Seduced Civilization*. New York: Grove Press, 2001.

Gates, Paul W. *The Farmer's Age: Agriculture 1815–1860*. Vol. 3 of *The Economic History of the United States*. Repr. 1989, M. E. Sharpe, Armonk, N. Y. New York: Holt, Rinehart, and Winston, 1960.

Goodwin, Jason. *Greenback: The Almighty Dollar and the Invention of America*. New York: Henry Holt, 2003.

Gordon, John Steele. *The Great Game: The Emergence of Wall Street as a World Power, 1653–2000*. New York: Scribner, 1999.

————. *Hamilton's Blessing: The Extraordinary Life and Times of Our National Debt.* New York: Walker, 1997.

————. *The Scarlet Woman of Wall Street.* New York: Wiedenfeld and Nicolson, 1988.

————. "When Our Ancestors Became Us," in *American Heritage,* December 1989.

Greenfeld, Liah. *The Spirit of Capitalism: Nationalism and Economic Growth.* Cambridge, Mass.: Harvard University Press, 2001.

Greider, William. *Secrets of the Temple: How the Federal Reserve Runs the Country.* New York: Simon and Schuster, 1987.

Hadley, Arthur T. *Railroad Transportation—Its History and Its Laws.* New York: G. P. Putnam's Sons, 1886.

Hallahan, William H. *The Day the American Revolution Began: 19 April 1775.* New York: William Morrow, 2000.

Hamilton, Alexander. *Papers on Public Credit, Commerce and Finance,* edited by Samuel McKee, Jr. New York: Columbia University Press, 1934.

Harris, Charles Townsend. *Memories of Manhattan in the Sixties and Seventies.* New York: Derrydale Press, 1928.

Hobhouse, Henry. *Seeds of Change: Five Plants That Transformed Mankind.* New York: Harper & Row, 1986.

Holbrook, Stewart H. *The Age of the Moguls: The Story of the Robber Barons and the Great Tycoons.* Garden City, N.Y.: Doubleday, 1954.

Hounsell, David A. *From the American System to Mass Production, 1800–1932.* Baltimore: Johns Hopkins University Press, 1984.

House of Representatives. *House Report Number 31, 41st Congress, 2nd Session.* Washington, D.C., 1871.

Hunter, Louis C. *Steamboats on the Western Rivers: An Economic and Technological Survey.* Cambridge, Mass.: Harvard University Press, 1949.

Jackson, Kenneth T., ed. *The Encyclopedia of New York City.* New Haven, Conn.: Yale University Press, 1995.

Johnson, Paul. *A History of the American People.* New York: HarperCollins, 1998.

Johnson, Richard R. *John Nelson Merchant Adventurer: A Life Between Empires.* New York: Oxford University Press, 1991.

Joseph, Alvin M., Jr., ed. *America in 1492: The World of the Indian Peoples Before the Arrival of Columbus.* New York: Alfred A. Knopf, 1992.

Kanigel, Robert. *The One Best Way: Frederick Winslow Taylor and the Enigma of Efficiency.* New York: Viking, 1997.

Kennedy, David M. *Freedom from Fear: The American People in Depression and War, 1929–1945.* Vol. 9 of *The Oxford History of the United States.* New York: Oxford University Press, 1999.

Kessner, Thomas. *Capital City: New York and the Men Behind America's Rise to Economic Dominance.* New York: Simon and Schuster, 2003.

King, Mary L. *The Great American Banking Snafu.* Lexington, Mass.: Lexington Books, 1985.

Kirkland, Edward C. *Industry Comes of Age: Business, Labor, and Public Policy 1860–1897.* Vol. 6 of *The Economic History of the United States.* New York: Holt Rinehart and Winston, 1961.

Klein, Maury. *The Life and Legend of Jay Gould.* Baltimore: Johns Hopkins University Press, 1986.

Klein, Milton, ed. *The Empire State: A History of New York.* Ithaca, N.Y.: Cornell University Press, 2001.

Krass, Peter. *Carnegie.* New York: John Wiley and Sons, 2002.

Kulikoff, Allan. *From British Peasants to Colonial American Farmers.* Chapel Hill University of North Carolina Press, 2000.

Landes, David S. *The Wealth and Poverty of Nations: Why Some Are So Rich and Some Are So Poor.* New York: W. W. Norton, 1998.

Lane, Wheaton J. *Commodore Vanderbilt, an Epic of the Steam Age.* New York: Alfred A. Knopf, 1942.

Larkin, Jack. *The Reshaping of Everyday Life.* Vol. 2 of *The Everyday Life in America.* New York: HarperPerennial, 1988.

Lee, Susan. *Hands Off: Why the Government Is a Menace to Economic Health.* New York: Simon and Schuster, 1996.

Lockwood, Charles. *Manhattan Moves Uptown: An Illustrated History.* Boston: Houghton Mifflin, 1976.

Marks, Paula Mitchell. *Precious Dust: The American Gold Rush Era: 1848–1900.* New York: William Morrow and Company, 1994.

McCrady, Edward. *The History of South Carolina.* New York: Macmillan, 1897.

McCullough, David. *John Adams.* New York: Simon and Schuster, 2001.

McCusker, John J. *How Much Is That in Real Money? A Historical Commodity Price Index for Use as a Deflator of Money Values in the Economy of the United States.* 2nd ed. Worcester, Mass.: American Antiquarian Society, 2001.

McCusker, John J., and Russell R. Menard. *The Economy of British North America, 1607–1789*. Chapel Hill: University of North Carolina Press, 1991.

McPherson, James M. *Battle Cry of Freedom: The Civil War Era*. New York: Oxford University Press, 1988.

Malabre, Alfred L., Jr. *Beyond Our Means: How America's Long Years of Debt, Deficits, and Reckless Borrowing Now Threatens to Overwhelm Us*. New York: Random House, 1987.

Malone, Dumas. *Jefferson and His Time*. 6 vols. Boston: Little, Brown, 1948–1981.

Martin, Albro. *Railroads Triumphant: The Growth, Rejection and Rebirth of a Vital American Force*. New York: Oxford University Press, 1992.

Medbery, James K. *Men and Mysteries of Wall Street*. Boston: Fields, Osgood, 1870.

Middlekauff, Robert. *The Glorious Cause: The American Revolution 1763–1789*. Vol. 2 of *The Oxford History of the United States*. New York: Oxford University Press, 1982.

Miller, John C. *Alexander Hamilton: Portrait in Paradox*. New York: Harper & Row, 1959.

Miller, Nathan. *Stealing from America: A History of Corruption from Jamestown to Reagan*. New York: Paragon Books, 1992.

Misa, Thomas J. *A Nation of Steel: The Making of Modern America 1865–1925*. Baltimore: Johns Hopkins University Press, 1995.

Mitchell, Broadus. *Depression Decade: From New Era Through New Deal 1929–1941*. Vol. 9 of *The Economic History of the United States*. New York: Holt, Rinehart and Winston, 1947.

Moran, William. *The Belles of New England: The Women of the Textile Mills and the Families Whose Wealth They Wove*. New York: St. Martin's Press, 2002.

Morison, Elting E. *Men, Machines, and Modern Times*. Cambridge, Mass.: M.I.T. Press 1966.

Morris, Edmund. *Theodore Rex*. New York: Random House, 2001.

Moss, David A. *When All Else Fails: Government as Ultimate Risk Manager*. Cambridge, Mass.: Harvard University Press, 2002.

Nettels, Curtis R. *The Emergence of a National Economy, 1775–1815*. Vol. 2 of *The Economic History of the United States*. New York: Holt, Rinehart and Winston, 1962.

Nevins, Allan, ed. *The Diaries of Philip Hone*. New York: Dodd, Mead, 1927.

Nevins, Allan, and Milton Thomas Halsey, eds. *The Diary of George Templeton Strong*. New York: Macmillan, 1952.

Norman, Bruce. *The Inventing of America*. New York: Taplinger, 1972.

Oberholtzer, Ellis Paxson. *Jay Cooke, Financier of the Civil War*. New York: Burt Franklin, 1970.

Parton, James. *Famous Americans of Recent Times*. Boston: Ticknor and Fields, 1866.

Patterson, James T. *Grand Expectations: The United States, 1945–1974*. Vol. 10 of *The Oxford History of the United States*. New York: Oxford University Press, 1996.

Paul, Randolph E. *Taxation in the United States*. Boston: Little, Brown, 1954.

Perlin, John. *A Forest Journey: The Role of Wood in the Development of Civilization*. New York: W. W. Norton, 1989.

Phillips, Cabell. *The 1940s: Decade of Triumph and Trouble*. New York: Macmillan, 1975.

Previts, Gary John, and Barbara Dubis Merino. *A History of Accounting in America*. New York: John Wiley and Sons, 1979.

Randall, Willard Sterne. *Thomas Jefferson: A Life*. New York: Henry Holt, 1993.

Ratner, Sidney, James H. Soltow, and Richard Sylla. *The Evolution of the American Economy: Growth, Welfare, and Decision Making*. 2nd ed. New York: Macmillan, 1993.

Remini, Robert V. *Andrew Jackson and the Course of American Empire, 1767–1821*. Vol. 1. New York: Harper & Row, 1977.

———. *Andrew Jackson and the Course of American Freedom, 1822–1832*. Vol. 2. New York: Harper & Row, 1981.

———. *Andrew Jackson and the Course of American Democracy, 1833–1845*. Vol. 3. New York: Harper & Row, 1984.

Richardson, Heather Cox. *The Greatest Nation of the Earth: Republican Economic Policies During the Civil War*. Cambridge, Mass.: Harvard University Press, 1997.

Richter, Daniel K. *Facing East from Indian Country: A Native History of Early America*. Cambridge, Mass.: Harvard University Press, 2001.

Rosenbaum, David E. "A Financial Disaster with Many Culprits," *New York Times*, June 6, 1990.

Rothschild, Michael. *Bionomics: The Inevitability of Capitalism*. New York: Henry Holt, 1990.

Roy, William G. *Socializing Capital: The Rise of the Large Industrial Corporation in America.* Princeton, N.J.: Princeton University Press, 1997.

Serrin, William. *Homestead: The Glory and Tragedy of an American Steel Town.* New York: Times Books, 1992.

Shannon, Fred A. *The Farmer's Last Frontier: Agriculture, 1860–1897.* Vol. 5 of *The Economic History of the United States.* New York: Holt, Rinehart and Winston, 1945. Repr. Harper Torchbooks, 1968.

Safire, William, ed. *Lend Me Your Ears: Great Speeches in History.* 2nd ed. New York: W. W. Norton, 1992.

Satterlee, Herbert L. *J. Pierpont Morgan: An Intimate Portrait.* New York: Macmillan, 1939.

Silverman, Kenneth. *Lightning Man: The Accursed Life of Samuel F. B. Morse.* New York: Alfred A. Knopf, 2003.

Simon, James F. *What Kind of Nation: Thomas Jefferson, John Marshall, and the Epic Struggle to Create a United States.* New York: Simon & Schuster, 2002.

Smith, Matthew Hale. *Twenty Years Among the Bulls and Bears of Wall Street.* Hartford, Conn.: J. B. Burr, 1870.

Smith, Page. *The Shaping of America.* Vol. 3 of *A People's History of the Young Republic.* New York: McGraw-Hill, 1980.

Sobel, Robert. *The Big Board: A History of the New York Stock Exchange.* New York: Free Press, 1965.

———. *The Great Boom 1950–2000: How a Generation of Americans Created the World's Most Prosperous Society.* New York: Truman Talley Books, St. Martin's Press, 2000.

———. *NYSE: A History of the New York Stock Exchange, 1935–1975.* New York: Weybright and Talley, 1975.

———. *Panic on Wall Street: A History of America's Financial Disasters.* New York: Macmillan, 1968.

Soule, George. *Prosperity Decade: From War to Depression: 1917–1929.* Vol. 3 of *The Economic History of the United States.* Repr. 1989, M. E. Sharpe, Armonk, N.Y. New York: Holt, Rinehart, and Winston, 1947.

Stamp, Kenneth. *America in 1857: A Nation on the Brink.* New York: Oxford University Press, 1990.

Stedman, Edmund Clarence. *The New York Stock Exchange.* New York: Stock Exchange Historical, 1905.

Stover, John F. *American Railroads*. 2nd ed. Chicago: University of Chicago Press, 1997.

Strouse, Jean. *Morgan: American Financier*. New York: Random House, 1999.

Tanner, Hudson C. *"The Lobby," and Public Men from Thurlow Weed's Time*. Albany, N.Y.: George MacDonald, 1888.

Taylor, Alan. *American Colonies*. New York: Viking, 2001.

Thomas, Emory M. *The Confederate Nation: 1861–1865*. New York: Harper & Row, 1979.

Tobin, James. *Great Projects*. New York: Free Press, 2001.

Trescott, Paul B. *Financing American Enterprise: The Story of Commercial Banking*. New York: Harper & Row, 1963.

Wall, Joseph Frazier. *Andrew Carnegie*. New York: Oxford University Press, 1970.

Warren, Charles. *The Supreme Court in United States History*. Rev. ed. Boston: Little, Brown, 1926.

Weightman, Gavin. *The Frozen-Water Trade: A True Story*. New York: Hyperion, 2003.

Wik, Reynold M. *Steam Power on the American Farm*. Philadelphia: University of Pennsylvania Press, 1953.

Wilson, George. *Stephen Girard: The Life and Times of America's First Tycoon*. Conshohocken, Pa.: Combined Books, 1995.

Yergin, Daniel. *The Prize: The Epic Quest for Oil, Money & Power*. New York: Simon and Schuster, 1991.

Yergin, Daniel, and Joseph Stanislaw. *The Commanding Heights: The Battle for the World Economy*. New York: Touchstone, 2002.

INDEX

labor relations and, 250
slums and, 243–44
income tax, 272–77, 388
  on capital gains, 394
  in Civil War, 194–95, 272–73,
    274
  in Constitution, 274, 276
  on corporations, 276–77
  Kemp-Roth proposals and, 390–91
  progressive, 390
  reform of, 394–96
  in World War I, 292–93
  in World War II, 358–59
indentured servants, 11–12, 16,
  18–19, 50
India, 7, 40, 54, 83, 85, 416
Indiana, 122, 141
Indians, 4–6, 8, 11, 13, 15, 16–17, 22,
  25, 54
indigo trade, 26, 82–83
Industrial Revolution, 26, 32, 49, 82,
  88, 92, 102, 141, 149–50, 152,
  162, 241, 243, 246, 409
Industrial Workers of the World
  (Wobblies), 250
inflation, 43–44, 47, 113, 267, 357, 379,
  381, 390–91, 396, 405
  in American Revolution, 60–61
  Civil War and, 196
  gold standard and, 265–66
  in 1970s, 383–86
  in postwar economy, 370–71
  after World War I, 295–96
Insull, Samuel, 304–6
Intel, 408
International Monetary Fund, 378
International Typographical Union,
  249–50
Internet, xiv, 12, 410–13
Interstate Commerce Commission,
  238, 392
Iran hostage crisis, 391

iron industry, 32–36, 184, 246
  see also steel industry
isolationism, 349, 353, 362
Italy, 9, 311, 353

Jackson, Andrew, xviii, 92, 122–31,
  151, 278, 281, 337, 389
  national debt and, 125–26
  nullification crisis and, 97
  pet banks and, 129–30
Jackson, Charles Thomas, 155
Jackson, Howell, 274
James I, king of England, 10, 15, 17
James II, king of England, 38, 52–53
Jamestown colony, 8, 12–17
Japan, 204, 278, 311, 352, 353, 361,
  363, 416, 417
Jefferson, Thomas, 52, 121, 122–23,
  171, 173, 223, 326, 389, 398
  and adoption of dollar, 68–69
  assumption debate and, 74–75
  central bank opposed by, 116
  decimal dollar system and, 69–70
  Embargo Act and, 94–95
  Erie Canal opposed by, 106
  national bank opposed by, 77–78,
    80–81
Jenks, Joseph, 35
Johnson, Andrew, 215, 221
Johnson, Anthony, 19
Johnson, Hiram, 349
Johnson, Lyndon B., 382, 390
joint-stock company, 9–10
Jones, William, 126
Jones, W. M., 248
Jonson, Ben, 27
junk bonds, 397

Kaiser, Henry J., 354
Kay, John, 88
Keats, John, 366
Kellogg-Briand Pact, 311

About the author

About the book

Read on

Insights,
Interviews
& More ...

*

Meet **John Steele Gordon**

Marcell Vandehaar

JOHN STEELE GORDON was born in New York City in 1944, into a family long associated with the city and its financial community. Both his grandfathers held seats on the New York Stock Exchange. He was educated at Millbrook School and Vanderbilt University, graduating with a bachelor of arts degree in history in 1966.

After college he worked for six years as a production editor for Harper & Row (now HarperCollins) before leaving to drive a Land Rover from New York to Tierra del Fuego—a nine-month journey of thirty-nine thousand miles and approximately forty flat tires. ("Ever had to pump a tire by hand at fourteen thousand feet?") This trek resulted in his first book, *Overlanding*. Altogether he has driven through forty-seven countries on five continents.

After returning to New York he served on the staffs of Congressmen Herman Badillo and Robert Garcia. He has been a full-time writer for the last twenty years. His second book, *The Scarlet Woman of Wall Street*, a history of Wall Street in the 1860s, was published in 1988. His third book, *Hamilton's Blessing: the Extraordinary Life and Times of Our National Debt,* was published in 1997. *The Great Game: The Emergence of Wall Street as a World Power: 1653–2000,* was published in 1999. A two-hour special based on *The Great Game* aired on CNBC on April 24, 2000. A collection of his columns from *American Heritage* magazine, entitled *The Business of America,* was published in 2001, followed by a history of the laying of the Atlantic Cable, *A Thread Across the Ocean,* in 2002.

He specializes in business and financial history, and has had articles published in, among others, *Forbes, Forbes ASAP, Worth,* the *New York Times* and the *Wall Street Journal* op-ed pages, *Washington Post Book World,* and *Outlook.* He is a contributing editor at *American Heritage,* where he has written the "Business of America" column since 1989.

In 1991 he traveled to Europe, Africa, North and South America, and Japan with the photographer Bruce Davidson for Schlumberger, Ltd. to create a photo essay called "Schlumberger People" for the company's annual report.

In 1992 he was the cowriter, with Timothy C. Forbes and Steve Forbes, of "Happily Ever After?", a video produced by *Forbes* in honor of the seventy-fifth anniversary of the magazine.

He is a frequent commentator on *Marketplace,* the daily public radio business-news program heard throughout the country. He has appeared on numerous other radio and television shows, including *New York: A Documentary Film* by Ric Burns, *Business Center* and *Squawk Box* on CNBC, and ▶

Meet John Steele Gordon *(continued)*

The News Hour with Jim Lehrer on PBS. He was a guest in 2001 on a live, two-hour edition of *Booknotes* with Brian Lamb on C-SPAN.

Mr. Gordon lives in North Salem, New York. ∽

66 He is a frequent commentator on *Marketplace*, the daily public radio business-news program. 99

John Steele Gordon on Writing *An Empire of Wealth*

I CAN'T IMAGINE why the subject came up, but I remember my mother many years ago giving me a piece of advice on, of all things, decorating. She told me that if a room is too familiar—such as your own living room—it can be hard to see it simply as a space. Too many memories clutter the picture, familiarity makes it, in a way, invisible. Her suggestion was to look at such a room in a mirror. By seeing the room reversed, she thought, one could look at it afresh.

No one has ever accused me of being very concerned with home decoration, but for some reason that notion stuck in my head. And a few years ago it occurred to me that the technique might be applied to an often all-too-familiar history, that of the United States.

The history of this country is one of the great epics of mankind. As with Greece, Rome, and England, an inconsequential country on the periphery of world concerns came, through the happenstance of history, to dominate those concerns and to spread its culture far and wide. American history, of course, is taught in every school in the country and cable and public television are chockablock with stories from the American past.

So I wondered if there might be another way to tell the story of America's rise to become the center of the world, a way to look at it afresh. ▶

66 My mother told me that if a room is too familiar it can be hard to see it simply as a space. Her suggestion was to look at such a room in a mirror. 99

5

> **``** In recent decades many historians have tried to write 'people's history.' . . . This has, I think, been less successful on the whole. **"**

John Steele Gordon on Writing *(continued)*

The standard histories of a country—any country—look at events through political eyes, recounting the rise and fall of kings, presidents, and generals as they struggled with the problems of their day. This is, of course, an entirely valid approach. In recent decades many historians have tried to write "people's history," telling the story from the point of view not of the high and the mighty but of the ordinary people. This has, I think, been less successful on the whole, if only because there is no real way (or at least no one has yet discovered it) to tell a dramatically satisfying story from the point of view of the chorus rather than the principals. The world doesn't really work that way.

But underlying much of politics and warfare (war, as von Clausewitz noted, being nothing more than politics carried on by other means) are always economic forces. Wars, as the Chinese proverb has it, are fought with silver bullets. The laws of supply and demand and of self-interest are every bit as universal and ineluctable as those of gravity and inertia, and they affect the destiny of nations in fundamental ways. But they often get short shrift in the writing of history.

When they are utilized, the result is often stupefyingly boring. The reason is simple enough. Thomas Carlyle dubbed economics the "dismal science" because economists so often reduce it to numbers and formulas, and thus eliminate the human element from what is, after all, a uniquely human activity. Much economic history is full of charts and graphs and empty of human genius. Little wonder that it's boring. Other economic histories have a political agenda at their core and are thus unavoidably tendentious and often deeply dishonest.

I decided to write what I hoped would be an economic history of the United States that was without charts, graphs, and political purposes but rich in human drama and that would give the reader a fresh slant on the extraordinary story that is American history.

The extent to which I have succeeded at this, of course, is up to the reader to decide. I know that I had a great deal of fun trying. ❧

> 66 I decided to write what I hoped would be an economic history of the United States that was without charts, graphs, and political purposes. 99

John Steele Gordon on
Books About Business

IT WAS KARL MARX who coined the word
capitalism, so it is perhaps not surprising that
its history, until recently, has been largely
written by its enemies. Indeed many of the
"classics" of American business and economic
history suffer from a profound animus on
the part of the authors toward the economic
system of the United States. But for all its
messiness and need for constant reform,
this system has made the country a fount of
wealth such as the world has never seen, and
distributed that wealth more widely through
the population than any system tried
elsewhere. Its history is a worthy subject.

Among the books in this tradition are
Gustavus Myers's *History of the Great
American Fortunes* (1910), Ida Tarbell's
History of the Standard Oil Company (1902),
and Charles A. and Mary Beard's *The Rise of
American Civilization* (1927, last revised and
enlarged in 1956). These books often contain
very useful information and references, but
must be used with caution, especially with
regard to the authors' interpretations and
basic economic point of view.

One of the most famous books of this ilk
is *The Robber Barons* by Matthew Josephson,
first published in 1934 (not a good year for
capitalism) and still very much in print
after nearly seventy years. It is, to be sure,
a great read and it was the book that firmly
established the previously obscure phrase
"robber barons" in the American lexicon as
the collective term for the capitalists of the
Gilded Age.

But the very title hints at how intellectually dishonest this book is. The supposed medieval robber barons charged merchants a fee to pass their castles on the Rhine unmolested. In other words they created no wealth, just used the threat of force to enrich themselves, like modern-day muggers. But the capitalists of the late nineteenth century such as Vanderbilt, Rockefeller, Carnegie, and Morgan, however ruthless they may have been, created enormous wealth in building and financing the railroads, oil refineries, and steel mills that made them so fabulously rich and established the United States as the world's greatest industrial economy.

One passage from *The Robber Barons* illustrates Josephson's point of view perfectly. Regarding Commodore Cornelius Vanderbilt he wrote, "In seeking quickened activity, great volume, and lower prices—instead of honest but limited services at high tariffs—he gave intimations of a new personal departure from the older bourgeois order." [1] Isn't that neat? Without actually saying so, Josephson implies that Vanderbilt—who was, in fact, an honest man to his fingertips—seemed somehow dishonest for offering greater service at lower prices on his steamboats. A far better popular history of the same subject is Stewart H. Holbrook's *The Age of the Moguls: The Story of the Robber Barons and the Great Tycoons* (1954).

The many books by Frederick Lewis Allen, a contemporary of Josephson and Holbrook, are also still well worth reading, especially his classic *Only Yesterday: An Informal History* ▶

> **❝ The supposed medieval robber barons charged merchants a fee to pass their castles on the Rhine unmolested. ❞**

[1] P. 16 of the Harcourt Brace Jovanovich 1962 paperback edition.

> 66 Inexplicably, there is no modern biography of one of the earliest and most interesting of the so-called robber barons, Cornelius Vanderbilt. 99

of the Nineteen-Twenties (1931), *The Lords of Creation* (1935), *Since Yesterday: The Nineteen-Thirties in America* (1940), and *The Great Pierpont Morgan* (1949).

Fortunately, as socialism, the great dream of so many twentieth-century intellectuals, has collapsed in country after country, newer books of less partisan scholarship have been appearing in the last twenty years or so. Excellent biographies, of impeccable scholarship, of many of the major business figures of the Gilded Age are now available. Among these are Jean Strouse's *Morgan: American Financier* (1999), Ron Chernow's *Titan: The Life of John D. Rockefeller, Sr.* (1998), Peter Krass's *Carnegie* (2002—but don't neglect Joseph Wall's 1970 bio, *Andrew Carnegie*), Maury Klein's *The Life and Legend of Jay Gould* (1986) and *The Life and Legend of E. H. Harriman* (2000), and David Nasaw's *The Chief: The Life of William Randolph Hearst* (2000). Inexplicably, there is no modern biography of one of the earliest and most interesting of the so-called robber barons, Cornelius Vanderbilt, but Wheaton J. Lane's *Commodore Vanderbilt: An Epic of the Steam Age* (1942) is excellent if dated.

The recently published Douglas Brinkley book, *Wheels for the World: Henry Ford, His Company, and a Century of Progress* (2003), is an excellent biography both of Henry Ford and the company he founded. But don't neglect Allan Nevins and Frank Ernest Hill's majestic three-volume work, *Ford: The Times, the Man, the Company* (1954), *Ford: Expansion and Challenge, 1915–1933* (1957), and *Ford: Decline and Rebirth, 1933–1962* (1963).

Nevins, one of the great twentieth-century American historians, also wrote two worthy books on Rockefeller, *John D. Rockefeller: The Heroic Age of American Enterprise* (1954) and *Study in Power: John D. Rockefeller, Industrialist and Philanthropist* (1953).

For general reference in economic history, the best and most accessible is the brand-new *Oxford Encyclopedia of Economic History* (2003) in five volumes. It covers the whole world and its articles have excellent bibliographies. (Full disclosure: I wrote one of the essays, on J. P. Morgan.) For a narrative history of the world economy, try Rondo Cameron's *A Concise Economic History of the World: From Paleolithic Times to the Present* (1993, second edition).

For raw statistics, while the Internet is increasingly chockablock with them, there is still no substitute for the *Historical Statistical Abstract of the United States* published by the Department of Commerce in 1976. The annual *Statistical Abstract of the United States* is the best source for recent statistics. The Federal Reserve also issues statistics, mostly of a financial nature, by the ton. Every good-sized library has these publications available.

For a primer on economics itself, one can still do no better than *The Wealth of Nations* by Adam Smith (1776). It revolutionized the discipline of economics and has been in print for 229 years. Like Shakespeare, the Bible, and Darwin's *Origin of Species,* it always will be in print and for the same reasons. While not exactly light reading, Smith wrote in elegant, Augustan prose. I wish someone would write a modern version, as was recently done with ▶

66 For a primer on economics itself, one can still do no better than *The Wealth of Nations* by Adam Smith (1776). 99

**John Steele Gordon on Books
About Business** *(continued)*

The Origin of Species, to make it more accessible to the modern reader.

Economists since John Maynard Keynes can seldom be accused of writing Augustan prose. Often it is hardly even English. And they have usually been so concerned with macroeconomics (aggregate economic effects at the national and international levels) that they have sadly neglected the microeconomics of the marketplace, without which the macroeconomics make no sense to the lay reader. One of the best modern introductions to economics is Thomas Sowell's *Basic Economics: A Citizen's Guide to the Economy* (2000). There's not a single chart in the entire book and it is mercifully free of the gobbledygook jargon of so many academic economists.

However, don't neglect the books by Harvard professor (emeritus) Alfred D. Chandler, Jr., the dean of American business historians, especially *The Visible Hand: The Managerial Revolution in American Business* (1977) and *Scale and Scope: The Dynamics of Industrial Capitalism* (1990), which, of course, cover topics that Adam Smith never dreamed of. They are not easy reading, but are basic literature. Also consider David Hackett Fischer's *Great Wave: Price Revolution and the Rhythm of History* (1996) and David S. Landis's *Wealth and Poverty of Nations: Why Some are So Rich and Some So Poor* (1998).

For Wall Street history, the many books on the subject by the late Robert Sobel are indispensable. His *Pursuit of Wealth: The Incredible Story of Money Throughout the Ages* (2000) is excellent as well. Another classic of Wall Street history is John Brooks's *Once* in *Golconda: A True Drama of Wall Street, 1920–1938* (1969). Brooks's many writings on business history, many of which first appeared in *The New Yorker,* are all well worth reading.

Let me add a few personal favorites. *How the West Grew Rich: The Economic Transformation of the Industrial World* (1986), by Nathan Rosenberg and L. E. Birdzell, Jr., is highly readable and eye-opening on why capitalism works and other economic systems do not.

James Grant's *Money of the Mind: Borrowing and Lending in America from the Civil War to Michael Milken* (1992) is beautifully written, highly entertaining, and wise.

Next is Daniel Yergin's *Prize: The Epic Quest for Oil, Money, and Power* (1991). Economic history boasts few page-turners of this size— nine hundred pages—but Yergin's tome earns my recommendation. I also recommend *The Commanding Heights: The Battle Between Government and the Marketplace That Is Remaking the Modern World* (1998) by Yergin and Joseph Stanislaw.

The Company: A Short History of a Revolutionary Idea (2003), by John Micklethwait and Adrian Wooldridge, is a history of the corporation, something we take entirely for granted but without which neither this country (Virginia and Plymouth were both founded by corporations, not the English government) nor the modern world would be possible.

Michael Rothschild's *Bionomics: The Inevitability of Capitalism* (1990) is the only book I know that deals with the biological ▶

> 66 *How the West Grew Rich: The Economic Transformation of the Industrial World* is highly readable and eye-opening on why capitalism works and other economic systems do not. 99

66 *Bionomics: The Inevitability of Capitalism* is an important and very readable book. 99

origins of all economic phenomena. Most economists treat economics as an abstraction to be reduced to mathematical formulas as though the basic units of an economy were analogous to atoms in the physical universe. But the ineluctable fact remains that the basic units of the economic universe are human beings, in all our quirky complexity. And that economic universe is an ecosystem, not a machine. The most powerful force operating in it is self-interest. This is an important and very readable book.

Finally, let me point out a long-forgotten gem of a book that, curiously, applies to today's economy as much as to that of the late nineteenth century when it was written, Arthur T. Hadley's *Railroad Transportation— Its History and Its Laws* (1886). A work of economics by a major economist of the day (Hadley would later be president of Yale), it is beautifully written and ideology free. And, if you would like to understand why the airline industry is a feast-and-famine business and why price wars frequently break out, Hadley will explain it to you, for railroads and airlines have very similar economic constraints. ⌒

Have You **Read?**

A THREAD ACROSS THE OCEAN

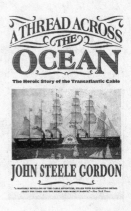

Today, in a world in which news flashes around the globe in an instant, time lags are inconceivable. In the mid nineteenth century, however, they were a fact of life. The United States was remote from Europe, the center of world affairs, and communication was only as quick as the fastest ship could cross the Atlantic. Instant contact seemed as unlikely then as walking on the moon did in the 1950s.

The Civil War had barely ended, however, when the Old and New Worlds had been united by the successful laying of the telegraph cable that spanned the Atlantic in 1866. *A Thread Across the Ocean* chronicles this extraordinary achievement, one of the greatest engineering feats of that century—and perhaps of all time. It was an epic struggle, requiring a decade of effort, numerous failed attempts, millions of dollars in capital, a near disaster at sea, the overcoming of seemingly insurmountable technological problems (many of them entirely unforeseen before work commenced), and uncommon physical, financial, and intellectual courage. In the end, it literally changed the world.

By bringing to life a dramatic and overlooked story in the annals of history ▶

Have You Read? *(continued)*

and technology, John Steele Gordon sheds fascinating new light on the American saga.

"We should not be surprised that this book is so joyously readable. . . . [A] sprightly and oddly engrossing mix of technological, financial, and human history."
—*Chicago Sun-Times*

Don't miss the next book by your favorite author. Sign up now for AuthorTracker by visiting www.AuthorTracker.com.